# TABLES

AND

# FORMULÆ

FOR

FINDING THE DISCHARGE OF WATER FROM ORIFICES, NOTCHES, WEIRS, PIPES, AND RIVERS.

BY

JOHN NEVILLE, C.E., M.R.I.A.,

COUNTY SURVEYOR OF LOUTH, AND OF THE COUNTY OF THE TOWN OF DROGHEDA.

---

"It ought to be more generally known, that theory is nothing more than the conclusions of reason from numerous and accurately-observed phenomena, and the deduction of the laws which connect causes with effects; that practice is the application of those general truths and principles to the common affairs and purposes of life; and that Science is the recorded experience and discoveries of mankind, or, as it has been well defined, 'the knowledge of many, orderly and methodically digested and arranged, so as to become attainable by one.'"— *American Quarterly Review.*

---

LONDON:
JOHN WEALE, 59, HIGH HOLBORN.

1853.

SAN FRANCISCO E. J. MUYGRIDGE.

LONDON:
WOODFALL AND KINDER, PRINTERS,
ANGEL COURT, SKINNER STREET.

# CONTENTS.

|  | PAGE. |
|---|---|
| INTRODUCTION | vii |

SECT.
I. Application and use of the Tables, Formulæ, &c. . . 1
II. Formulæ for the Velocity and Discharge from Orifices, Weirs, and Notches.—Coefficients of Velocity, Contraction, and Discharge.—Practical Remarks on the Use of the Formulæ . . . . . . . . . 17
III. Experimental Results.—Coefficients of Discharge . . 31
IV. Variations in the Coefficients from the Position of the Orifice.—General and Partial Contraction.—Velocity of Approach.—Central and Mean Velocities . . . 51
V. Submerged Orifices and Weirs.—Contracted River Channels . . . . . . . . . . 71
VI. Short Tubes, Mouth-pieces, and Approaches.—Coefficients of Discharge for Simple and Compound Short Tubes.—Shoots . . . . . . . . . . 78
VII. Lateral Contact of the Water and Tube.—Atmospheric Pressure.—Head measured to the Discharging Orifice. Coefficient of Resistance.—Formula for the Discharge from a Short Tube.—Diaphragms.—Oblique Junctions. —Formula for the Time of the Surface sinking a Given Depth . . . . . . . . . . 89
VIII. Flow of Water in uniform Channels.—Mean Velocity.—Mean Radii and Hydraulic Mean Depths.—Border.—Train.—Hydraulic Inclination.—Effects of Friction.—Formulæ for calculating the Mean Velocity.—Application of the Formulæ and Tables to the Solutions of three useful Problems . . . . . . . 99
IX. Best forms of the Channel.—Regimen . . . . 127
X. Effects of Enlargements and Contractions.—Backwater Weir Case.—Long and Short Weirs . . . . 136
XI. Bends and Curves.—Branch Pipes.—Different Losses of Head.—General Equation for finding the Velocity.—Hydrostatic and Hydraulic Pressure.—Piëzometer.—Catchment Basins.—Rain Fall per annum.—Water Power . . . . . . . . . . 150

A 2

## CONTENTS.

NOTE A. Discharge from one Vessel or Chamber into another.—
Lock-chambers . . . . . . . . 169

TABLE.
I. Coefficients of Discharge from Square and Differently Proportioned Rectangular Lateral Orifices in Thin Vertical Plates . . . . . . . . . . 174
II. For Finding the Velocities from the Altitudes and the Altitudes from the Velocities . . . . . 176
III. Square Roots for finding the Effects of the Velocity of Approach when the Orifice is small in proportion to the Head. Also for finding the Increase in the Discharge from an Increase of Head. See p. 55 . . . . 186
IV. For Finding the Discharge through Rectangular Orifices; in which $n = \dfrac{h}{d}$. Also for Finding the Effects of the Velocity of Approach to Weirs, and the Depression on the Crest. See p. 55 . . . . . . . 188
V. Coefficients of Discharge for different Ratios of the Channel to the Orifice . . . . . . . 192
VI. The Discharge over Weirs or Notches of One Foot in Length in Cubic Feet per Minute . . . . 198
VII. For Finding the Mean Velocity from the Maximum Velocity at the Surface, in Mill Races, Streams, and Rivers with uniform Channels; and the Maximum Velocity from the Mean Velocity. See p. 101 . . . . 204
VIII. For Finding the Mean Velocities of Water flowing in Pipes, Drains, Streams, and Rivers . . . . 205
IX. For Finding the Discharge in Cubic Feet, per Minute, when the Diameter of a Pipe, or Orifice, and the Velocity of Discharge, are known; and *vice versâ* . . 218
X. For Finding the Depths on Weirs of different Lengths, the Quantity discharged over each being supposed constant. See pages 149 and 150 . . . . . . . 220
XI. Relative Dimensions of equal Discharging Trapezoidal Channels, with slopes from 0 to 1, up to 2 to 1 . . 221
XII. Discharges from the Primary Channel in the first column of TABLE XI. . . . . . . . . 222
XIII. The Square Roots of the Fifth Powers of Numbers for finding the Diameter of a Pipe, or Dimensions of a Channel from the Discharge, or the reverse; showing the relative discharging powers of Pipes of different Diameters, and of any similar Channels whatever, closed or open. (See pages 13, 123, 124, &c.) . . . 224

# ERRATA.

Page 4, line 11, for "3807" read "·3807."
Page 6, line 10, for "1·497" read "14·97."
,, line 15, for "1·3089" read "13·089."
,, line 16, for "·1745" read "1·745."
,, line 17, for "·0131" read "·131."
,, line 18, for "·0004" read "·004."
,, line 19, for "1·4969" read "14·969."
,, line 22, for "1·497" read "14·97."
,, ,, for "1·485" read "14·85."
Page 9, line 24, for "1732·2" read "17322."
Page 13, line 15, for "TABLE IV." read "TABLE XIII."
,, lines 30 and 31, for "EXAMPLE 9, page 6" read "EXAMPLE IX., page 109."
Page 14, line 5 from the bottom, for "formulæ" read "formula."
Page 20, line 13, for "$or = ·617$ OR" read "$or = ·7854$ OR."
Page 24, line 2, for "$\frac{2}{-}$" read "$\frac{2}{5}$".
Page 25, line 8, for "3·4478" read "·34478."
Page 29, line 15, for "$\frac{2}{-}\sqrt{2y}$" read "$\frac{2}{3}\sqrt{2y}$."
Page 36, in note, for "·637" read "·636."
Page 142, line 9, for "proposition" read "proportion."

# INTRODUCTION.

In preparing the following little work, we had three objects in view: first, a collection of useful hydraulic formulæ; secondly, a collection of experimental results, and coefficients; and, thirdly, a collection of useful practical tables, some calculated entirely from the formulæ and experiments, and others for the purpose of rendering the calculations more easy.

The TABLES at the end of the volume are all original, with the exception of TABLE I., which contains the well-known coefficients of PONCELET and LESBROS; but those are newly arranged, the heads reduced to English inches, and the coefficients for heads measured over and back from the orifice, placed side by side, for more ready comparison. The coefficients in the small Tables throughout the work have been all calculated by us from the original experiments; the formulæ have been carefully examined, and the continental ones reduced to English measures—some of them, as will be seen, for the first time. No labour has been spared in preparing the TABLES, and they are all purely hydraulic, though some of them are capable of being otherwise applied. We have filled no gap by the introduction of Tables applicable to other subjects, and in every-day use.

The correction of some of the experimental formulæ, particularly the continental ones, as printed in some English books, cost us some labour. Even Du Buât's well-known formula is frequently misprinted; and in a late hydraulic work, $\sqrt{d} - \cdot 1$, one of the factors, is printed $\sqrt{d - \cdot 1}$ in every page where it is quoted. It is not always that such mistakes can be avoided, but experimental formulæ are so often copied from one work into another without sufficient examination, that an error of this kind frequently becomes fixed; and when applied to practical purposes erroneous formulæ get the correct ones into disrepute. See note to formula (91), p. 113.

The TABLES of velocities and discharges over weirs and notches have been calculated for a great number of coefficients to meet different circumstances of approach and overfall, and for various heads from $\frac{1}{4}$th of an inch up to 6 feet. TABLE II. embodies the velocities acquired by falling bodies under the head of "theoretical velocity," and the velocities, suited to various coefficients, for heads up to 40 feet.

The formulæ for calculating the effects of the velocity of approach to orifices and weirs, and the necessary corrections for the ratio of the channel to the orifice, as well as TABLE V., we believe to be original. They will be found of much value in determining the proper coefficients suited to various ratios. The remarks throughout SECTION IV. are particularly applicable to the proper use of this TABLE.

TABLE VII. of surface and mean velocities will be found to vary from those generally in use, and to be much more correct, and better suited for practical purposes, particularly as applied to finding the mean velocities in rivers.

We have extended TABLE VIII. so as to make it directly available for hydraulic mean depths, from $\frac{1}{16}$th of an inch to 12 feet, and for various hydraulic inclinations, even up to vertical, for pipes. The fall in rivers seldom exceeds 2 or 3 feet per mile, or the velocity 5 or 6 feet per second. The extension of the table for great inclinations, and consequently great velocities, was made for purposes of calculation, and to include pipes. It must be understood throughout this TABLE that the velocities are those which continue unchanged for any length of channel, viz., when the resistance of friction is equal to the acceleration of gravity, the moving water and channel being then *in train*. Several of Du Buât's experiments were made with small vertical pipes. This TABLE is equally applicable to pipes and rivers, and gives directly either the hydraulic inclination, the hydraulic mean depth, or the velocity when any two of them are known.

## INTRODUCTION.

Hydraulic formulæ have been frequently rendered unnecessarily complex, and unsuited for practical application, by combining them with those of mere mensuration in order to find the discharge. We have therefore given formulæ for finding the mean velocity principally,—unless in a few instances, as in orifices near the surface, where the discharge itself is first necessary to find the mean velocity: this once determined, the calculation of the discharge becomes one of simple mensuration.

We have preferred giving the mean velocity to the discharge itself in TABLE VIII., because, while an infinite number of channels having the same hydraulic inclination ($s$) and the same hydraulic mean depth ($r$) must have the same velocity ($v$), yet the sectional areas, and consequently the discharges, may vary upwards from $6\cdot2832\,r^2$, the area of a semicircular channel, to any extent; and the operation of multiplying the area by the mean velocity, to find the discharge, is so very simple that any tabulation for that purpose is unnecessary. Besides this, the banks of rivers, unless artificially protected, remain very seldom at a constant slope, and therefore any TABLES of discharge for particular side slopes are only of use so far as they apply to hypothetical cases. Indeed we have seen, in new river cuts, the banks, cut first to a given slope, alter very considerably in a few months; while the necessary regimen between the velocity of the water and the channel was in the course of being established. The velocity suited to the permanency of any proposed river channels, though too often entirely neglected, is the very first element to be considered.

For circular pipes, however, TABLE IX. gives the discharge in cubic feet per minute when the velocity in inches per second is known, or found from TABLE VIII., and is calculated for pipes from $\frac{1}{4}$th of an inch up to 12 inches in diameter. TABLE XII. gives also the discharges in cubic feet per minute from the different equivalent river channels in TABLE XI.

TABLE X., for finding the heads on weirs of different lengths, TABLE XI., of equal discharging river channels, and TABLE XII., of the actual discharges from the equivalents in

TABLE XI., will be found of great practical value when new weirs and water-cuts have to be made. TABLES XI. and XII. are equally applicable to channels having side slopes, the widths being then the mean or central widths : see pages 124, 125, and 126.

When the discharge and fall are known, and the hydraulic mean depth and the dimensions of *any* channel have to be determined, Problem III., page 122, as illustrated in EXAMPLE 17, page 13, gives a new and perhaps the most practically useful solution yet published. Tables XI., XII., and XIII. are particularly applicable to this problem.

A uniform notation is preserved throughout the work, so that the different experimental formulæ can be compared without any further reduction. The letter $h$ is used in every instance for the head, $c$ for the coefficient, $r$ for the mean radius or hydraulic mean depth, and $s$ for the sine of the hydraulic inclination, unless it be otherwise stated. In order to designate particular values, the primary letters have *deponent or initial* letters below to explain them. Thus $h_t$ is the head to the *t*op of an orifice, $h_b$ the head at the *b*ottom, $h_w$ the head on a *w*eir, $h_f$ the head due to *f*riction, $c_d$ the coefficient of *d*ischarge, $c_v$ the coefficient of *v*elocity, $c_c$ the coefficient of *c*ontraction, &c. When the whole head is made up of different elements, such as the portions due to friction, velocity, contractions, bends, &c., it is expressed by the capital letter H.

Some writers and engineers appear to confound the inclination of a pipe, simply so called, or the head divided by the length, with the hydraulic inclination; and consequently have fallen into error in applying such of the known formulæ as take into consideration only the head due to the resistance of friction. When pipes are of considerable length, and the water is supplied from a reservoir at one end, the inclination, found as above, and the hydraulic inclination, may be taken equal to each other without sensible error; but for shorter pipes, of say up to 100 feet long, or even longer, the greater number of formulæ, as Du Buât's and others, do not directly apply; and it is necessary to take into consideration the head due to the orifice of entry, the velocity in the tube, and also to the impulse of supply when there are junctions. These separate elements, and their effects, will be considered in the following pages; but it will be of use to refer here to some late experiments, and the imperfect application of formulæ to them, first premising that a *pipe may be horizontal, or even turn upwards, and yet have a considerable hydraulic inclination.*

Mr. Provis's valuable experiments* with $1\frac{1}{2}$-inch pipes, from 20 to 100 feet long, have been used in a recent work for the purpose of testing the accuracy of Du Buât's and some other formulæ; but the head divided by the length is assumed to be the hydraulic inclination throughout, and no allowance is made for the head due to the orifice of entry and velocity in the pipe. Of course the writer's conclusions are erroneous. We have shown, SECTION I., page 12, how nearly the formulæ and experiments agree.

The formulæ appear to have been also misunderstood by the surveyor who experimented for the General Board of Health; for the inclination of the pipe in itself is assumed to be the hydraulic inclination, and no allowance is made for the head due to the impulse of supply. We quote from the CIVIL ENGINEER AND ARCHITECT'S JOURNAL, Vol. XV., page 366, in which it is stated that "the chief results as respect the house drains are thus described in the examination of the surveyor appointed to make the trials."†

"What quantity of water would be discharged through a 3-inch pipe on an inclination of 1 in 120?—Full at the head, it would discharge 100 gallons in three minutes, the pipe being 50 feet in length. This is with stone-ware pipe manufactured at Lambeth. This applies to a pipe receiving water only at the inlet, the water not being higher than the head of the pipe.

"What water was this?—Sewage-water of the full consistency, and it was discharged so completely that the pipe was perfectly clean.

---

\* Transactions of the Institution of Civil Engineers, Vol. II., pp. 201—210.

† Minutes of Information with reference to Works for the removal of Soil, Water, or Drainage, &c., &c. Presented to both Houses of Parliament, 1852.

"At the same inclination what would a 4-inch pipe discharge with the same distances?—Twice the amount (that I found from experiment); or, in other words, 100 gallons would be discharged in half the time. This likewise applies to a pipe receiving water only at the inlet, and of not greater height than the head. In these cases the section of the stream is diminished at the outlet to about half the area of the pipe.

"Before these experiments were made, were there not various hypothetical formulæ* proposed for general use?—Yes.

"What would these formulæ have given with a 3-inch pipe, and at an inclination of 1 in 100? and what was the result of your experiments with the 3-inch pipe?—The formulæ would give 7 cubic feet, the actual experiment gave 11¼ cubic feet; converting it into time, the discharge, according to the formulæ, compared with the discharge found by actual practice, would be as 2 to 3.

"How would it be with a 4-inch pipe?—The formulæ would give about 14·7 cubic feet per minute, whereas practice gave 23 cubic feet per minute.

"Take the case of a 6-inch pipe of the same inclination?—The results, according to Mr. Hawkesley's formula, would be 40¼ cubic feet per minute; from experiment it was found to be 63½ cubic feet per minute.

"Then with respect to mains and drainage over a flat surface, the result of course becomes of much more value, as the difference proved by actual practice increases with the diminution of the inclination?—Certainly, to a very great extent. For example, the tables give only 14·2 cubic feet per minute as the discharge from a pipe 6 inches diameter, with a fall of 1 in 800; practice shows that, under the same conditions, 47·2 cubic feet will be discharged.

"Will you give an example of the practical value of this when it is required to carry out drainage works over a very flat surface?—An inclination of 1 in 800 gives only 14 cubic feet per minute, according to theory, while, according to actual experiment, and with the same inclination, 47 cubic feet are given.

"Then this difference may be converted either into a saving of water to effect the same object, or into power of water to remove feculent matter from beneath the site of any houses or town?—It may be so.

"And also the power of small inclinations properly managed?—Yes;

---

* It is a mistake to call those formulæ hypothetical, unless so far as the hypothesis is founded on facts. Every formula with which we are acquainted is founded on experiments, and has been deduced from them, but those formulæ are too often hypothetically applied to short tubes without the necessary corrections. It will be seen from SECTION VIII. that the experiments from which the formulæ there given were derived, were in every way greatly more extensive than those made by the directions of the Board of Health.

for example, if it was required to construct a water course that should discharge, say 200 feet per minute, the formula would require an inclination of 1 in 60 = 2 inches in 10 feet; whereas, experiment has shown that the same would be discharged at an inclination of 1 in 200 = ⅔ inch in 10 feet, thus effecting a considerable saving in excavation, or a smaller drain would suffice at the greater inclination."

We have extracted and tabulated the results given above, in the following Table, and also eight of the experiments made for the Metropolitan Commissioners of Sewers*; and assuming for the present, with the surveyor examined by the Commissioners, that the inclinations of the pipes and hydraulic inclinations of the formulæ are the same, *which is incorrect*, we give the calculated discharges, found by means of TABLES VIII. and IX., in the last column of the Table.

| Diameter of pipe in inches. | Inclination of pipe. | Discharge in cubic feet per minute by experiment. | Hypothetical discharge by Du Buât's formula. |
|---|---|---|---|
| 3 | 1 in 120 | 5·3 | 6·6 |
| 4 | 1 in 120 | 10·7 | 14 |
| 3 | 1 in 100 | 11·2 | 7·5 |
| 4 | 1 in 100 | 23 | 15·6 |
| 6 | 1 in 100 | 63·5 | 43·8 |
| 6 | 1 in 800 | 47·2 | 13·3 |
| 6 | 1 in 60 | 75 | 59·3 |
| 6 | 1 in 100 | 63 | 43 8 |
| 6 | 1 in 160 | 54 | 33·4 |
| 6 | 1 in 200 | 52 | 29·2 |
| 6 | 1 in 320 | 49 | 21·8 |
| 6 | 1 in 400 | 48·5 | 19·6 |
| 6 | 1 in 800 | 47·2 | 13·3 |
| 6 | Level | 46 | 00 |

Du Buât's formula, therefore, gives larger results than the experiments in the two first cases, because the water received at one end only barely filled it, and the pipe was not full at

---

* Adcock's Engineer's Pocket Book, 1852, pp. 261 and 262.

xiv                           INTRODUCTION.

the lower end; but less in the others. If in these the head due to the impulse of entry, at the upper end, and at the side junctions, were known, and the proper hydraulic inclination determined by the experiments, the formulæ would be found to give larger approximate results in every case, as might have been expected from the sewage-water used. In the last eight experiments it is stated*, that "the water was admitted at the head of the pipe, and at *five junctions or tributary pipes* on each side, so regulated as to keep the main pipe full," and that "without the addition of junctions the transverse sectional area of the stream of water near the discharging end was reduced to one-fifth of the corresponding area of the pipe, *and that it required a simple head of water of about 22 inches to give the same result as that accruing under the circumstances of the junctions.*" It is also stated, that "in the case of the 6-inch pipe, which discharged 75 cubic feet per minute, the lateral streams had a velocity of a few feet per minute."

Now, the head of "about 22 inches" is wholly neglected in the foregoing calculations, *though in a pipe* 100 *feet long it would be equal to an inclination of* 1 *in* 55! It however includes three elements at least, viz. the portion due to the orifice of entry, the portion due to the velocity in the pipe, and the portion due to friction. Let us assume the case of *the horizontal pipe, which discharged* 46 *cubic feet per minute*†. This is equal to a mean velocity of 46·9 inches per second; with this velocity, we find from TABLE VIII. the hydraulic inclination of a 6-inch pipe to be 1 in 94, and, therefore, the head due to friction in a pipe 100 feet long is 12·7 inches. Assuming the coefficient for the orifice of entry and velocity to be ·815, we also find from TABLE II. a head of $4\frac{1}{4}$ inches due to these. We then have,

---

\* Adcock's Engineer's Pocket Book, 1852, pp. 261 and 262.

† The horizontal pipe would discharge equally at both ends, unless there was a head of water at either, or an equivalent in the velocity of approach. Of course, a smaller pipe with a fall, must be better than the larger one with none at all, in preventing deposits.

## INTRODUCTION.

> Head due to the velocity and orifice of entry    4·25 inches
> Head due to the resistance of friction    .   . 12·70   „
> Radius of pipe   .   .   .   .   .   .   . 3·00   „
>
>                                        Total   .   .   . 19·95

which is about 2 inches less than the observed head: this, however, is not stated definitely. *It is therefore evident, that the formula gives, if anything, larger results than these experiments\*, as might have been expected, instead of less in the ratio of 2 to 3, as is stated in the Report.*

Wherever junctions are applied, as in the examples above referred to, the formulæ in general use require correction; for the quantity of water then flowing below each junction is increased. A certain amount of error is, perhaps, inseparable from every calculation of this kind; but before we condemn formulæ deduced from experiment by men every way qualified for the task, it would be well that we should learn to understand and properly apply them.

The diameter of a short pipe gives in itself the means of increasing very considerably the surface inclination of the fluid stream, by reducing the section at the lower end. If we assume a horizontal pipe 50 feet long and 6 inches in diameter, we perceive, that if the receiving end be full, and the discharging end one-third full, this inclination will be $\dfrac{6-2}{50 \times 12} = \dfrac{1}{150}$; and that the discharging end cannot be kept full unless a head of several inches be maintained at the receiving end, or an equivalent from a lateral supply. When the pipe is about two diameters long it becomes a short tube; and when the length vanishes, the transverse section becomes, simply, a discharging orifice.

We have been led into the foregoing remarks, not from any desire to find fault with a Report containing so much valuable information as the one referred to, but for the pur-

---

\* This is also true of the other formulæ, for finding the discharge from pipes, given in this work.

pose of defending from unmerited reproach, in a *Blue Book*, the researches in this department, of

> "Those dead but sceptred sovereigns who still rule
> Our science from their urns—"

Du Buât, Young, Eytelwein, Prony, and others.

We do not pretend to any particular accuracy in the sketches scattered throughout the work; they are only intended to illustrate the text, and were sketched while writing it, without further aim; neither do we pretend to have entered fully into the principles or practice of hydraulics, our object being to select, construct, and arrange useful hydraulic formulæ, experiments, and Tables for the use of all classes of engineers. We make, however, no apology for preferring formulæ, in their simplicity, to any written rules which may be deduced from them, as being in every way more general, concise, and elegant. In conclusion, it is hoped that any errors of consequence in the work, will be found corrected in the errata.

*Roden Place, Dundalk,*
*January,* 1853.

ON THE

# DISCHARGE OF WATER

FROM

# ORIFICES, WEIRS, PIPES, AND RIVERS.

## SECTION I.

APPLICATION AND USE OF THE TABLES, FORMULÆ, &c.

*To find the velocity of a falling body from the height fallen, or the height fallen from the velocity.*

RULE.—MULTIPLY THE SQUARE ROOT OF THE HEIGHT IN INCHES BY 27·8, AND THE PRODUCT WILL BE THE VELOCITY IN INCHES. TO FIND THE HEIGHT FROM THE VELOCITY, SQUARE THE VELOCITY IN INCHES AND DIVIDE THE SQUARE BY 772·84, THE QUOTIENT WILL BE THE HEIGHT IN INCHES. See equation (1). TABLE II., column 1, will give the velocity from the height, found in the column of "altitudes," or the height from the velocity, directly.

EXAMPLE 1.—*What is the velocity acquired by a heavy body falling $\frac{1}{8}$th of an inch?* In the Table opposite to $\frac{1}{8}$th of an inch, found in the column headed "altitudes $h$," we find 9·829 in column 1, for the required velocity in inches per second.

EXAMPLE 2.—*What is the velocity acquired by a fall of 11 feet 3 inches?* Opposite to 11 feet 3 inches, as before, we find 323·007 inches, for the velocity required.

EXAMPLE 3.—*What height must a heavy body fall from to acquire a velocity of* $40\frac{1}{2}$ *feet per second?* Here $40\frac{1}{2}$ feet equal 486 inches, opposite to the nearest number to which, found in column 1, we find 25 feet 6 inches for the required fall. In this example, the nearest number to 486

B

found in the Table is 486·301. The difference ·301 corresponds, very nearly, to ⅜ths of an inch in altitude, and, therefore, the true head according to the rule would be 25′ 6⅜″; but for all practical purposes the difference is immaterial.

By means of TABLE II. we can find directly, or by simple interpolation, the velocity due to all heights from $\frac{1}{100}$ part of an inch up to 40 feet, and the heights from the velocities. For a greater height than 40 feet it may be divided by 4, 9, or some square number $s^2$, and the velocity found for the quotient, from the Table, multiplied by 2, 3, or $s$, the square root of the divisor, will give the velocity required.

EXAMPLE 4.—*What is the velocity acquired by a fall of 45 feet?* $\frac{45}{4} = 11′\ 3″$, the velocity corresponding to which, found from the Table, is 323·007. Hence,

323·007 × $\sqrt{4}$ = 323·007 × 2 = 646″·014 = 53′ 10″·014

is the velocity per second required. The reverse of this example is equally simple.

Columns 3, 4, 5, 6, 7, 8, 9, 10, 11, and 12 in the Table, give the values of $\sqrt{2gh}$ multiplied by the coefficients therein stated. These columns will be found of great practical use in finding the mean velocities in the *vena-contracta*, in the orifice, and in short tubes; and consequently also in finding the mechanical force and discharge. An examination of the coefficients in the small Tables in SECTION III., and also those in TABLES I. and V., at the end of the work, will show how much they vary; but those most generally useful and their products by the theoretical velocity due to different heads, up to 40 feet, are given in the columns referred to.

EXAMPLE 5.—*What is the discharge from an orifice 4 inches by 8 inches, the centre sunk 20 feet below the surface of a reservoir?* From TABLE II., we find 430·676 inches equal 35·89 feet for the theoretical velocity of discharge: hence, $\frac{8 \times 4}{144} \times 35·89 = \frac{2}{9} \times 35·89 = 7·976$ cubic feet per second is the theoretical discharge. If the discharge takes place through a thin plate, or if the inner arrisses next the water in the reservoir be *perfectly square*, and the water in flowing out does not fill the passage so as to convert the orifice into a short tube, the coefficient will be found from TABLE I. to be ·603. The true discharge then is 7·976 × ·603 = 4·809 cubic feet per second.

For the determination of the coefficient suited to any particular orifice, and the circumstances of its position, we must refer generally to the following pages. If in the example just given, the arrisses next the reservoir were rounded into the form of the contracted vein, see Fig. 4, the coefficient would increase from ·603 to ·974 or ·956, for a passage not exceeding a couple of feet in length. With the former the discharge would be 7·976 × ·974 = 7·769 cubic feet, and with the latter 7·976 × ·956 = 7·625 cubic feet. We may find from TABLE II. the latter results otherwise. With a head of 20 feet and the coefficient ·974, the velocity is 419·48 inches = 34·957 feet; hence, the discharge is $\frac{2}{9}$ × 34·957 = 7·768 cubic feet. With a coefficient of ·956, the velocity is 411·73 inches = 34·31 feet, and $\frac{2}{9}$ × 34·31 = 7·624 cubic feet. These results are the same, practically, as those previously found.

If the inner arrisses be square, and the passage out be from 18 inches to 2 feet long, the orifice will be converted into a short tube, the coefficient for which is ·815. With this coefficient, and a head of 20 feet, we find as before, from TABLE II., the mean velocity of discharge equals 351 inches = 29·25 feet; hence, the discharge now is $\frac{2}{9}$ × 29·25 = 6·5 cubic feet per second.

*The velocities in inches per second, given in* TABLES II. *and* VIII., *or elsewhere in the following pages, may be converted into velocities in feet per minute, by multiplying by 5, equal* $\frac{60}{12}$.

EXAMPLE 6.—*The discharge from a small orifice having its centre placed 10 feet below the surface of a reservoir is 18 feet per minute, what will be the discharge from the same orifice at a depth of 17 feet?* The discharges will be to each other as $\sqrt{10} : \sqrt{17}$, or as $1 : \sqrt{1\cdot 7}$; or, from TABLE III., as $1 : 1\cdot 3038$, whence we get the discharge sought equal $1\cdot 3038 \times 18 = 23\cdot 4684$ cubic feet.

EXAMPLE 7.—*What is the value of the expression* $c_d \left\{ 1 + \dfrac{c^2_d}{m^2 - c^2_d} \right\}^{\frac{1}{2}}$ *in equation* (45), *when* $c_d = \cdot 617$, *and* $m = 2$? Here we have—
$$\frac{c^2_d}{m^2 - c^2_d} = \frac{\cdot 617^2}{4 - \cdot 617^2} = \frac{3807}{3\cdot 6193} = \cdot 1052;$$
whence the expression becomes equal to $\cdot 617\,(1\cdot 1052)^{\frac{1}{2}}$ equal, from TABLE III., $\cdot 617 \times 1\cdot 0513 = \cdot 649$, the value sought. TABLE V. contains the values of this expression for various values of $c_d$ and $m$, which latter, $m$, stands for the ratio of the channel to an orifice; and we can immediately find from it, opposite 2 in the first column, and under the coefficient $\cdot 617$ in the sixth column, $\cdot 649$, the value sought. When the head due to the pressure, and the velocity of approach, are both known, we can determine the new coefficient of discharge by the above expression, and thence the discharge itself. The coefficient suited to the velocity of approach is, however, to be found directly in TABLE V.

EXAMPLE 8.—*What is the discharge from an orifice 17 inches long and 9 inches deep, having the upper edge placed 4 inches below the surface, and the lower edge 13 inches?* The expression for the discharge is
$$\tfrac{2}{3} \times \text{A} \sqrt{2gd} \times c_d \left\{ \left(1 + \frac{h_t}{d}\right)^{\frac{3}{2}} - \left(\frac{h_t}{d}\right)^{\frac{3}{2}} \right\}$$
equation (43), in which we must take $d = 9$ inches; $h_t = 9$ inches; $\text{A} = 17 \times 9 = 153$ square inches; and $\sqrt{2gd}$, found from TABLE II. $= 83\cdot 4$ inches. We have, also, $\dfrac{h_t}{d} = \dfrac{4}{9} = \cdot 444$, and hence the value of
$$(1\cdot 444)^{\frac{3}{2}} - (\cdot 444)^{\frac{3}{2}} = \text{(from TABLE IV.) } 1\cdot 44.$$

Assuming the coefficient of discharge to be ·617, we then have the discharge in cubic inches per second equal to

$$\frac{2}{3} \times 153 \times 83 \cdot 4 \times \cdot 617 \times 1 \cdot 44 =$$

$$\frac{2}{3} \times 12760 \cdot 2 \times \cdot 88848 = 7558.$$

Consequently, $\frac{7558}{1728} = 4 \cdot 374$ is the discharge in cubic feet per second. From equation (6), we get the discharge equal to

$$\frac{2}{3} \times \cdot 617 \times 27 \cdot 8 \times 17 \times \{13^{\frac{3}{2}} - 4^{\frac{3}{2}}\}.$$

But $13^{\frac{3}{2}} - 4^{\frac{3}{2}} = 46 \cdot 872 - 8$, from TABLE IV., equal to $38 \cdot 872$, whence the discharge is

$$\frac{2}{3} \times \cdot 617 \times 27 \cdot 8 \times 17 \times 38 \cdot 872 = 11 \cdot 4351 \times 17 \times 38 \cdot 872 = 194 \cdot 3967 \times 38 \cdot 872 = 7557 \text{ cubic inches} = 4 \cdot 374 \text{ cubic feet},$$

the same as before.

It is shown in equation (31), that by using the mean depth for orifices near the surface, the discharge will approximate very closely to the true discharge, and that even for weirs the error will not exceed 6 per cent. The discharge is then expressed by $\cdot 617 \sqrt{2g \times 8\frac{1}{2}} \times 9 \times 17 =$ (from TABLE II.) $50 \cdot 01 \times 153 = 7651 \cdot 53$ cubic inches $= 4 \cdot 427$ cubic feet per second. The head to the centre of the orifice is here $8\frac{1}{2}$ inches, and the depth of the orifice 9 inches, therefore, in equation (31), $h = d$ very nearly; and, therefore, this result must be multiplied by ·989, as shown in that equation; then $\cdot 989 \times 4 \cdot 427 = 4 \cdot 378$ cubic feet, which gives a result differing from those otherwise found, by a very small quantity, which, practically, is of no value. By means of TABLE VI. the discharge from rectangular orifices near the surface can be found with very great facility.

*We may always find the discharge from an orifice near the surface with sufficient accuracy by taking the head to the centre, as if it were sunk to a considerable depth, and* then applying the corrections given in equation (31); or if the orifice be circular, those given in equation (28).

EXAMPLE 9.—*What is the discharge from a circular orifice* 4 *inches in diameter, having its centre placed* 4 *inches below the surface, when the coefficient of discharge is* ·617? The area of the orifice is $4 \times 4 \times ·7854 = 12·566$ square inches. The velocity at the mean depth of 4 inches, with a coefficient of ·617, is 34·31 inches, whence the discharge is $12·566 \times 34·31 = 431·139$ cubic inches $= ·2496$ cubic feet per second, or 1·497 cubic feet per minute. By means of TABLE IX. the discharge in cubic feet per minute can be found very readily when the velocity, 34·31 inches per second, is known. Thus,

|  | Inches. |  | Cubic feet. |
|---|---|---|---|
| For a velocity of 30·00 | the discharge is | 1·3089 |
| ,, ,, 4·00 | ,, ,, | ·1745 |
| ,, ,, ·30 | ,, ,, | ·0131 |
| ,, ,, ·01 | ,, ,, | ·0004 |
| ,, ,, 34·31 | ,, ,, | 1·4969 |

By applying the coefficient found from equation (28), which is ·992, when the depth at the centre is twice the radius, as it is in this example, we get $·992 \times 1·497 = 1·485$ for the discharge in cubic feet per minute.

The application of TABLE VI. will enable us to find the discharge from rectangular orifices near the surface very quickly. Resuming "EXAMPLE 8," the discharge may be found from this Table for each foot in length of the orifice, as follows. The discharge in cubic feet per minute, when the coefficient is ·617 for a notch 1 foot long and 13 inches deep, is 223·323; and for a notch of 4 inches deep, 38·116; therefore, the discharge from an orifice 9 inches deep, with the upper edge 4 inches below the surface, is $223·323 - 38·116 = 185·207$ cubic feet per minute. But as the length of the orifice is 17 inches, this must be multiplied by $\frac{17}{12}$, and the product 262·377 is the discharge in cubic feet per minute; this is equal to a discharge of 4·373 cubic feet per second, which agrees with that before found. This is the simplest way of finding the discharge from rectangular orifices near the surface.

EXAMPLE 10.—*What is the discharge in cubic feet per minute, from an orifice 2 feet 6 inches long and 7 inches deep, the upper edge being 3 inches below the surface, and the coefficient of discharge ·628?* From TABLE VI. we find the discharge from a notch 1 foot long and 10 inches deep to be 153·353, and for a notch 3 inches deep, 25·199. The difference, or 128·154, multiplied by $2\frac{1}{2}$, will be the discharge required; viz. $2\frac{1}{2} \times 128\cdot154 = 320\cdot385$ cubic feet per minute.

EXAMPLE 11.—*The size of a channel is 2·75 times the size of an orifice, what is the coefficient of discharge when that for a very large channel in proportion to the orifice is ·628?* We find from TABLE V. the coefficient sought to be ·645, when the approaching water suffers full contraction. By attending to the auxiliary Tables in the text, we find, for this case, $\dfrac{\text{orifice}}{\text{channel}} = \dfrac{1}{2\cdot75} = \cdot36$. We must, therefore, multiply 2·75 by ·857, which gives 2·36 for the ratio of the mean velocities in the orifice and approaching it. With this new value of the ratio of the channel to the orifice, we find, as before, the value of the coefficient from TABLE V. to be ·651. The remarks throughout the work, with the auxiliary tables, will be found of much use in determining the coefficients for different ratios of the channel to the orifice, notch, or weir, and the corrections suited to each. If in this example we were considering, other things being the same, the alteration in the coefficient for a notch, or weir, it would be found from the Table, column 4, to be ·672 instead of ·645 found in column 3, for an orifice sunk some depth below the surface. For the corrections suited to mean and central velocity, and to the nature of the approaches, we must refer to the body of this work and the auxiliary tables.

EXAMPLE 12.—*What is the discharge over a weir 50 feet long; the circumstances of the overfall, crest, and approaches, being such that the coefficient of discharge is ·617,* when the head measured from the water in the weir basin, 6 feet above the crest, is 17½ inches? TABLE VI. will give the discharge in cubic feet per minute, over each foot in length of weir, for various depths up to 6 feet. It is divided into two parts; the first for "greater coefficients," viz. ·667 to ·617; and the second for "lesser coefficients," viz. ·606 to ·518. The coefficient assumed being ·617, we find the discharge over 1 foot in length, with a head of 17½ inches, to be 348·799 cubic feet per minute; hence the required discharge is 50 × 348·799 = 17439·95 cubic feet.

The determination of the coefficient suited to the circumstances of the overfall, crest, approaches, and approaching section, will be found discussed elsewhere through this work. The valuable Table derived from Mr. Blackwell's experiments will also be of use; but the heads being taken at a much greater distance back from the crest than is generally usual, the coefficients taken from it for heads greater than 5 or 6 inches, will be found under the true ones for heads measured immediately at or about 6 feet, above the crest. For heads measured *on* the crest, the small Table of coefficients in SECTION III., applicable to the purpose, will be of use.

EXAMPLE 13.—*What is the mean velocity in a large channel, when the maximum velocity along the central line of the surface is 31 inches per second?* TABLE VII. gives 25·89 inches for the required velocity, and for smaller channels 24·86 inches. In order to find the mean velocity at the surface from the maximum central velocity, the latter must be multiplied by ·914.

The velocity at the surface is best found by means of a floating hollow ball, which just rises out of the water. The velocity at a given depth is best found by means of two hollow balls connected with a link, the lower being made heavier than the upper, and both so weighted by the admission of a certain quantity of water that they shall float along the current, the upper one being in advance but nearly vertical over the other. The velocity of both will then be the velocity at half the depth between them. The velocity at the surface, found by means of a single ball, being also found, the velocity lost at the half depth is had by subtracting the common velocity due to the linked balls from that of the single ball at the surface. The velocity at any given depth is then easily found by a simple proportion; but the result will be most accurate when the given depth is nearly half the distance between the balls, which distance can never exceed the depth of the channel. *Pitot's tube, Woltmann's tachometer,* the *hydrometric pendulum,* the *rheometer,* and several other hydrometers, have been used for finding the velocity; but these instruments require certain corrections suited to each separate instrument, as well as kind of instrument, and are not so correct or simple for measuring the velocity in open channels as a ball and linked balls.

EXAMPLE 14.—*What is the discharge from a river having a surface inclination of* 18 *inches per mile, or* 1 *in* 3520, 40 *feet wide, with nearly vertical banks, and* 3 *feet deep?* The area is $40 \times 3 = 120$ feet, and the border $40 + 2 \times 3 = 46$ feet; therefore the hydraulic mean depth is $\frac{120}{46} = 2\cdot61$ feet $= 2$ feet $7\cdot3$ inches*. With this and the inclination we find from TABLE VIII. $28\cdot27 + 2\cdot75 \times \frac{1\cdot3}{6} = 28\cdot87$ inches per second $= 28\cdot87 \times 5 = 144\cdot35$ feet per minute for the mean velocity; hence we get $144\cdot35 \times 120 = 1732\cdot2$ cubic feet per minute for the required discharge. For channels with sloping banks we have only to divide the border, which is always known, into the area for the hydraulic mean depth, with which, and the surface inclination, we can always find the velocity by TABLE VIII., and thence the discharge. Unless the banks of rivers be protected by stone pavement or otherwise, the slopes will not continue permanent; it is therefore almost useless to give the discharges for channels of particular widths and side slopes. When the mean velocity is once known, the remaining calculations are

---

* For greater hydraulic depths than 144 inches, the extent of the TABLE, divide by 9, and find the corresponding velocity. This multiplied by 3 will be the velocity sought.

those of mere mensuration, and they should be made separately.

EXAMPLE 15.—*The diameter of a very long pipe is $1\frac{1}{2}$ inch, and the rate of inclination, or whole length of the pipe divided by the whole fall, is 1 in $71\frac{1}{2}$; what is the discharge in cubic feet per minute?* The hydraulic mean depth, or mean radius, is $\frac{1\cdot5}{4} = \cdot 375$ inches $= \frac{3}{8}$ inch. Consequently we find from TABLE VIII. the velocity in inches per second equal to $25\cdot09 - 1\cdot92 \times \frac{1\cdot5}{10} = 25\cdot09 - \cdot 29 = 24\cdot80$. The discharge in cubic feet per minute for a $1\frac{1}{2}$-inch pipe is now found most readily by means of TABLE IX., as follows:—

|  | Inches. |  | Cubic feet. |
|---|---|---|---|
| For a velocity of | 20·0 | the discharge is | 1·227 |
| ,, ,, | 4·0 | ,, ,, | ·245 |
| ,, ,, | ·8 | ,, ,, | ·049 |
| ,, ,, | 24·8 | ,, ,, | 1·521 |

Whence the discharge in cubic feet per minute is 1·521.

For short pipes, of 100 or 200 feet in length and under, the height due to the velocity and orifice of entry must be deducted from the whole height to find the proper hydraulic inclination, and also the height due to bends, curves, cocks, slides, and erogation. The neglect of these corrections has led some writers into mistakes in applying certain formulæ, and in testing them by experimental results obtained with short pipes. We shall now apply the TABLES to the determination of the discharge from short pipes, and compare the results with experiment, referring generally to equation (153) and the remarks preceding it for a correct and direct solution.

EXAMPLE 16.—*What is the discharge in cubic feet per minute from a pipe 100 feet long, with a fall or head of 35 inches at the lower end, when the diameter is $1\frac{1}{2}$ inch? Find also the discharge from pipes 80 feet, 60 feet, 40 feet, and 20 feet, of the same diameter and having the same head.* If the water be admitted by a stop-cock at the upper end, the coefficient due to the orifice of entry will probably be about ·75 or less, ·815 being that for a clear entry to a short cylindrical tube. The approximate inclination is $\frac{100 \times 12}{35} = 1$ in $34\cdot3$; but as a portion of the fall must be absorbed by the velocity and orifice of entry, we may assume for the present that the inclination is 1 in 35. With this inclination and the mean radius $1\frac{1}{2} = \frac{3}{8}$ inches, we find the mean velocity from TABLE VIII. to be 38·06 inches. Now when the coefficient due to the orifice of entry and velocity is ·75, we find from TABLE II. the head due to this velocity to be $3\frac{3}{8}$ inches nearly, whence $35 - 3\frac{3}{8} = 31\frac{5}{8} = 31\cdot625$ inches is the height due to friction, and $\frac{100 \times 12}{31\cdot625}$ equals 1 in 37·9, the inclination, very nearly. With this new inclination we find, as before, from TABLE VIII. the mean velocity of discharge to be now 36·35 inches; and by repeating the operation we shall find the velocity to any degree of accuracy in accordance with the table, and the shorter the pipe is, the oftener must it be repeated. The height due to 36·35 inches taken from TABLE II. as before, with a coefficient of ·750, is $3\frac{1}{8} = 3\cdot125$ inches. The corrected fall due to the friction is now $35 - 3\cdot125 = 31\cdot875$, and $\frac{1200}{31\cdot875}$ equal 1 in 37·6, the corrected inclination. With this inclination we find the corrected velocity to be now 36·53 inches per second. It is not necessary to repeat the operation again. The discharge determined from TABLE IX. is as follows:—

|  | Inches. |  | Cubic feet. |
|---|---|---|---|
| For a velocity of | 30·00 | the discharge is | 1·841 |
| ,, ,, | 6·00 | ,, ,, | ·368 |
| ,, ,, | ·50 | ,, ,, | ·031 |
| ,, ,, | ·03 | ,, ,, | ·002 |
| ,, ,, | 36·53 | ,, ,, | 2·242 |

The experimental discharge found by Mr. Provis was 2·264 cubic feet per minute in one experiment, and 2·285 in another. The discharge from the shorter pipes may be found in a similar manner, and we place the results alongside

the experimental ones given in the work referred to below*
in the following short table:—

EXPERIMENTAL AND CALCULATED DISCHARGES FROM SHORT PIPES.

| Length of pipe, in feet. | Head, in inches. | Observed discharge, in cubic feet. | Velocity per second. | Head due to the orifice and velocity. | Head due to friction. | Hydraulic inclinations. | Calculated velocity. | Calculated discharge. |
|---|---|---|---|---|---|---|---|---|
| 100 | 35 | 2·275 | 37·082 | 3⅛ | 31⅞ | 37·6 | 36·53 | 2·242 |
| 80 | 35 | 2·500 | 40·750 | 3¾ | 31¼ | 30·8 | 41·18 | 2·521 |
| 60 | 35 | 2·874 | 46·846 | 5 | 30 | 24·0 | 48·02 | 2·946 |
| 40 | 35 | 3·504 | 57·115 | 7½ | 27½ | 17·5 | 58·50 | 3·590 |
| 20 | 35 | 4·528 | 73·801 | 12½ | 22½ | 10·7 | 78·61 | 4·824 |

The velocities in the fourth column have been calculated by the writer from the observed quantities discharged, from which the height due to the orifice of entry and velocity in column 5 is determined, and thence the quantities in the other columns as above shown. The differences between the experimental and calculated results are not large, and had we used a lesser coefficient than ·750 for the velocity, stop-cock, and orifice of entry, say ·715, the calculated results, and those in all of Mr. Provis's experiments in the work referred to, would be nearly identical†.

EXAMPLE 17.—*It is proposed to supply a reservoir near the town of Drogheda with water by a long pipe, having an inclination of 1 in 480, the daily supply to be 80,000 cubic feet; what must the diameter of the pipe be?* The discharge

* "Transactions of the Institution of Civil Engineers," vol. ii. p. 203. "Experiments on the Flow of Water through small Pipes." By W. A. Provis.

† In a late work, "Researches in Hydraulics," the author is led into a series of mistakes as to the accuracy of Du Buât's and several other formulæ, from neglecting to take into consideration the head due to the velocity and orifice of entry when testing them by the experiments above referred to.

per minute must be $\frac{80,000}{1440} = 56$ cubic feet, nearly. Assume a pipe whose "mean radius" is 1 inch, or diameter 4 inches, and the velocity per second found from TABLE VIII. will be 14·41 inches. We then have from TABLE IX.,

|  |  | Inches. |  |  | Cubic feet. |
|---|---|---|---|---|---|
| For a velocity of | | 10·00 | a discharge of | | 4·363 |
| " | " | 4·00 | " | " | 1·745 |
| " | " | ·40 | " | " | ·175 |
| " | " | ·01 | " | " | ·004 |
| " | " | 14·41 | " | " | 6·287 |

The discharge from a pipe 4 inches in diameter would be therefore 6·287 cubic feet per minute. We then have

$4^{\frac{5}{2}} : d^{\frac{5}{2}} :: 6·287 : 56$, or $1 : d^{\frac{5}{2}} :: ·196 : 56 :: 1 : 286$;

therefore $d^{\frac{5}{2}} = 286$, and $d = 9·61$ inches, nearly, as found from TABLE IV., &c. This is the required diameter. It is to be observed that this diameter will not agree exactly with Du Buât's formula, because by it the discharges are not strictly as $d^{\frac{5}{2}}$, but in practice the difference is immaterial.

EXAMPLE 18.—*The area of a channel is 50 square feet, and the border 20·6 feet; the surface has an inclination of 4 inches in a mile; what is the mean velocity of discharge?*

$\frac{50}{20·6} = 2·427$ feet $= 29·124$ inches is the hydraulic mean depth; and we get from TABLE VIII. $12·03 - \frac{1·30 \times ·876}{6}$

$= 12·03 - ·19 = 11·84$ inches per second for the required velocity. Though this velocity will be found under the true value for straight clear channels, it will yet be more correct for ordinary river courses, with bends and turns, of the dimensions given, than the velocity found from equation (114). For a straight clear channel of these dimensions, Watt found the mean velocity to be 13·35 inches. See EXAMPLE 9, page 6.

EXAMPLE 19.—*A pipe 5 inches in diameter, 14637 feet in length, has a fall of 44 feet; what is the discharge in cubic feet per minute?* The inclination is $\frac{14637}{44} = 332\cdot7$, and mean radius $\frac{5}{4} = 1\frac{1}{4}$. We then find from TABLE VIII. the velocity equal to $19\cdot81 + \frac{\cdot41 \times 4\cdot8}{12\cdot5} = 19\cdot81 + \cdot16 = 19\cdot97$, or 20 inches per second very nearly; and by TABLE IX. the discharge in cubic feet per minute is, as before found to be, 13·635.

EXAMPLE 20.—*What is the velocity of discharge from a pipe or culvert 4 feet in diameter, having a fall of 1 foot to a mile?* Here $s = \frac{1}{5280}$, and $r = 1$ foot. We then find the velocity of discharge from TABLE VIII. to be 14·09 inches equal 1·174 feet per second. By calculating from the different formulæ referred to below, we shall find the velocities as follows:—

|  |  |  | Velocity in feet. |
|---|---|---|---|
| Reduction of Du Buât's formula . . . . . . . equation | (81) | 1·174 |
| „ Girard's do. (Canals with aquatic plants) | „ | (86) | ·521 |
| „ Prony's do. (Canals) . . . . . . . | „ | (88) | 1·201 |
| „ Prony's do. (Pipes) . . . . . . . | „ | (90) | 1·257 |
| „ Prony's do. (Pipes and Canals) . . . | „ | (92) | 1·229 |
| „ Eytelwein's do. (Rivers) . . . . . | „ | (94) | 1·200 |
| „ Eytelwein's do. (Rivers) . . . . . | „ | (96) | 1·285 |
| „ Eytelwein's do. (Pipes) . . . . . | „ | (98) | 1·364 |
| „ Eytelwein's do. (Pipes) . . . . . | „ | (99) | 1·350 |
| „ Dr. Young's do. . . . . . . . . | „ | (104) | 1·120 |
| „ D'Aùbuisson's do. (Pipes) . . . . . | „ | (109) | 1·259 |
| „ D'Aubuisson's do. (Rivers) . . . . | „ | (111) | 1·199 |
| „ The writer's do. (Clear straight Channels) | „ | (114) | 1·268 |
| „ Weïsbach's do. (Pipes) . . . . . . | „ | (119) | 1·285 |

We have calculated this example from the several formulæ above referred to, whether for pipes or rivers, in order that the results may be more readily compared. The formulæ from which the velocity and tables for the discharge in rivers are usually calculated is, for measures in feet, $v = 94\cdot17 \sqrt{rs}$. This gives the mean velocity, in the foregoing example, equal to 1·295 feet per second, which is certainly much too high if bends and curves be not allowed for separately. For all practical purposes the result of Du Buât's general formula, equation (81), should be adopted; and we have, accordingly, preferred calculating the results in TABLE VIII. from it, notwithstanding the greater additional labour required in the calculations, than from any other.

Dr. Young's formula gives less results for rivers and large pipes than Du Buât's, but they are too small unless where the curves and bends are numerous and sudden. Girard's formula (86) is only suited for canals containing aquatic plants, and must be increased by about ·7 foot per second to find the velocity in clear channels. A knowledge of various formulæ, and their comparative results, applied to any particular case, will be found of great value to the hydraulic engineer, and the differences in the results show the probable amount of error.

EXAMPLE 21.—*Water flowing down a river rises to a height of* $10\frac{1}{2}$ *inches on a weir 62 feet long; to what height will the same quantity of water rise, on a weir similarly circumstanced,* 120 *feet long?* $\frac{62}{120} = ·517$, nearly. In TABLE X. we find, by inspection, opposite to ·517, the ratio of the lengths, the coefficient ·644, rejecting the fourth place of decimals; whence $10\frac{1}{2} \times ·644 = 6·76$ inches, the height required. When the height is given in inches it is not necessary to take out the coefficient to further than two places of decimals.

EXAMPLE 22.—*The head on a weir* 220 *feet long is* 6 *inches; what will the head be on a weir* 60 *feet long, similarly circumstanced, the same quantity of water flowing over each?* $\frac{60}{220} = ·273$. As this lies between ·27 and ·28, we find from TABLE X. the coefficient ·4208; hence $\frac{6}{·4208} = 14·26$ inches, the head required.

TABLE X. will be found equally applicable in finding the head at the pass into weir basins and in contracted water channels. See SECTION X.

EXAMPLE 23.—*A river channel* 40 *feet wide and* 4·5 *feet deep is to be altered and widened to* 70 *feet; what must the depth of the new channel be so that the surface inclination and discharge shall remain unaltered?* In "TABLE XI., OF EQUAL DISCHARGING RECTANGULAR CHANNELS," we find opposite to 4·54, in the column of 40 feet widths, 3 in the column of 70 feet widths, which is the depth required.

EXAMPLE 24.—*It is necessary to unwater a river channel* 70 *feet wide and* 1 *foot deep, by a rectangular side cut* 10 *feet wide; what must the depth of the side cut be, the surface inclination remaining the same as in the old channel?* In TABLE XI. we find 4·5 feet for the required depth. When the width of a channel remains constant, the discharge varies as $\sqrt{rs} \times d$, in which $d$ is the depth; and when the width is very large compared with the depth, the hydraulic mean depth $r$ approximates very closely to the depth $d$, and $d = r$; consequently the discharge varies as $d^{\frac{3}{2}} \times s^{\frac{1}{2}}$, and when it is given $d^{\frac{3}{2}}$ must vary inversely as $s^{\frac{1}{2}}$; or more generally $d\,r^{\frac{1}{2}}$ must vary inversely, as $s^{\frac{1}{2}}$, when the width and discharge remain constant.

In narrow cuts for unwatering, it is prudent to make the depth of the water half the width of the cut very nearly, when local circumstances will admit of these proportions\*; for then a maximum effect will be obtained with the least possible quantity of excavation; but for rivers and permanent channels the proper relation of the depth to the width must be regulated by the principles referred to in SECTION IX.

TABLE XI. is equally applicable, whether the measures be in feet, yards, or any other standards.

\* Section IX.

## SECTION II.

FORMULÆ FOR THE VELOCITY AND DISCHARGE FROM ORIFICES, WEIRS, AND NOTCHES.—COEFFICIENTS OF VELOCITY, CONTRACTION, AND DISCHARGE.—PRACTICAL REMARKS ON THE USE OF THE FORMULÆ.

The quantity of water discharged in a given time through an aperture of a given area in the side or bottom of a vessel, is modified by different circumstances, and varies more or less with the form, position, and depth of the orifice; but the discharge may be easily found, when we have determined the velocity and the contraction of the fluid vein.

### VELOCITY.

If $g$ be the velocity acquired by a heavy body falling from a state of rest for one second, *in vacuo*, then it has been shown by writers on mechanics, that the velocity $v$ per second acquired by falling from a height $h$, will be

(1.) $$v = \sqrt{2gh}.$$

The numerical value of $g$ varies with the latitude; we shall assume $2g = 772.84$ inches $= 64.403$ feet. These will give for measures in inches,

$$v = 27.8 \sqrt{h},^* \text{ and } h = \frac{v^2}{772.84};$$

and for measures in feet,

$$v = 8.025 \sqrt{h}, \text{ and } h = \frac{v^2}{64.403}.\dagger$$

---

\* The velocities for different heights are given in the column number 1, TABLE II.

† The force of gravity increases with the latitude, and decreases with the altitude above the level of the sea, but not to any considerable extent. If $\lambda$ be the latitude, and $h$ the altitude, in feet, above the sea level, then we may, generally, take

$$g = 32.17 (1 - .0029 \cos 2\lambda) \times \left(1 - \frac{2h}{R}\right),$$

in which R, the radius of the earth at the given latitude is equal to
$$20887600 (1 + .0016 \cos 2\lambda).$$

## COEFFICIENT OF VELOCITY.

Let the vessel A B C D, Fig. 1, be filled with water to the level E F: then it has been found by experiment that the velocity of discharge through a small orifice o, in a thin plate, at the distance of half the diameter outside it, will be very nearly that due to a heavy body falling freely from the height $h$, of the surface of the water E F, above the centre

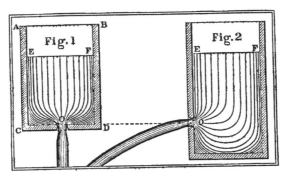

of the orifice. The velocity of discharge determined by the equation $v = \sqrt{2gh}$, for falling bodies, is, therefore, called the *theoretical velocity*. If we now put $v_d$ for the actual mean velocity of discharge, and $c_v$ for its ratio to the theoretical velocity $v$, we shall get $v_d = c_v v$; and by substituting for $v$, its value $\sqrt{2gh}$.

(2.) $$v_d = c_v \sqrt{2gh},$$

$c_v$ is termed *the coefficient of velocity;* its numerical value at half the diameter from the orifice is about ·974, and, consequently,

$$v_d = ·974 \sqrt{2gh}.$$

This for measures in inches becomes

$$v_d = 27·077 \sqrt{h},*$$

and for measures in feet

$$v_d = 7·816 \sqrt{h}.$$

\* The velocities for different heights calculated from this formula, are given in the column numbered 2, TABLE II.

ORIFICES, WEIRS, PIPES, AND RIVERS.         19

The orifice o, is termed an *horizontal orifice* in Fig. 1, and in Fig. 2 a *vertical* or *lateral orifice*. When small, each is found to have practically the same velocity of discharge, when the centres of the contracted sections are at the same depth, $h$, below the surface; but when lateral orifices are large, or rather deep, the velocity at the centre is not, even practically, the mean velocity; and in thick plates and modified forms of adjutage, the mean velocities are found to vary.

### VENA-CONTRACTA AND CONTRACTION.

It has been found that the diameter of a column issuing from a circular orifice in a thin plate, is contracted to very nearly eight-tenths of the whole diameter at the distance of the radius from it, and that at this distance the contraction is greatest. The ratio of the diameter of the orifice to that of the contracted vein, *vena-contracta*, is not always found constant by the same or different experimentalists.

| | | | | | | |
|---|---|---|---|---|---|---|
| Newton makes it | 1 : | ·841, | and, therefore, that of the areas as | 1 : | ·707 | |
| Poleni | „ | 1 : {·846 / ·788} | „ | „ | 1 : {·7156 / ·622} | |
| Borda | „ | 1 : ·802 | „ | „ | 1 : ·6432 | |
| Michellotti | „ | 1 : ·8 | „ | „ | 1 : ·64 | |
| Bossut | „ | 1 : {·81 / ·818} | „ | „ | 1 : {·656 / ·669} | |
| Du Buât | „ | 1 : ·816 | „ | „ | 1 : ·667 | |
| Venturi | „ | 1 : ·798 | „ | „ | 1 : ·637 | |
| Eytelwein | „ | 1 : ·8 | „ | „ | 1 : ·64 | |
| Bayer | „ | 1 : ·7854 | „ | „ | 1 : ·617 | |

Bayer's value for the contraction has been determined on the hypothesis, that the velocities of the particles of water as they approach the orifice from all sides, are inversely as the squares of their distances from its centre; and the calculations made of the discharge from circular, square, and rectangular orifices, on this hypothesis, coincide pretty closely with experiments.

### FORM OF THE CONTRACTED VEIN.

Let $o\,R = d$, Fig. 3, be the diameter of an orifice; then at the distance $R\,s = \dfrac{d}{2}$ the contraction is found to be greatest;

we shall assume the contracted diameter $or = \cdot7854\,d$. If we suppose the fluid column between OR and $or$ to be so reduced, that the curve lines R$r$ and O$o$ shall become arcs

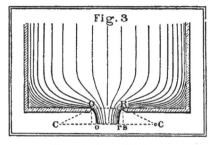

of circles, then it is easy to show from the properties of the circle, that the radius C$r$ must be equal to $1\cdot22\,d$. The mean velocity in the orifice, OR, is to that in the *vena-contracta*, $or$, as $\cdot617:1$; and the mouth piece, R$r$OO, Fig. 4, in which O$p = \frac{1}{2}$OR, and $or = \cdot617$ OR, will give for the velocity of discharge at $o, r$,

$$v_{\mathrm{d}} = \cdot974\,\sqrt{2gh} = 7\cdot816\,\sqrt{h},$$

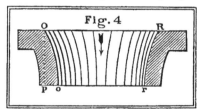

in feet very nearly. In speaking of the velocity of discharge from orifices in thin plates, we always take it to be the velocity in the *vena-contracta*, and not that in the orifice itself, unless in TABLE II., where the mean velocity in the latter, as representing $c_{\mathrm{d}}\,\sqrt{2gh}$, is also given.

### COEFFICIENTS OF CONTRACTION AND DISCHARGE.

If we put A for the area of the orifice OR, Fig. 3, and $c_{\mathrm{c}} \times$ A for that of the contracted section at $or$, then $c_{\mathrm{c}}$ is called the *coefficient of contraction*. The velocity of discharge $v_{\mathrm{d}}$ is equal to $c_{\mathrm{v}}\,\sqrt{2gh}$, equation (2). If we multiply this by the area of the contracted section $c_{\mathrm{c}} \times$ A, we shall get for the discharge

$$\mathrm{D} = c_{\mathrm{v}} \times c_{\mathrm{c}} \times \mathrm{A}\,\sqrt{2gh}.*$$

---

\* The expression $c_{\mathrm{v}}\,c_{\mathrm{c}}\,\sqrt{2gh} = c_{\mathrm{d}}\,\sqrt{2gh}$ is the coefficient of the area A, and, consequently, represents the mean velocity in the orifice, the coefficient of which is, therefore, equal to $c_{\mathrm{d}}$. The values of $c_{\mathrm{d}}\,\sqrt{2gh}$ for different heights and coefficients, are given in TABLE II.

It is evident $A\sqrt{2gh}$ would be the discharge if there were no contraction and no change of velocity due to the height $h$; $c_v \times c_c$ is therefore equal to the coefficient of discharge. If we call the latter $c_d$, we shall have the equation

(3.) $$c_d = c_v \times c_c,$$

and hence we perceive that the *coefficient of discharge is equal to the product of the coefficients of velocity and contraction.* In the foregoing expression for the discharge D, $h$ must be so taken, that the velocity at that depth shall be the mean velocity in the orifice A. *In full prismatic tubes the coefficients of velocity and discharge are equal to each other.*

### MEAN AND CENTRAL VELOCITY.

In order to find the mean velocity of discharge from an orifice, it is, in the first instance, necessary to determine the velocity due to each point in its surface, and the discharge itself; after which, the mean velocity is found by simply dividing the area of the orifice into the discharge. The velocity due to the height of water at the centre of a circular, square, or rectangular orifice, is not the mean velocity, nor is the latter in these, or other figures, that at the centre of gravity. When, however, an orifice is small in proportion to its depth in the water, the velocity of efflux determined for the centre approaches very closely to the mean velocity; and, indeed, at depths exceeding four times the depth of the orifice, the error in assuming the mean velocity to be that at the centre of the orifice is so small as to be of little or no practical consequence. It is, therefore, for greater simplicity, the practice to determine the velocity from the depth $h$ of the centre of the orifice; and the coefficients of discharge and velocity in the following pages have been calculated from experiments on this assumption, unless it be otherwise stated.

### DISCHARGES THROUGH ORIFICES OF DIFFERENT FORMS IN THIN PLATES.

The orifices which we have to deal with in practice are square, rectangular, or circular; and sometimes, perhaps, triangular or quadrangular in form. It will be necessary to give here the theoretical expressions for the discharge and velocity for each kind of form, but as the demonstrations are unsuited to our present purposes we shall omit them.

### TRAPEZOIDAL ORIFICES WITH TWO HORIZONTAL SIDES.

Put $d$ for the vertical depth of an orifice, $h_t$ for the altitude of pressure above the upper side, and $h_b$ for the altitude above the lower side, we then get

$$h_b - h_t = d.$$

Let us also represent the upper side of the orifice A or C, Fig. 5, by $l_t$, and the lower side by $l_b$, and put $\dfrac{l_t + l_b}{2} = l.$

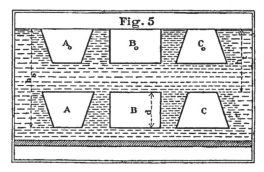

Now, when $l_t = l_b$, the trapezoid becomes a parallelogram whose length is $l$ and depth $d$; and putting $h$ for the depth to the centre of gravity, we get the equation

$$h_t + \frac{d}{2} = h_b - \frac{d}{2} = h.$$

The general expression for the discharge, D, through a trapezoidal orifice, A, is

(4.) $\quad \text{D} = c_d \sqrt{2g} \times \dfrac{2}{3} \left\{ l_b h_b^{\frac{3}{2}} - l_t h_t^{\frac{3}{2}} + \dfrac{2}{5}(l_t - l_b) \dfrac{h_b^{\frac{5}{2}} - h_t^{\frac{5}{2}}}{d} \right\},$

in which $c_d$ is the coefficient of discharge; and when the smaller side is uppermost as at C,

(5.) $\quad \text{D} = c_d \sqrt{2g} \times \dfrac{2}{3} \left\{ l_b h_b^{\frac{3}{2}} - l_t h_t^{\frac{3}{2}} - \dfrac{2}{5}(l_b - l_t) \dfrac{h_b^{\frac{5}{2}} - h_t^{\frac{5}{2}}}{d} \right\}.$

## PARALLELOGRAMIC AND RECTANGULAR ORIFICES.

When $l_t = l_b = l$, the orifice becomes a parallelogram, or a rectangle, B, and we have for the discharge

(6.) $\quad D = c_d \sqrt{2g} \times \frac{2}{3} l \{h_b^{\frac{3}{2}} - h_t^{\frac{3}{2}}\}.$

### NOTCHES.

When the upper sides of the orifices A, B, and C, rise to the surface as at $A_0$, $B_0$, and $C_0$, $h_t$ becomes nothing, and we get — as $h_b = d$ — : for the trapezoidal notch $A_0$ with the larger side up,

(7.) $\quad D = c_d \sqrt{2g} \times \frac{2}{3} d^{\frac{3}{2}} \left\{ l_b + \frac{2}{5}(l_t - l_b) \right\}$

$\qquad\qquad = \frac{2}{15} c_d \sqrt{2g} \, d^{\frac{3}{2}} (2 l_t + 3 l_b);$

for the trapezoidal notch, $C_0$, with the smaller side up,

(8.) $\quad D = c_d \sqrt{2g} \times \frac{2}{3} d^{\frac{3}{2}} \left\{ l_b - \frac{2}{5}(l_b - l_t) \right\}$

$\qquad\qquad = \frac{2}{15} c_d \sqrt{2g} \, d^{\frac{3}{2}} (2 l_t + 3 l_b);$

and for a parallelogramic or rectangular notch, $B_0$,

(9.) $\quad D = c_d \sqrt{2g} \times \frac{2}{3} l d^{\frac{3}{2}} = \frac{2}{3} c_d l d^{\frac{3}{2}} \sqrt{2g}.$

It is easy to perceive that the forms of equations (4) and (5), and also of equations (7) and (8), are identical. The values for the discharge in equations (6) and (9) are equally applicable, whether the form of the orifice be a parallelogram or a rectangle, the only difference being in the value of the coefficient of discharge, $c_d$, which becomes slightly modified for each form of orifice.

## TRIANGULAR ORIFICES WITH HORIZONTAL BASES AND RECTILINEAL ORIFICES IN GENERAL.

When the length of the lower side, $l_b = 0$, the orifice becomes a triangle, D, Fig. 6, with the base upwards.

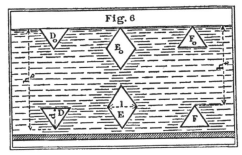

In this case, equation (4) becomes

(10.) $\quad \text{D} = c_\text{d} \sqrt{2g} \times \frac{2}{3} l_t \left\{ \frac{2}{-} \times \frac{h_b^{\frac{5}{2}} - h_t^{\frac{5}{2}}}{d} - h_t^{\frac{3}{2}} \right\};$

which gives the discharge through the triangular orifice, D.

When $l_t = 0$, in equation (5), the orifice becomes a triangle F, with the base downwards; in this case, we find for the value of the discharge,

(11.) $\quad \text{D} = c_\text{d} \sqrt{2g} \times \frac{2}{3} l_b \left\{ h_b^{\frac{3}{2}} - \frac{2}{5} \times \frac{h_b^{\frac{5}{2}} - h_t^{\frac{5}{2}}}{d} \right\}.$

As any triangular orifice whatever can be divided into two others by a line of division through one of the angles parallel to the horizon; and as the discharge from the triangular orifice D or F is the same as for any other on the same base and between the same parallels, we can easily find, by such a division, the discharge from any triangle not having one side parallel to the horizon.

If the triangle F be raised so that the base shall be on the same level with the upper side of the triangular orifice D; if, also, the bases be equal, and also the depths, we shall find, by adding equations (10) and (11), and making the necessary changes indicated by the diagram,

(12.) $\quad \text{D} = c_\text{d} \sqrt{2g} \times \frac{4}{15} \frac{l}{d} \left\{ h_b^{\frac{5}{2}} + h_t^{\frac{5}{2}} - 2 \times \overline{h_t + d}^{\frac{5}{2}} \right\}$

for the discharge from a parallelogram E with one diagonal horizontal. If the orifices D, F, and E rise to the surface of the water, as at $\text{D}_\text{o}$, $\text{E}_\text{o}$, and $\text{F}_\text{o}$, we shall then have for the discharge from the notch $\text{D}_\text{o}$,

(13.) $$D = c_d \sqrt{2g} \times \tfrac{4}{15} l_t d^{\frac{3}{2}}:$$

for the discharge from the notch $F_o$,

(14.) $$D = c_d \sqrt{2g} \times \tfrac{6}{15} l_b d^{\frac{3}{2}}:$$

and for the discharge through the notch $E_o$,

(15.) $$D = c_d \sqrt{2g} \times \tfrac{4}{15}\tfrac{l}{d}\{h_b^{\frac{5}{2}} - 2d^{\frac{5}{2}}\} = c_d \sqrt{2g} \times \cdot 9752\, l d^{\frac{3}{2}}.$$

When the parallelogram $E_o$ becomes a square $l = 2d$, and hence,

(16.) $$D = c_d \sqrt{2g} \times \cdot 9752\, l^{\frac{3}{2}} \times \sqrt{\tfrac{1}{8}} = c_d \sqrt{2gl} \times 3\cdot 4478\, l^2.$$

The foregoing equations will enable us to find an expression for the discharge from any rectilineal orifice whatever, as it can be divided into triangles, the discharge from each of which can be determined, as already shown in the remark following equation (11). The examples which we have given will be found to comprehend every form of rectilineal orifice which occurs in practice; but for the greater number of orifices, sunk to any depth below the surface, the discharge will be found with sufficient accuracy by multiplying the area by the velocity due to the centre.

## CIRCULAR AND SEMICIRCULAR ORIFICES.

The discharge through circular and semicircular orifices in thin plates can only be represented by means of infinite series. Let us represent by $s_1$ the sum of the series

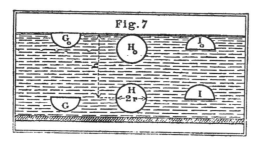

Fig. 7

$$\left\{\frac{1}{2} - \left(\frac{1}{2}\cdot\frac{1}{4}\right)\left(\frac{1}{2}\cdot\frac{1}{4}\right)\frac{r^2}{h^2} - \left(\frac{1}{2}\cdot\frac{1}{4}\cdot\frac{3}{6}\cdot\frac{5}{8}\right)\left(\frac{1}{2}\cdot\frac{1}{4}\cdot\frac{3}{6}\right)\frac{r^4}{h^4}\right.$$
$$\left. - \left(\frac{1}{2}\cdot\frac{1}{4}\cdot\frac{3}{6}\cdot\frac{5}{8}\cdot\frac{7}{10}\cdot\frac{9}{12}\right)\left(\frac{1}{2}\cdot\frac{1}{4}\cdot\frac{3}{6}\cdot\frac{5}{8}\right)\frac{r^6}{h^6} - \&c. \right\}:$$

Let us also represent by $s_2$ the sum of the series

$$\frac{2}{3\cdot 1416}\left\{\left(\frac{1}{2}\cdot\frac{1}{3}\right)\frac{r}{h} + \left(\frac{1}{2}\cdot\frac{1}{4}\cdot\frac{3}{6}\right)\left(\frac{1}{3}\cdot\frac{2}{5}\right)\frac{r^3}{h^3}\right.$$
$$\left. + \left(\frac{1}{2}\cdot\frac{1}{4}\cdot\frac{3}{6}\cdot\frac{5}{8}\cdot\frac{7}{10}\right)\left(\frac{1}{3}\cdot\frac{2}{5}\cdot\frac{4}{7}\right)\frac{r^5}{h^5} + \&c. \right\}:$$

then the discharge from the semicircle G, Fig. 7, with the diameter upwards and horizontal, is

(17.) $\quad \text{D} = c_\text{d}\sqrt{2gh} \times 3\cdot 1416\, r^2 (s_1 + s_2).$

And the discharge from the semicircle I, with the diameter downwards and horizontal, is

(18.) $\quad \text{D} = c_\text{d}\sqrt{2gh} \times 3\cdot 1416\, r^2 (s_1 - s_2).$

If we put A for the area, we shall also have for the discharge from a circle H,

(19.) $\quad \text{D} = c_\text{d}\sqrt{2gh} \times 2\text{A}\, s_1.$

In each of these three equations (17), (18), and (19), $h$ is the depth of the centre of the circumference below the surface, and $r$ the radius.

When the orifices rise to the surface, we have for the discharge from a semicircular notch $\text{G}_0$, with the diameter horizontal and at the surface,

(20.) $\quad \text{D} = c_\text{d}\sqrt{2gr} \times \cdot 9586\, r^2 = c_\text{d}\sqrt{2gr} \times \cdot 6103\,\text{A};$

when the circumference of the semicircle is at the surface, and the diameter horizontal, as at $\text{I}_0$,

(21.) $\quad \text{D} = c_\text{d}\sqrt{2gr} \times \frac{4}{15}(\sqrt{128} - 7)r^2 = c_\text{d}\sqrt{2gr} \times \cdot 7324\,\text{A};$

when the horizontal diameter of the semicircle is uppermost, and at the depth $r$ below the surface,

(22.) $\quad \text{D} = c_\text{d}\sqrt{2gr} \times 1\cdot 8667\, r^2 = c_\text{d}\sqrt{2gr} \times 1\cdot 1884\,\text{A};$

and when the circumference of the entire circle is at the surface, as at $H_o$,

(23.) $D = c_d \sqrt{2gr} \times 3.0171\, r^2 = c_d \sqrt{2gr} \times .9604\, A$.

If we desire to reduce equations (20), (21), and (22), to others in which the depth $h$ of the centre of gravity from the surface is contained, we have only to substitute $\dfrac{h}{.4244}$ for $r$ in equation (20), and we shall get, for the discharge from a semicircle with the diameter at the surface,

(24.) $\qquad D = c_d \sqrt{2gh} \times .9367\, A$ :

also, by substituting $\dfrac{h}{.5756}$ for $r$ in equation (21), we get, for the discharge from a semicircle when the circumference is at the surface and the diameter horizontal,

(25.) $\qquad D = c_d \sqrt{2gh} \times .9653\, A$ ;

and when the horizontal diameter is uppermost, and at the depth $r$ below the surface $r = \dfrac{h}{1.4244}$ and

(26.) $\qquad D = c_d \sqrt{2gh} \times .9957\, A$.

As A stands for the area of the particular orifice in each of the preceding expressions for the discharge, it must be taken of double the value—in equation (23), for instance, where it stands for the area of a circle—to what it is in equations (20), (21), or (23), where it represents only the area of a semicircle.

## MEAN VELOCITY.

The mean velocity is easily found by dividing the area into the discharge per second given in the preceding equations. For instance, the mean velocity in the example represented in equation (9) is equal $\frac{2}{3} c_d \sqrt{2gd}$, which is had by dividing the area $l\,d$ into the discharge; and in like manner the mean velocity in equation (23) is $.9604\, c_d \sqrt{2gr}$.

## THE DISCHARGE OF WATER FROM

### PRACTICAL REMARKS.—CIRCULAR ORIFICES.

It has been shown, equation (19), that, for the discharge from a circle, we have

$$\text{D} = c_\text{d} \sqrt{2gh} \times 2 \text{A} s_1,$$

in which $h$ is the depth of the centre, $\text{A}$ the area, and $s_1$ the sum of the series

$$\left\{ \frac{1}{2} - \left(\frac{1}{2}\cdot\frac{1}{4}\right)\left(\frac{1}{2}\cdot\frac{1}{4}\right)\frac{r^2}{h^2} - \left(\frac{1}{2}\cdot\frac{1}{4}\cdot\frac{3}{6}\cdot\frac{5}{8}\right)\left(\frac{1}{2}\cdot\frac{1}{4}\cdot\frac{3}{6}\right)\frac{r^4}{h^4} - \&\text{c.} \right\}:$$

and it has also been shown, equation (23), that, when the circumference touches the surface, this value becomes

$$\text{D} = c_\text{d} \sqrt{2gr} \times \cdot 9604 \text{ A}.$$

Now when $h$ is very large compared with $r$, it is easy to perceive that $2 s_1 = 1$, and hence

(27.) $$\text{D} = c_\text{d} \sqrt{2gh} \times \text{A}.$$

As this is the formula commonly used for finding the discharge, it is clear, if the coefficient $c_\text{d}$ remain constant, that the result obtained from it for $\text{D}$ would be too large. The differences, however, for depths greater than three times the diameter, or $6r$, are practically of no importance; for, by calculating the values of the discharge at different depths, we shall find, when

(28.) $$\begin{cases} h = r, & \text{that } \text{D} = c_\text{d} \sqrt{2gh} \times \cdot 960 \text{ A}; \\ h = \frac{5r}{4}, & \text{,, } \text{D} = c_\text{d} \sqrt{2gh} \times \cdot 978 \text{ A}; \\ h = \frac{3r}{2}, & \text{,, } \text{D} = c_\text{d} \sqrt{2gh} \times \cdot 985 \text{ A}; \\ h = \frac{7r}{4}, & \text{,, } \text{D} = c_\text{d} \sqrt{2gh} \times \cdot 989 \text{ A}; \\ h = 2r, & \text{,, } \text{D} = c_\text{d} \sqrt{2gh} \times \cdot 992 \text{ A}; \\ h = 3r, & \text{,, } \text{D} = c_\text{d} \sqrt{2gh} \times \cdot 996 \text{ A}; \\ h = 4r, & \text{,, } \text{D} = c_\text{d} \sqrt{2gh} \times \cdot 998 \text{ A}; \\ h = 5r, & \text{,, } \text{D} = c_\text{d} \sqrt{2gh} \times \cdot 9987 \text{ A}; \\ h = 6r, & \text{,, } \text{D} = c_\text{d} \sqrt{2gh} \times \cdot 9991 \text{ A}. \end{cases}$$

These results show very clearly that, for circular orifices, the common expression for the discharge $c_d \sqrt{2gh}$ A is abundantly correct for all depths exceeding three times the diameter, and that for lesser depths the extreme error cannot exceed four per cent. in reduction of the quantity found by this formula. We shall show, hereafter, when discussing the value of $c_d$, that from the sinking of the surface, and perhaps other causes, the discharge at lesser depths is even larger than that exhibited by the expression $c_d \sqrt{2gh}$ A, the value of $c_d$ being found to increase as the depths $h$ decrease.

### RECTANGULAR ORIFICES.

It has been shown, equation (6), that the discharge from rectangular orifices with two sides parallel to the surface is expressed by the equation

$$D = c_d \times \frac{2}{3} \sqrt{2g} \times l \{h_b^{\frac{3}{2}} - h_t^{\frac{3}{2}}\},$$

in which $l$ is the horizontal length of the orifice, $h_b$ the depth of water on the lower, and $h_t$ the depth on the upper, side. As it is desirable in practice to change this form into a more simple one, in which the height $h$ of the centre and depth $d$ of the orifice only are included, we then have $h_b = h + \frac{d}{2}$ and $h_t = h - \frac{d}{2}$. By substituting these values of $h_b$ and $h_t$ in the foregoing equations, and developing the result into a series, the terms of which, after the third, may be neglected, and putting A for the area $ld$, we shall find

(29.) $$D = c_d \sqrt{2gh} \times A \left\{1 - \frac{1}{96} \frac{d^2}{h^2}\right\} \text{very nearly.}$$

We have therefore for the accurate theoretical discharge

(30.) $$D = c_d \sqrt{2gh} \times \frac{2}{3} A \left\{\frac{(h + \frac{1}{2}d)^{\frac{3}{2}} - (h - \frac{1}{2}d)^{\frac{3}{2}}}{d h^{\frac{1}{2}}}\right\};$$

for the approximate discharge

$$\text{D} = c_\text{d} \sqrt{2gh} \times \text{A} \left\{ 1 - \frac{1}{96} \frac{d^2}{h^2} \right\};$$

and for the discharge by the common formula

$$\text{D} = c_\text{d} \sqrt{2gh} \times \text{A}.$$

When the head ($h$) is large compared with ($d$) the height of the orifice, each of the three last equations gives the same value for the discharge; but as the common expression $c_\text{d} \sqrt{2gh}$ A is the most simple; and as the greatest possible error in using it for lesser depths does not exceed six per cent., viz. when the orifice rises to the surface and becomes a notch, it is evidently that formula best suited for practical purposes. The following Table will show more clearly the differences in the results as obtained from the true, the approximate and the common formulas, applied to lesser heads.

(31.)

| 1 | 2 | 3 |
|---|---|---|
| $h = \frac{d}{2}$, | $\text{D} = c_\text{d}\sqrt{2gh} \times \cdot9428\,\text{A}.$ | $\text{D} = c_\text{d}\sqrt{2gh} \times \cdot9583\,\text{A}.$ |
| $h = \frac{5d}{8}$, | ,, ,, ,, $\times \cdot9693\,\text{A}.$ | ,, ,, ,, $\times \cdot9733\,\text{A}.$ |
| $h = \frac{3d}{4}$, | ,, ,, ,, $\times \cdot9796\,\text{A}.$ | ,, ,, ,, $\times \cdot9815\,\text{A}.$ |
| $h = \frac{7d}{8}$, | ,, ,, ,, $\times \cdot9854\,\text{A}.$ | ,, ,, ,, $\times \cdot9864\,\text{A}.$ |
| $h = d$, | ,, ,, ,, $\times \cdot9890\,\text{A}.$ | ,, ,, ,, $\times \cdot9896\,\text{A}.$ |
| $h = \frac{3d}{2}$, | ,, ,, ,, $\times \cdot9953\,\text{A}.$ | ,, ,, ,, $\times \cdot9954\,\text{A}.$ |
| $h = 2d$, | ,, ,, ,, $\times \cdot9974\,\text{A}.$ | ,, ,, ,, $\times \cdot9974\,\text{A}.$ |
| $h = \frac{5d}{2}$, | ,, ,, ,, $\times \cdot9983\,\text{A}.$ | ,, ,, ,, $\times \cdot9983\,\text{A}.$ |
| $h = 3d$, | ,, ,, ,, $\times \cdot9988\,\text{A}.$ | ,, ,, ,, $\times \cdot9988\,\text{A}.$ |
| $h = \frac{7d}{2}$, | ,, ,, ,, $\times \cdot9991\,\text{A}.$ | ,, ,, ,, $\times \cdot9991\,\text{A}.$ |
| $h = 4d$, | ,, ,, ,, $\times \cdot9994\,\text{A}.$ | ,, ,, ,, $\times \cdot9994\,\text{A}.$ |
| $h = 10d$, | ,, ,, ,, $\times \cdot9999\,\text{A}.$ | ,, ,, ,, $\times \cdot9999\,\text{A}.$ |

In the foregoing Table the first column contains the head at the centre of the orifice expressed in parts of its height $d$; the second contains the values of the discharges according to equation (30); and the third column contains the approximate values determined from equation (29), the results in which are something larger than those in column 2, derived from the correct formula. The numerical coefficients of A, at every depth, are less in both than unity, the constant coefficient according to the common formula. The latter, therefore (as in circular orifices), gives results exceeding the true ones, but the excess is inappreciable at greater depths than $h = 3\,d$, and for lesser depths than this the error cannot exceed six per cent. It may be useful to remark here, that when the orifice rises to the surface the centre of mean velocity is at four-ninths of the depth, and the centre of gravity at two-thirds of the depth from the surface. The former fraction is the square of the latter.

## SECTION III.

EXPERIMENTAL RESULTS.—COEFFICIENTS OF DISCHARGE.

We have heretofore dwelt but very partially on the numerical values of the general coefficient of discharge $c_d$. In order to determine its value under different circumstances more particularly, it will be now necessary to consider some of the experiments which have been made from time to time. These do not always give the same results, even when conducted under the same circumstances and by the same parties, and there appears to exist a certain amount of error, more or less, inseparable from the subject. The experiments with orifices in thin plates afford the most consistent results; but even here the differences are sometimes greater than might be expected. In many of the earlier experiments the value of the coefficient $c_d$ comes out too large, which arises, very probably, from the orifices experimented with not being in thin plates, and partaking, more or less, of the nature of a short tube or mouth-piece with rounded arrisses, which we shall see gives larger coefficients than simple orifices. When an orifice is in the bottom of a vessel, it would appear more correct to measure the head from the surface to the *vena-contracta* than from the surface to the orifice itself; and as any error in measuring the head in any experiment must affect the value of the coefficient derived from such experiment, so as to increase it when the error is to make the head less, and *vice versâ*, it appears that heads measured to an orifice in the bottom of a vessel, and not to the *vena-contracta*, must give larger coefficients from the experimental results than, perhaps, the true ones. The coefficients in the following pages have been almost all arranged and calculated, by the writer, from the original experiments.

In 1739 Dr. Bryan Robinson made some experiments on the discharge through small circular orifices, from one-tenth to eight-tenths of an inch in diameter, with heads of two and four feet*, which give the following coefficients.

COEFFICIENTS FROM DR. B. ROBINSON'S EXPERIMENTS.

| Heads. | $\frac{1}{10}$ inch diameter. | $\frac{4}{10}$ inch diameter. | $\frac{5}{10}$ inch diameter. | $\frac{8}{10}$ inch diameter. |
|---|---|---|---|---|
| 2 feet head | ·768 | ·767 | ·761 | ·728 |
| 4 feet head | ·768 | ·774 | ·765 | ·742 |

These results are pretty uniform, and the values from which they are derived are said to be " means taken from five or six experiments;" as values of $c_d$ they are, however, too high. The apparatus made use of is not described; but it is probable, from the results, that the plate containing the hole or orifice was of some thickness, and that the inner arriss was slightly rounded. There is here, however, a very perceptible increase in the coefficients for the smaller orifices, but none for the smaller depth.

* Helsham's Lectures, p. 390. Dublin, 1739.

In a paper in the Transactions of the Royal Irish Academy, vol. ii. p. 81, read March 1st, 1788, Dr. Mathew Young determines the value of the coefficient for an orifice $\frac{2}{10}$ inch in diameter, with a mean head of 14 inches, to be ·623. The manner in which this value is determined is very elegant, viz. by comparing the observed with the theoretical time of the water in the vessel sinking from 16 inches to 12 inches.

The following experiments by Michelotti, with circular orifices from 1 to about 3 inches diameter, and with from 6 to 23 feet heads, give for the mean value $c_d =$ ·613; and for square orifices of from 1 to 9 square inches in area, at like depths, the mean value of $c_d =$ ·628. The experiments are given in French feet and inches, according to which standard we have, in feet, $D = 7·77 \, A \sqrt{h} \times t$, $t$ being the time in seconds*. As the time of discharge in these experiments varies from ten minutes to an hour, and as the depths are considerable, the results must be looked upon as pretty accurate; and it is worthy of note that here the coefficients are larger for square than for circular orifices.

* The value of $\sqrt{2gh}$, equation (1), for measures in French feet, is 7·77 $\sqrt{h}$, and for measures in French inches, 26·9 $\sqrt{h}$, $g$ being equal to 30·2 feet, or 362·4 inches, French measure. One French foot is equal to 1·06578 English feet, and the inches preserve the same proportion. The resulting coefficients must be the same, whatever standards we make our calculations from. Many of the most valuable formulæ and experiments in hydraulics are given in French measures of the old style. As our object, however, in the present section is simply to determine from experiment the relation of the experimental to the theoretical discharge, it is not necessary to reduce the experiments to other measures than those in the original; but the value of the force of gravity, $g$, must of course be taken in those measures with which the experiments were made. In the French decimal, or modern style, the metre is equal to 3·2809 English feet, or 39·371 inches. The tenth part of a metre is the decimetre, and the tenth part of the decimetre is the centimetre, as the names imply.

THE DISCHARGE OF WATER FROM

COEFFICIENTS FROM MICHELOTTI'S EXPERIMENTS.

| Description and size of orifice, in French inches. | Depth of the centre of the orifice, in French feet. | Quantity discharged, in cubic feet. | Time of discharge, in seconds. | Theoretical time, calculated from $t = \dfrac{D}{7.77 \text{ A} \sqrt{h}}$. | Resulting coefficients of discharge. |
|---|---|---|---|---|---|
| Square orifice, 3″ × 3″ | 6·613 | 463·604 | 600 | 371·3 | ·619 |
|  | 6·852 | 566·458 | 720 | 445·6 | ·619 |
|  | 11·676 | 516·785 | 510 | 311·4 | ·610 |
|  | 11·818 | 612·118 | 600 | 366·6 | ·611 |
|  | 21·691 | 415·437 | 300 | 183·7 | ·612 |
|  | 21·715 | 499·222 | 360 | 220·6 | ·613 |
| Mean value of the coefficient; square orifice 3″ × 3″ ............... | | | | | ·614 |
| Square orifice, 2″ × 2″ | 6·625 | 329·806 | 900 | 594· | ·660 |
|  | 11·426 | 423·465 | 900 | 580·4 | ·645 |
|  | 21·442 | 385·333 | 600 | 385·7 | ·643 |
| Mean value of the coefficient; square orifice 2″ × 2″ ............... | | | | | ·649 |
| Square orifice, 1″ × 1″ | 6·757 | 158·549 | 1800 | 1585· | ·628 |
|  | 11·889 | 163·792 | 1440 | 880·6 | ·612 |
|  | 21·507 | 562·944 | 3600 | 2249·9 | ·625 |
| Mean value of the coefficient; square orifice 1″ × 1″ ............... | | | | | ·621 |
| Circular orifice, 3″ diameter | 6·694 | 542·85 | 900 | 550·1 | ·611 |
|  | 11·590 | 570·972 | 720 | 439·6 | ·610 |
|  | 21·611 | 521·299 | 480 | 293·8 | ·612 |
| Mean value of the coefficient; circular orifice 3″ diameter ......... | | | | | ·611 |
| Circular orifice, 2″ diameter | 6·785 | 488·687 | 1800 | 1108·1 | ·616 |
|  | 11·722 | 589·535 | 1680 | 1016·4 | ·605 |
|  | 21·903 | 575·486 | 1200 | 725·9 | ·605 |
| Mean value of the coefficient; circular orifice 2″ diameter ......... | | | | | ·609 |
| Circular orifice, 1″ diameter | 6·875 | 247·354 | 3600 | 2227· | ·619 |
|  | 11·743 | 324·11 | 3600 | 2233· | ·620 |
|  | 22·014 | 444·535 | 3600 | 2237·2 | ·621 |
| Mean value of the coefficient; circular orifice 1″ diameter ......... | | | | | ·620 |

The experiments made by the Abbé Bossut, contained in the following table, give the mean value of $c_d$, for both circular and square orifices, equal to ·616 nearly; and it may be perceived that, for the small depth in the last experiment, the coefficient rises so high as ·649. These and other experiments

COEFFICIENTS FROM BOSSUT'S EXPERIMENTS.

| Description, position, and size of orifice, in French inches. | Depth of the centre of the orifice, in French inches. | Number of French cubical inches discharged per minute. | Theoretical discharge per minute, $D = 1614 \, a \sqrt{h}$. | Resulting coefficients. |
|---|---|---|---|---|
| Horizontal and circular, ¼" diameter | 140·832 | 2311 | 3760·8 | ·614 |
| Horizontal and circular, 1" diameter | 140·832 | 9281 | 15043·3 | ·617 |
| Horizontal and circular, 2" diameter | 140·832 | 37203 | 60173·1 | ·618 |
| Horizontal and rectangular, 1" × ¼" | 140·832 | 2933 | 4788·4 | ·613 |
| Horizontal and square, 1" × 1" ... | 140·832 | 11817 | 19153·7 | ·617 |
| Horizontal and square, 2" × 2" ... | 140·832 | 47361 | 76614·6 | ·618 |
| Lateral and circular, ¼" diameter... | 108· | 2018 | 3293·3 | ·613 |
| Lateral and circular, 1" diameter... | 108· | 8135 | 13173·3 | ·617 |
| Lateral and circular, ¼" diameter... | 48· | 1353 | 2195·5 | ·616 |
| Lateral and circular, 1" diameter... | 48· | 5436 | 8782·2 | ·619 |
| Lateral and circular, 1" diameter... | 0·5833 | 628 | 968· | ·649 |

led the Abbé to construct a table of the discharges, at different depths, from a circular orifice 1 inch in diameter, from which we have determined the following table of coefficients.

COEFFICIENTS DEDUCED FROM BOSSUT'S EXPERIMENTS.

| Heads, in feet. | Coefficients. | Heads, in feet. | Coefficients. | Heads, in feet. | Coefficients. |
|---|---|---|---|---|---|
| 1 | ·621 | 6 | ·620 | 11 | ·619 |
| 2 | ·621 | 7 | ·620 | 12 | ·618 |
| 3 | ·621 | 8 | ·619 | 13 | ·618 |
| 4 | ·620 | 9 | ·619 | 14 | ·618 |
| 5 | ·620 | 10 | ·619 | 15 | ·617 |

These increase, as the orifice approaches the surface, from ·617 to ·621; and at lesser depths than 1 foot other experiments show an increase in the coefficient up to ·650. The experiments of Poncelet and Lesbros show, however, a reduction in the coefficients for square orifices 8″ × 8″ as they approach the surface from ·601 to ·572.

Brindley and Smeaton's experiments, with an orifice 1 inch square placed at different depths, give a mean value for $c_d$ of

COEFFICIENTS CALCULATED FROM BRINDLEY AND SMEATON'S EXPERIMENTS.

1 foot head : orifice 1″ × 1″ : coefficient ·639 ⎫
2 feet head : orifice 1″ × 1″ : coefficient ·635 ⎪
3 feet head : orifice 1″ × 1″ : coefficient ·648 ⎬ mean ·637.
4 feet head : orifice 1″ × 1″ : coefficient ·632 ⎪
5 feet head : orifice 1″ × 1″ : coefficient ·632 ⎭
6 feet head : orifice ½″ × ½″ : coefficient ·577

·637. The last experiment, with an orifice only ½ inch by ½ inch, gives so small a coefficient as ·557 placed at a depth of 6 feet!

For notches 6 inches wide and from 1 to 6½ inches deep, Brindley and Smeaton's experiments give the mean value of

COEFFICIENTS FOR NOTCHES, CALCULATED FROM BRINDLEY AND SMEATON'S EXPERIMENTS.

| Ratio of the length to the depth. | Size of notches in inches. | Coefficients. | Ratio of the length to the depth. | Size of notches in inches. | Coefficients. |
|---|---|---|---|---|---|
| ·92 to 1 | 6 × 6½ | ·633 | 3·7 to 1 | 6 × 1⅝ | ·638 |
| 1·07 to 1 | 6 × 5⅝ | ·571 | 4·4 to 1 | 6 × 1⅜ | ·654 |
| 1·2 to 1 | 6 × 5 | ·609 | 4·8 to 1 | 6 × 1¼ | ·681 |
| 1·92 to 1 | 6 × 3⅛ | ·602 | 6· to 1 | 6 × 1 | ·713 |
| 2·4 to 1 | 6 × 2⁵⁄₁₆* | ·636 | Mean value. | | ·637 |

$c_d = $ ·637. The coefficient of discharge for notches and orifices appear to differ as little from each other as those for either do in themselves. The results also show a general though not uniform increase in the coefficients for smaller depths.

Du Buât's experiments with notches 18·4 inches long,

* This depth is misprinted 2⅝ inches in the Encyclopædias, the resulting coefficient for which would be ·568 instead of ·637, as above, for a depth of 2⁵⁄₁₆ inches.

give the mean value of $c_d = \cdot 632$, which differs very little from the mean value determined from Brindley and Smeaton's experiments.

COEFFICIENTS FOR NOTCHES, CALCULATED FROM DU BUAT'S EXPERIMENTS.

| Ratio of the length to the depth. | Size of notches in inches. | Coefficients. | Ratio of the length to the depth. | Size of notches in inches. | Coefficients. |
|---|---|---|---|---|---|
| 2·72 to 1 | 18·4 × 6·753 | ·630 | 5·75 to 1 | 18·4 × 3·199 | ·624 |
| 3·94 to 1 | 18·4 × 4·665 | ·627 | 10·3 to 1 | 18·4 × 1·778 | ·648 |

Poncelet and Lesbros' experiments give the coefficients in the following table, for notches 8 inches wide; the mean

COEFFICIENTS FOR NOTCHES, BY PONCELET AND LESBROS.

| Ratio of the length to the depth. | Size of notches in inches. | Coefficients. | Ratio of the length to the depth. | Size of notches in inches. | Coefficients. |
|---|---|---|---|---|---|
| ·9 to 1 | 8 × 9 | ·577 | 3·33 to 1 | 8 × 2·4 | ·601 |
| 1 to 1 | 8 × 8 | ·585 | 5 to 1 | 8 × 1·6 | ·611 |
| 1·3 to 1 | 8 × 6 | ·590 | 6·7 to 1 | 8 × 1·2 | ·618 |
| 2 to 1 | 8 × 4 | ·592 | 10 to 1 | 8 × 0·8 | ·625 |
| 2·5 to 1 | 8 × 3·2 | ·595 | 20 to 1 | 8 × 0·4 | ·636 |

value of all the coefficient in these experiments is ·603. Here the coefficients increase in every instance as the depth decreases, or as the ratio of the length of the notch to its depth increases. We shall have to refer to the valuable experiments made at Metz, on the discharge from differently-proportioned orifices immediately.

Rennie's experiments for circular orifices at depths from

COEFFICIENTS FOR CIRCULAR ORIFICES, FROM RENNIE'S EXPERIMENTS.

| Heads at the centre of the orifice in feet. | ¼ inch diameter. | ½ inch diameter. | ¾ inch diameter. | 1 inch diameter. | Mean values. |
|---|---|---|---|---|---|
| 1 | ·671 | ·634 | ·644 | ·633 | ·645 |
| 2 | ·653 | ·621 | ·652 | ·619 | ·636 |
| 3 | ·660 | ·636 | ·632 | ·628 | ·639 |
| 4 | ·662 | ·626 | ·614 | ·584 | ·621 |
| Means | ·661 | ·629 | ·635 | ·616 | ·635 |

1 foot to 4 feet, and of diameters from ¼ inch to 1 inch, give the following coefficients. Here the increase in the coefficients for lesser orifices and at lesser depths exhibits itself very clearly, notwithstanding a few instances to the contrary. The mean value of the coefficient $c_d$ derived from the whole, is ·635. For small rectilineal orifices the coefficients were as follows:—

COEFFICIENTS FOR RECTANGULAR ORIFICES, FROM RENNIE'S EXPERIMENTS.

| Heads at the centre of gravity, in feet. | Square orifice, 1 inch × 1 inch. | Rectangular orifice, longer side horizontal, 2″ × ½″. | Rectangular orifice, longer side horizontal, 1¼″ × ⅜″. | Equilateral triangle of 1 square inch, with base down. | Equilateral triangle of 1 square inch, with base up. |
|---|---|---|---|---|---|
| 1 | ·617 | ·617 | ·663 | ,, | ·596 |
| 2 | ·635 | ·635 | ·668 | ,, | ·577 |
| 3 | ·606 | ·606 | ·606 | ,, | ·572 |
| 4 | ·593 | ·593 | ·593 | ·593 | ·593 |
| Means | ·613 | ·613 | ·632 | ·593 | ·585 |

The most valuable series of experiments of which we are possessed are those made at Metz, by Poncelet and Lesbros. These were made with orifices eight inches wide, nearly, and of different vertical dimensions placed at various depths down to 10 feet. The discrepancies as to any general law in the relation of the different values of the coefficient of discharge $c_d$ to the size and depth of the orifice in the preceding experiments, have been remedied to a great extent by these. They give an increase of the coefficients for the smaller and very oblong orifices as they approach the surface, and a decrease under the same circumstances in those for the larger square and oblong orifices. There are a few depths where maximum and minimum values are obtained: we use the terms "maximum and minimum values" for those which are greater in the one case and less in the other than the coefficients immediately before and after them, and not as being numerically the greatest or least values in the column. We have marked with a *, in the arrangement of the coefficients, TABLE I., these maximum and minimum values. The heads given in this table are measured to the upper side of the orifices, and by adding half the depth ($d$) to any particular head, we obtain the head at the centre.

As a perceptible sinking of the surface takes place in heads less than from five to three times the depth of the orifice, the coefficients are arranged in pairs, the first column containing the coefficients for heads measured from the still water surface some distance back from the orifice, and the second those obtained when the lesser heads, measured directly at the orifice, were used. A very considerable increase in the value of the coefficients for very oblong and shallow small orifices, may be perceived as they approach the surface, and the mean value for all rectilinear orifices at considerable depths, seems to approach to ·605 or ·606.

We have shown, equation (29), that the discharge is

$$D = c_d \times \left\{1 - \frac{d^2}{96 h^2}\right\} \times A \sqrt{2gh},$$

approximately, in which $d$ is the depth of the orifice, and $h$ the head at its centre. Now it is to be observed, that it is not the value of $c_d$ simply, which is given in TABLE I., but the value of $c_d \left\{1 - \frac{d^2}{96 h^2}\right\}$, the coefficient of $A \sqrt{2gh}$. The coefficients in the table are, therefore, less than the coefficient of discharge, strictly so called, by a quantity equal to $\frac{c_d d^2}{96 h^2}$. The value of this expression is in general very small, and it is easy to perceive from the first of the expressions in equation (31), that it can never exceed $4\frac{1}{6}$ per cent., or ·0417 in unity. If we wish to know the discharge from an orifice 4 inches square $= 4'' \times 4''$, with its centre 4 feet below the surface, which is equivalent to a head of 3 feet 10 inches at the upper side, we find from the table the value of $c_d \left\{1 - \frac{d}{96 h^2}\right\} = ·601$; hence we get

$$D = \cdot 601 \times A\sqrt{2gh} = \cdot 601 \times \frac{1}{9} \times 8\cdot025 \times 2 =$$
$$\cdot 601 \times \frac{1}{9} \times 16\cdot05 = \frac{1}{9} \times 9\cdot646 = 1\cdot072$$

cubic feet per second. In the absence of any experiments with larger orifices, we must, when they occur, use the co-efficients given in this table; and, in order to do so with judgment, it is only necessary to observe the relation of the sides and heads. For example, if the size of an orifice be 16″ × 4″, we must seek for the coefficient in that column where the ratio of the sides is as four to one, and if the head at the upper side be five times the length of the orifice, we shall find the coefficient ·626, which in this case is the same for depths measured behind, or at the orifice. For lesser orifices, the results obtained from the experiments of *Michelotti* and *Bossut*, pages 34 and 35, are most applicable; and also the coefficients of *Rennie*, pages 37 and 38. It is almost needless to observe, that all these coefficients are only applicable to orifices in thin plates, or those having the outside arrises chamfered, as in Fig. 8. Very little de-pendance can be placed on calculations of the quantities of water dis-charged from other ori-fices, unless where the coefficients have been al-ready obtained by ex-periment for them. If the inner arris next the water be rounded, the coefficient will be increased.

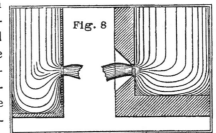

Fig. 8

### NOTCHES AND WEIRS.

We have already given some coefficients, pages 36 and 37, derived from the experiments of Du Buât, Brindley and Smeaton, and Poncelet and Lesbros, for finding the dis-charge over notches in the sides of large vessels; and it does not appear that there is any difference of importance between these and those for orifices sunk some depth below the surface, when the proper formula for finding the discharge for each is used. If we compare Poncelet and Lesbros' coefficients for notches, page 37, with those for an orifice at the surface, TABLE I., we perceive little practical difference in the results, the heads being measured back from the orifice, unless in the very shallow depths, and where the ratio of the length to the depth exceeds five to one. The depths being in these examples less than an inch, it is probable that the larger coefficients found for the *orifice* at the surface, arise from the upper edge attracting the fluid to it and lessening the effects of contraction. Indeed, the results obtained from experiments with very shallow weirs, or notches, have not been at all uniform, and at small depths the discharge must proportionably be more affected by movements of the air and external circumstances than when the depths are considerable. We shall see that in Mr. Blackwell's experiments the coefficient obtained for depths of 1 and 2 inches was ·676 for a thin plate 3 feet long, while for a thin plate 10 feet long it increased up to ·805.

The experiments of Castel, with weirs up to about 30 inches long, and with variable heads of from 1 to 8 inches, lead to the coefficient ·597 for notches extending over one-fourth of the side of a reservoir; and to the coefficient ·664 when they extend for the whole width. For lesser widths than one-fourth, the coefficients decrease down to ·584; and for those extending between one-third of, and the whole width, they increase from ·600 to ·665 and ·680. Bidone finds $c_d = ·620$, and Eytelwein $c_d = ·635$. It will be perceived from these and the foregoing results, that the third place of decimals in the value of $c_d$, and even sometimes the second, is very uncertain; that the coefficient varies with the head and ratio of the notch to the side in which it is placed; and we shall soon show that the form and size of the weir, weir-basin, and approaches, still further modify its value.

When the sides and edge of a notch increase in thickness, or are extended into a shoot, the coefficients are found to reduce very considerably; and for small heads, to an extent beyond what the increase of resistance, from friction alone, indicates. Poncelet and Lesbros found, *for orifices*, that the addition of a horizontal shoot, 21 inches long, reduced the coefficient from ·604 to ·601, with a head of 4 feet; but for a head of only 4½ inches, the coefficient fell from ·572 to ·483, the orifice being 8" × 8". For *notches* 8 inches wide, with a horizontal shoot 9 feet 10 inches long, the coefficient fell from ·582 to ·479, for a head of 8 inches; and from ·622 to ·340, for a head of only 1 inch. Castel found also, for a notch 8 inches wide with a shoot 8 inches long attached and inclined at an angle of 4° 18', that the mean coefficient for heads from 2 to 4½ inches was only ·527*.

We have obtained the following table of coefficients from some experiments made by Mr. Ballard, on the river Severn, near Worcester, "with a weir 2 feet long, formed by a board standing perpendicularly across a trough."† *The*

COEFFICIENTS FOR SHORT WEIRS OVER BOARDS.
*Heads measured on the crest.*

| Depths in inches. | Coefficients. | Depths in inches. | Coefficients. | Depths in inches. | Coefficients. |
|---|---|---|---|---|---|
| 1 | ·762 | 3 | ·801 | 5 | ·733 |
| 1¼ | ·662 | 3¼ | ·765 | 5¼ | ·713 |
| 1½ | ·673 | 3½ | ·748 | 5½ | ·735 |
| 1¾ | ·692 | 3¾ | ·740 | 5¾ | ·729 |
| 2 | ·684 | 4 | ·759 | 6 | ·727 |
| 2¼ | ·702 | 4¼ | ·731 | 7 | ·716 |
| 2½ | ·756 | 4½ | ·744 | 8 | ·726 |
| 2¾ | ·786 | 4¾ | ·745 | Mean | ·732 |

*heads or depths were here measured on the weir*, and hence the coefficients are larger than those found from heads measured back to the surface of still water.

Experiments made at Chew Magna, in Somersetshire, by Messrs. Blackwell and Simpson, in 1850‡, give the following coefficients.

* Traité Hydraulique, par D'Aubuisson, pp. 46, 94 et 95.
† Civil Engineer and Architect's Journal for 1851, p. 647.
‡ Civil Engineer and Architect's Journal for 1851, pp. 642 and 645

ORIFICES, WEIRS, PIPES, AND RIVERS. 43

COEFFICIENTS DERIVED FROM THE EXPERIMENTS OF BLACKWELL AND SIMPSON.

| Heads in inches. | Coefficients. | Heads in inches. | Coefficients. | Heads in inches. | Coefficients. |
|---|---|---|---|---|---|
| 1 to $\frac{7}{8}$ | ·591 | $4\frac{1}{4}$ | ·743 | 6 | ·749 |
| 1 to $1\frac{1}{16}$ | ·626 | $4\frac{5}{16}$ | ·760 | $6\frac{3}{16}$ | ·748 |
| $2\frac{3}{16}$ to $2\frac{1}{4}$ | ·682 | $4\frac{3}{8}$ | ·741 | $6\frac{3}{16}$ to $6\frac{1}{4}$ | ·747 |
| $2\frac{3}{4}$ | ·665 | $4\frac{7}{16}$ | ·750 | $6\frac{1}{8}$ | ·772 |
| $2\frac{13}{16}$ | ·670 | $4\frac{1}{2}$ | ·725 | $7\frac{21}{32}$ | ·717 |
| $2\frac{7}{8}$ | ·655 | 5 | ·780 | 8 | ·802 |
| $2\frac{29}{32}$ | ·653 | $5\frac{5}{16}$ | ·781 | 8 to $8\frac{13}{16}$ | ·737 |
| $2\frac{15}{16}$ | ·654 | $5\frac{13}{32}$ | ·749 | $8\frac{15}{16}$ | ·750 |
| 3 to $3\frac{1}{16}$ | ·725 | $5\frac{7}{16}$ to $5\frac{1}{2}$ | ·751 | 9 | ·781 |
| 4 | ·745 | $5\frac{15}{16}$ | ·728 | Mean | ·723 |

"The overfall bar was a cast-iron plate 2 inches thick, with a square top." The length of the overfall was 10 feet. The heads were measured from still water at the side of the reservoir, and at some distance up in it. The area of the reservoir was 21 statute perches, of an irregular figure, and nearly 4 feet deep on an average. It was supplied from an upper reservoir, by a pipe 2 feet in diameter and of 19 feet fall; the distance between the supply and the weir was about 100 feet. The width of the reservoir as it ap-

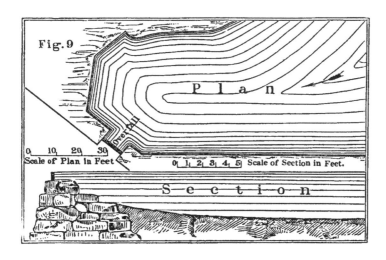

proached the overfall was about 50 feet, and the plan and section, Fig. 9, of the weir and overfall in connection with it, will give a fair idea of the circumstances attending the experiments. For heads over 5 inches the velocity of approach to the weir was "perceptible to the eye," though its amount was not determined. We perceive that the coefficient (derived from two experiments) for a depth of 8 inches is ·802, while the coefficient (derived from three experiments) for a depth of $7\frac{9}{32}$ inches is ·717, and for depths from 8 to $8\frac{15}{16}$ inches the mean coefficient is ·743 : as all the attendant circumstances appear the same, these discrepances and others must arise from the circumstances of the case: perhaps the supply, and, consequently, the velocity of approach, was increased while making one set of experiments, without affecting the still water near the side where the heads appear to have been taken. By comparing the results with those obtained by one of the same experimenters, Mr. Blackwell, on the Kennet and Avon Canal, we shall immediately perceive that the velocity of approach, and every circumstance which tends to alter and modify it, has a very important effect on the amount of the discharge, and, consequently, on the coefficient.

The experiments made by Mr. Blackwell, on the Kennet and Avon Canal, in 1850[*], afford very valuable instruction as the form and width of the crest were varied, and brought to agree more closely with actual weirs in rivers than the thin plates or boards of earlier experimenters. We have calculated and arranged the coefficients in the following table from these experiments. The variations in the values for different widths of crest, other circumstances being the same, are very considerable; and the differences in the coefficients, at depths of 5 inches and under, for thin plates and crests 2 inches wide, are greater than mere friction can account for; and greater also than the differences at the same depths between the coefficients for crests 2 inches wide, and 3 feet wide.

---

[*] Civil Engineer and Architect's Journal, 1851, p. 642.

COEFFICIENTS FOR THE DISCHARGE OVER WEIRS, ARRANGED AND DERIVED FROM THE EXPERIMENTS OF MR. BLACKWELL.

*When more than one experiment was made with the same head, and the results were pretty uniform, the resulting coefficients are marked with a* \*. *The effect of the converging wing-boards is very strongly marked.*

| Heads in inches, measured from still water in the reservoir. | Thin plates. | | Planks 2 inches thick, square on crest. | | | 10 feet long, wing boards making an angle of 60°. | Crests 3 feet wide. | | | | | |
|---|---|---|---|---|---|---|---|---|---|---|---|---|
| | 3 feet long. | 10 feet long. | 3 feet long. | 6 feet long. | 10 feet long. | | 3 feet long, level. | 3 feet long, fall 1 in 18. | 3 feet long, fall 1 in 12. | 6 feet long, level. | 10 feet long, level. | 10 feet long, fall 1 in 18. |
| 1 | ·677 | ·809 | ·467 | ·459 | ·435† | ·754 | ·452 | ·545 | ·467 | ... | ·381 | ·467 |
| 2 | ·675 | ·803 | ·509* | ·561 | ·585* | ·675 | ·482 | ·546 | ·533 | ... | ·479* | ·495* |
| 3 | ·630 | ·642* | ·563* | ·597* | ·569* | ... | ·441 | ·537 | ·539 | ·492* | ... | ... |
| 4 | ·617 | ·656 | ·549 | ·575 | ·602* | ·656 | ·419 | ·431 | ·455 | ·497* | ... | ·515 |
| 5 | ·602 | ·650* | ·588 | ·601* | ·609* | ·671 | ·479 | ·516 | ... | ... | ·518 | ... |
| 6 | ·593 | ... | ·593* | ·608* | ·576* | ... | ·501* | ... | ·531 | ... | ·513 | ·543 |
| 7 | ... | ... | ·617* | ·608* | ·576* | ... | ·488 | ·513 | ·527 | ·507 | ... | ... |
| 8 | ... | ·581* | ·606* | ·590* | ·548* | ... | ·470 | ·491 | ... | ·497 | ·468 | ·507 |
| 9 | ... | ·530 | ·600 | ·569* | ·558* | ... | ·476 | ·492* | ·498 | ... | ·486 | ... |
| 10 | ... | ... | ·614* | ·539 | ·534* | ... | ... | ... | ... | ·480* | ·455 | ... |
| 12 | ... | ... | ... | ·525 | ·534* | ... | ... | ... | ... | ·465* | ... | ... |
| 14 | ... | ... | ... | ·549* | ... | ... | ... | ... | ... | ·467* | ... | ... |

† The discharge per second varied from ·461 to ·665 cubic feet in two experiments. The coefficient ·435 is derived from the mean value.

The plan and section, Fig. 10, will give a fair idea of the approach to, and nature of the overfall made use of in these

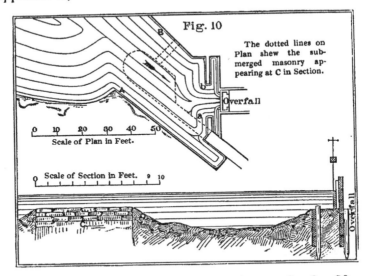

experiments. The area of the reservoir was 2A. 1R. 30P., and the head was measured from the surface of the still water in it, which remained unchanged between the beginning and end of each experiment. The width of the approach AB from the reservoir was about 32 feet; the width at $ab$ about 13 feet, below which the waterway widened suddenly, and again narrowed to the length of the overfall. The depth in front of the dam appears to have been about 3 feet; the depth on the dam, next the overfall, about 2 feet; and the depth on the sunk masonry in the channel of approach, about 18 inches. Altogether, the circumstances were such as to increase the amount of resistances between the reservoir, from which the head was measured, and the overfall, particularly for the larger heads, and we accordingly see that the coefficients become less for heads over six inches, with a few exceptions. The measurements of the quantities discharged appear to have been made very accurately, yet the discharges per second, with the same head and same length of overfall, sometimes vary;

for instance, with the plank 2 inches thick and 10 feet long, the discharge per second for 4 inches head varies from 6·098 cubic feet to 6·491 cubic feet, or by about one-sixteenth of the whole quantity. Most of the results, however, are means from several experiments. The quantities discharged varied from one-tenth of a cubic foot to 22 cubic feet per second, and the times from 24 to 420 seconds. If we compare the coefficients for a plank 10 feet long and 2 inches thick in the foregoing table with those for the same overfall at Chew Magna, we shall immediately perceive how much the form of the approaches affects the discharge. Indeed, were the area of the reservoir at Chew Magna even larger than that for the Kennet and Avon experiments, it would be found, notwithstanding, that the coefficients in the former would still continue the larger, though not fully as large as those found under the particular circumstances.

### HEAD, AND FROM WHENCE MEASURED.

By referring to TABLE I., we shall see that there is a difference in the coefficients as obtained from heads measured on or above the orifice. This difference is greater in notches, or weirs, than in orifices sunk below the surface; and when the crest of a weir is of some width, the depths upon it vary*. In the Kennet and Avon experiments, the heads measured from the surface of the water in the reservoir, and the depths at the "outer edge" (by which we understand the lower edge) of the crest were as follows:—

---

* There is a very important omission in almost all the preceding experiments on weirs and notches. In Fig. 10, for instance, it would have been necessary to have obtained the heads at $AB$ and $ab$ in each experiment, above the crest, and also the head on and a few feet above the crest itself. These would indicate the resistances at the different passages of approach, and enable us to calculate the coefficients correctly, and thereby render them more generally applicable to practical purposes. The coefficients in the table at page 45, are not as valuable as they otherwise would be from this omission. The level of still water near the banks is below that of the moving water in the current, therefore, heads measured from still water must give larger coefficients than if taken from the centre of the current. This may account, to some extent, for the larger coefficients in the table at p. 43.

DIFFERENCE OF HEADS MEASURED ON AND ABOVE WEIRS.

| Heads from reservoir to crest, in inches. | Heads on crests 2 inches thick. | | Heads on crests 3 feet wide. | | | | | |
|---|---|---|---|---|---|---|---|---|
| | 3 feet long. | 6 feet long. | 3 feet long, crest level. | 3 feet long, crest slope 1 in 18. | 3 feet long, crest slope 1 in 12. | 6 feet long, crest level. | 10 feet long, crest level. | 10 feet long, crest slope 1 in 18. |
| 1 | $\frac{5}{8}$ | ... | $\frac{7}{16}$ | ... | $\frac{1}{4}$ | $\frac{1}{2}$ | $\frac{7}{16}$ | $\frac{5}{16}$ |
| 2 | ... | ... | $\frac{7}{8}$ | ... | ... | ... | $\frac{7}{8}$ | $\frac{9}{16}$ |
| 3 | ... | $1\frac{15}{16}$ | ... | ... | ... | $1\frac{1}{8}$ to $1\frac{1}{4}$ | ... | ... |
| 4 | 3 to $2\frac{11}{16}$ | $3\frac{1}{2}$ | $1\frac{3}{8}$ | $1\frac{3}{8}$ | $1\frac{1}{2}$ | ... | ... | $1\frac{1}{2}$ |
| 5 | $3\frac{1}{2}$ | $3\frac{5}{8}$ | $2\frac{1}{4}$ | $1\frac{3}{4}$ | ... | ... | $1\frac{3}{4}$ | ... |
| 6 | $4\frac{3}{8}$ | $4\frac{3}{8}$ | $2\frac{3}{8}$ | ... | ... | $2\frac{3}{4}$ | $2\frac{5}{8}$ | $2\frac{1}{4}$ |
| 7 | ... | ... | ... | $2\frac{1}{8}$ | $2\frac{3}{8}$ | ... | ... | ... |
| 8 | $6\frac{3}{8}$ | ... | ... | ... | ... | ... | $3\frac{5}{8}$ | $3\frac{1}{2}$ |
| 9 | ... | ... | ... | ... | ... | $4\frac{1}{8}$ | $3\frac{1}{2}$ | ... |
| 10 | ... | ... | ... | ... | ... | ... | 4 | ... |

No intermediate heads are given, but those registered point out very clearly the great differences which often exist between the heads measured on a weir, or notch, and those measured from the still water above it; and how the form of the weir itself, as well as the nature of the approaches, alters the depth passing over. On a crest 2 feet wide, with $14\frac{1}{2}$ inches depth on the upper edge, we have found that the depth on the lower edge is reduced to $11\frac{1}{2}$ inches. The head taken from 3 to 20 feet above the crest, where the plane of the approaching water surface becomes curved, is that in general which is best suited for finding the discharge by means of the common coefficients, but a correct section of the channel and water-line, showing the different depths upon and for some distance above the crest, is necessary in all experiments for determining accurately by calculation the value of the coefficient of discharge $c_d$.

Du Buât, finding the theoretical expression for the discharge through an orifice of half the depth $h$,

$$D = \tfrac{2}{3}\sqrt{2g} \times l\left\{h^{\tfrac{3}{2}} - \left(\tfrac{h}{2}\right)^{\tfrac{3}{2}}\right\},$$

equation (6)

$$= \tfrac{2}{3} lh\sqrt{2gh} \times \left\{1 - \left(\tfrac{1}{2}\right)^{\tfrac{3}{2}}\right\} = \cdot 646 \times \tfrac{2}{3} lh\sqrt{2gh},$$

to agree pretty closely with his experiments, seems to have assumed that the head $h$ was reduced to $\dfrac{h}{2}$ in passing over. This is a reduction, however, which never takes place unless with a wide crest and at its lower edge, or where the head $h$ is measured at a considerable distance above the weir, and a loss of head due to the distance, and obstructions in channel takes place. When there is a clear weir basin immediately above the weir, we have found that, putting $h$ for the head measured from the surface in the weir basin and $h_w$ for the depth on the upper edge of the weir, that

(32.) $$h - h_w = \cdot 14 \sqrt{h},$$

for measures in feet, and

(33.) $$h - h_w = \cdot 48 \sqrt{h},$$

for measures in inches. The comparative values of $h$ and $h_w$ depend, however, a good deal on the particular circumstances of the case. Dr. Robinson found* $h = 1\cdot 111 h_w$, when $h$ was about 5 inches. The expressions we have given are founded on the hypothesis, that $h - h_w$ is as the velocity of discharge, or as the $\sqrt{h}$ nearly. For small depths, there is a practical difficulty in measuring with sufficient accuracy the relative values of $h$ and $h_w$. Unless for very small heads the sinking will be found in general to vary from $\dfrac{h}{10}$ to $\dfrac{h}{4}$, and in practice it will always be useful to observe the depths on the weir as well as the heads for some distances (and where the widths contract) above it.

In order to convey to our readers a more definite idea of the differences between the coefficients for heads measured

---

* Proceedings of the Royal Irish Academy, vol. iv. p. 212.

at the weir, or notch, and at some distance above it, we shall assume the difference of the heads $h - h_w = \dfrac{h_w}{r}$; hence $h = \dfrac{r+1}{r} h_w$ and $h_w = \dfrac{r}{r+1} h$.

Now the discharge may be considered as that through an orifice whose depth is $h_w$ with a head over the upper edge equal to $h - h_w = \dfrac{h_w}{r}$; hence from equation (6) the discharge is equal to

$$\tfrac{2}{3} l \sqrt{2g} \times c_d \left\{ h^{\frac{3}{2}} - \left(\dfrac{h_w}{r}\right)^{\frac{3}{2}} \right\},$$

and substituting for $h^{\frac{3}{2}}$ its value $\left(\dfrac{r+1}{r} h_w\right)^{\frac{3}{2}}$, we shall find the value of

(34.) $\quad D = \tfrac{2}{3} l h_w \sqrt{2 g h_w} \times c_d \left\{ \left(1 + \dfrac{1}{r}\right)^{\frac{3}{2}} - \left(\dfrac{1}{r}\right)^{\frac{3}{2}} \right\}.$

As the value of the discharge would be expressed by

$$\tfrac{2}{3} l h_w \sqrt{2 g h_w} \times c_d$$

if the head $h - h_w$ were neglected, it is evident the coefficient is increased, under the circumstances, from $c_d$ to

$$c_d \times \left\{ \left(1 + \dfrac{1}{r}\right)^{\frac{3}{2}} - \left(\dfrac{1}{r}\right)^{\frac{3}{2}} \right\};$$

or, more correctly, the common formula has to be multiplied by $\left(1 + \dfrac{1}{r}\right)^{\frac{3}{2}} - \left(\dfrac{1}{r}\right)^{\frac{3}{2}}$, to find the true discharge, and the value of this expression for different values of $\dfrac{1}{r} = n$ will be found in TABLE IV. If we suppose that

$$h - h_w = \dfrac{h_w}{10}, \text{ then } \dfrac{1}{r} = \dfrac{1}{10} = n;$$

and we find from the table $\left(1 + \dfrac{1}{r}\right)^{\frac{3}{2}} - \left(\dfrac{1}{r}\right)^{\frac{3}{2}} = 1\cdot1221$.

Now if we take the value of $c_d$ for the full head $h$ to be ·628, we shall find $1\cdot1221 \times ·628 = ·705$, rejecting the latter figures, for the coefficient when the head is measured at the orifice; and if $\dfrac{1}{r} = \dfrac{2}{10} = n$, we should find in the same manner the new coefficient to be $1\cdot2251 \times ·628 = ·769$ nearly. The increase of the coefficients determined, page 42, from Mr. Ballard's experiments is, therefore, evident from principle, as the heads were taken at the notch; and it is also pretty clear that, *in order to determine the true discharge, the heads both on, at, and above a weir should be taken.*

## SECTION IV.

VARIATIONS IN THE COEFFICIENTS FROM THE POSITION OF THE ORIFICE.—GENERAL AND PARTIAL CONTRACTION.—VELOCITY OF APPROACH.—CENTRAL AND MEAN VELOCITIES.

A glance at Table I. will show us that the coefficients increase as the orifices approach the surface, to a certain depth dependent on the ratio of the sides, and that this increase increases with the ratio of the length to the depth. Some experimenters have found the increase to continue uninterrupted for all orifices up to the surface, but this seems to hold only for depths taken at or near the orifice when it is square or nearly so. It has also been found that the coefficient increases as the orifice approaches to the sides or bottom of a vessel: as the contraction becomes imperfect the coefficient increases. The lateral orifices A, B, C, D, E, F, G, H, I, and K, Fig. 11, have coefficients differing more or less from each other. The coefficient for A is found to be larger than either of those for B, C, E, or D; that for G or K larger than that for H or I; that for H larger than that for I; and that for F, where the contraction is general, least of all. The contraction of the fluid on entering the orifice F removed from the bottom and sides is complete; it is termed, therefore,

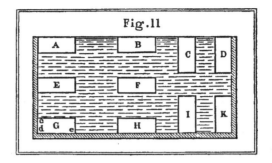

Fig. 11

*general contraction;* that at the orifices A, E, G, H, I, K, and D, is interfered with by the sides; it is therefore incomplete, and termed *partial contraction.* The increase in the co-efficients for the same-sized orifices at the same mean depths may be assumed as proportionate to the length of the perimeter at which the contraction is partial, or from which the lateral flow is shut off; for example, the increase for the orifice G is to that for H as $cd + de : de$; and in the same manner the increase for G is to that for E as $cd + de : cd$. If we put $n$ for the ratio of the contracted portion $cde$ to the entire perimeter, and, as before, $c_d$ for the coefficient of general contraction, we shall find the coefficient of partial contraction to be equal to

(35.) $\qquad c_d + \cdot 09\,n$

for rectangular orifices. The value of the second term $\cdot 09\,n$ is derived from experiments. If we assume $\cdot 617$ for the mean value of $c_d$, we may change the expression into the form $(1 + \cdot 146\,n)\,c_d$. When $n = \frac{1}{4}$, this becomes $1 \cdot 036\,c_d$; when $n = \frac{1}{2}$, it becomes $1 \cdot 073\,c_d$; and when $n = \frac{3}{4}$, contraction is prevented for three-fourths of the perimeter, and the coefficient for partial contraction becomes $1 \cdot 109\,c_d$. The form which we have given equation (35) is, however, the simplest; but the value of $n$ must not exceed $\frac{3}{4}$. If in this case $c_d = \cdot 617$, the coefficient for partial contraction becomes $\cdot 617 + \cdot 09 \times \frac{3}{4}$ $= \cdot 617 + \cdot 067 = \cdot 684$. Bidone's experiments give for the coefficient of partial contraction $(1 + \cdot 152\,n)\,c_d$; and Weisbach's $(1 + \cdot 132\,n)\,c_d$.

## VARIATION IN THE COEFFICIENTS FROM THE EFFECTS OF THE VELOCITY OF APPROACH.

Heretofore we have supposed the water in the vessel to be almost still, its surface level unchanged, and the vessel consequently large compared with the area of the orifice. When the water flows to the orifice with a perceptible velocity, the contracted vein and the discharge are both found to be increased, other circumstances being the same. If the area of the vessel or channel in front exceed thirty times that of the orifice, the discharge will not be perceptibly increased by the induced velocity in the conduit; but for lesser areas of the approaching channel corrections due to the velocity of approach become necessary. It is clear that this velocity may arise from either a surface inclination in the channel, an increase of head, or a small channel of approach.

We get equation (6) for the discharge from a rectangular

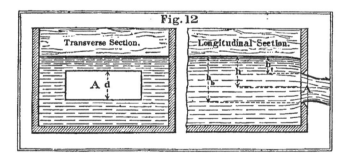

Fig. 12

orifice A, Fig. 12, of the length $l$, with a head measured from still water

$$D = \frac{2}{3} c_d \, l \sqrt{2g} \times \{h_b^{\frac{3}{2}} - h_t^{\frac{3}{2}}\},$$

in which $h_b$ and $h_t$ are measured at some distance back from the orifice, as shown in the section. The water here, however, must move along the channel towards the orifice with considerable velocity. If A be the area of the orifice, and c the area of the channel, we may suppose with tolerable accuracy that this velocity is equal to $\frac{A}{C}v_o$, in which $v_o$ represents the mean velocity in the orifice. If we also represent by $v_a$ the velocity of approach, we get the equation

(36.) $$v_a = \frac{A}{C} \times v_o,$$

and consequently the height ($h_a$) due to it is

(37.) $$h_a = \frac{A^2}{C^2} \times \frac{v_o^2}{2g}.$$

The height $h_a$ may be considered as an increase of head, converting $h_b$ into $h_b + h_a$, and $h_t$ into $h_t + h_a$. The discharge therefore now becomes

(38.) $$D = \frac{2}{3} c_d l \sqrt{2g} \left\{ (h_b + h_a)^{\frac{3}{2}} - (h_t + h_a)^{\frac{3}{2}} \right\};$$

which, for notches or weirs, is reduced to

(39.) $$D = \frac{2}{3} c_d l \sqrt{2g} \left\{ (h_b + h_a)^{\frac{3}{2}} - h_a^{\frac{3}{2}} \right\},$$

as $h_t$ then vanishes. As D is also equal to $A \times v_o$, equation (37) may be changed into

(40.) $$h_a = \frac{D^2}{C^2} \times \frac{1}{2g}.$$

If this value for $h_a$ be substituted in equations (38) and (39), the resulting equations will be of a high order and do not admit of a direct solution; and in (38) and (39), as they stand, $h_a$ involves implicitly the value of D, which we are seeking for. By finding at first an approximate value for the velocity of approach, the height $h_a$ due to it can be easily found, equation (37); this height, substituted in equation (38) or (39), will give a closer value of D, from which again a more correct value of $h_a$ can be determined; and by repeating the operation the values of D and $h_a$ can be had to any degree of accuracy. In general the values found at the second operation will be sufficiently correct for all practical purposes.

It has been already observed that, for orifices, it is advisable to find the discharge from a formula in which only one head, that at the centre, is made use of; and though TABLE IV., as we shall show, enables us to calculate the discharge with facility from either formula, it will be of use to reduce equation (38) to a form in which only the head ($h$) at the centre is used. The error in so doing can never exceed six per cent., even at small depths, equation (31), and this is more than balanced by the observed increase in the coefficients for smaller heads.

The formula for the discharge from an orifice, $h$ being the head at the centre, is

$$D = c_d \sqrt{2gh} \times A;$$

and when the additional head $h_a$ due to the velocity of approach is considered,

$$D = c_d \sqrt{2g(h + h_a)} \times A,$$

which may be changed into

(41.) $$D = A \sqrt{2gh} \times c_d \left\{1 + \frac{h_a}{h}\right\}^{\frac{1}{2}}.$$

Equation (39), for weirs, may be also changed to the form

(42.) $$D = \tfrac{2}{3} A \sqrt{2gh_b} \times c_d \left\{ \left(1 + \frac{h_a}{h_b}\right)^{\frac{3}{2}} - \left(\frac{h_a}{h_b}\right)^{\frac{3}{2}} \right\};$$

this is similar in every way to the equation

(43.) $$D = \tfrac{2}{3} A \sqrt{2gd} \times c_d \left\{ \left(1 + \frac{h_t}{d}\right)^{\frac{3}{2}} - \left(\frac{h_t}{d}\right)^{\frac{3}{2}} \right\},$$

for the discharge from a rectangular orifice whose depth is $d$, with the head $h_t$ at the upper edge. TABLE III. contains the values of $\left\{1 + \frac{h_a}{h}\right\}^{\frac{1}{2}}$ in equation (41), and TABLE IV. the values of $\left(1 + \frac{h_a}{h_b}\right)^{\frac{3}{2}} - \left(\frac{h_a}{h_b}\right)^{\frac{3}{2}}$ in equation (42), or the similar expression in (43), $\frac{h_a}{h_b}$ or $\frac{h_t}{d}$ being put equal to $\bar{n}$; and we perceive that the effect of the velocity of approach is such as to increase the coefficient $c_d$ to $c_d \left\{1 + \dfrac{h_a}{h}\right\}^{\frac{1}{2}}$ for orifices sunk some distance below the surface, and into

$$c_d \left\{\left(1 + \dfrac{h_a}{h_b}\right)^{\frac{3}{2}} - \left(\dfrac{h_a}{h_b}\right)^{\frac{3}{2}}\right\}$$

for weirs when $h_a$ is the height due to the velocity of approach, $h$ the depth of the centre of the orifice, and $h_b$ the head on the weir. A few examples, showing the application of the formulæ (41), (42), and (43), and the application of TABLES I., II., III., and IV. to them, will be of use. We shall suppose, for the present, the velocity of approach $v_a$ to be given.

EXAMPLE I. *A rectangular orifice, 12 inches wide by 4 inches deep, has its centre placed 4 feet below the surface, and the water approaches the head with a velocity of 28 inches per second; what is the discharge?* For an orifice of the given proportions, and sunk to a depth nearly four times its length, we shall find from TABLE I.

$$c_d = \dfrac{\cdot 616 + \cdot 627}{2} = \cdot 621 \text{ nearly.}$$

As the coefficient of velocity, equation (2), for water flowing in a channel is about ·956, we shall find, column No. 3, TABLE II., the height $h_a = 1\frac{1}{8} = 1\cdot 125$ inch, nearly corresponding to the velocity 28 inches. Equation (41),

$$D = A\sqrt{2gh} \times c_d \left\{1 + \dfrac{h_a}{h}\right\}^{\frac{1}{2}},$$

now becomes

$$D = 12 \times 4 \sqrt{2gh} \times \cdot 621 \left\{1 + \dfrac{1\cdot 125}{48}\right\}^{\frac{1}{2}}.$$

We also find $\sqrt{2gh} = 192\cdot 6$ inches when $h = 48$ inches, TABLE II.; therefore

$$D = 12 \times 4 \times 192\cdot 6 \times \cdot 621 \left\{1 + \dfrac{1\cdot 125}{48}\right\}^{\frac{1}{2}}$$

$$= 9244\cdot 8 \times \cdot 621 \{1 + \cdot 0234\}^{\frac{1}{2}} = 9244\cdot 8 \times \cdot 621 \times 1\cdot 0116,$$

ORIFICES, WEIRS, PIPES, AND RIVERS. 57

(as $\{1\cdot0234\}^{\frac{3}{2}} = 1\cdot0116$ from TABLE III.) $= 9244\cdot8 \times \cdot628$ nearly $= 5805\cdot7$ cubic inches $= 3\cdot306$ cubic feet per second. *Or thus:* The value of $\cdot621 \times (1\cdot0234)$ being found equal $\cdot628$, D $=$ A $\times \cdot628 \sqrt{2g \times 48}$. Now for the coefficient $\cdot628$, and $h = 48$ inches, TABLE II. gives us $\cdot628 \sqrt{2g \times 48} = 120\cdot96$ inches; hence we get D $= 12 \times 4 \times 120\cdot96 = 5806\cdot08$ cubic inches $= 3\cdot306$ cubic feet, the same as before, the difference $\cdot38$ in the cubic inches being of no practical value. If the centre of the orifice were within 1 foot of the surface, the effect of the velocity of approach would be much greater; for then

$$c_d \times \left\{1 + \frac{h_a}{h}\right\}^{\frac{3}{2}} = \text{(from TABLE I.)} \cdot623 \left\{1 + \frac{1\cdot125}{12}\right\}^{\frac{3}{2}}$$

$=$ (from TABLE III.) $\cdot623 \times 1\cdot047 = \cdot652$ instead of $\cdot628$. In this case the discharge is D $= 12 \times 4 \times \cdot652 \sqrt{2g \times 12} = 12 \times 4 \times \cdot652 \times 96\cdot3$ (from TABLE II.) $= 12 \times 4 \times 62\cdot8 = 3014\cdot4$ cubic inches $= 1\cdot744$ cubic feet per second. Or we may find the value of $\cdot652 \sqrt{2gh}$ directly from Table II. thus:

The value of $\cdot628 \sqrt{2g \times 12} = 60\cdot48$    $\cdot628$
The value of $\cdot666 \sqrt{2g \times 12} = 64\cdot14$    $\cdot652$
                        38      :        $3\cdot66$ ::   $24 : 2\cdot31$.

Hence $\cdot652 \sqrt{2gh} = 60\cdot48 + 2\cdot31 = 62\cdot79$, and the discharge $= 12 \times 4 \times 62\cdot79 = 3013\cdot92$ cubic inches $= 1\cdot744$ cubic feet per second, the same as before.

EXAMPLE II. *A rectangular notch, 7 feet long, has a head of 8 inches measured at about 4 feet above the orifice, and the water approaches the head with a velocity of* $16\frac{1}{4}$ *inches per second; what is the discharge?* For a still head we shall assume $c_d = \cdot628$ in this case, and we have from equation (42)

$$\text{D} = \tfrac{2}{3}\text{A}\sqrt{2gh_b} \times c_d \left\{\left(1 + \frac{h_a}{h_b}\right)^{\frac{3}{2}} - \left(\frac{h_a}{h_b}\right)^{\frac{3}{2}}\right\}.$$

As in the last example, we shall find from TABLE II. ($h_a$) the height due to the velocity of approach (16¼ inches) to be $\frac{3}{8} = 0.375$ inch, assuming the coefficient of velocity to be ·956. We have, therefore, $h_a = \cdot 375$, $h_b = 8$, $c_d = \cdot 628$, and $A = 7 \times 12 \times 8$, or for measures in feet $\frac{h_a}{h_b} = \cdot 047$, $h_b = \frac{2}{3}$, and $A = 7 \times \frac{2}{3}$; hence

$$D = \frac{2}{3} \times 7 \times \frac{2}{3} \sqrt{2g \times \frac{2}{3}} \times \cdot 628 \left\{ (1 \cdot 047)^{\frac{3}{2}} - (\cdot 047)^{\frac{3}{2}} \right\}.$$

The value of $(1 \cdot 047)^{\frac{3}{2}} - (\cdot 047)^{\frac{3}{2}}$ will be found from TABLE IV. equal to 1·0612; the value of $\sqrt{2g \times \frac{2}{3}}$ will be found from TABLE II. equal to 6·552, viz. by dividing the velocity 78·630, to be found opposite 8 inches, by 12; hence

$$D = \frac{2}{3} \times 7 \times \frac{2}{3} \times 6 \cdot 552 \times \cdot 628 \times 1 \cdot 0612$$
$$= \frac{2}{3} \times 7 \times 4 \cdot 368 \times \cdot 628 \times 1 \cdot 0612$$
$$= \frac{2}{3} \times 7 \times 4 \cdot 368 \times \cdot 666 \text{ nearly}$$
$$= \frac{2}{3} \times 7 \times 2 \cdot 909 = 7 \times 1 \cdot 939$$

$= 13 \cdot 573$ cubic feet per second $= 814 \cdot 38$ cubic feet per minute. *Or thus:* From TABLE VI. we find, when the coefficient is ·628, the discharge from a weir 1 foot long, with a head of 8 inches, to be 109·731 cubic feet per minute. The discharge for a weir 7 feet long, when $\frac{h_a}{h} = \cdot 047$ is therefore $109 \cdot 731 \times 7 \times 1 \cdot 0612 = 815 \cdot 12$ cubic feet per minute. The difference between this value and that before found, 814·38 cubic feet, is immaterial, and has arisen from not continuing all the products to a sufficient number of places of decimals.

We have, in equations (36) and (37), pointed out the relations between the channel, orifice, velocity of approach, and velocity in the orifice, viz.

$v_\mathrm{a} = \dfrac{\mathrm{A}}{\mathrm{C}} \times v_\mathrm{o}$, and $h_\mathrm{a} = \dfrac{\mathrm{A}^2}{\mathrm{C}^2} \times \dfrac{v_\mathrm{o}^2}{2g} = \dfrac{\mathrm{D}^2}{2g\,\mathrm{C}^2}$, in which $h_\mathrm{a} = \dfrac{v_\mathrm{a}^2}{2g}$ (neglecting the coefficient of velocity ·974 or ·956). As $v_\mathrm{o}$ is the velocity in the orifice, $\dfrac{v_\mathrm{o}}{c_\mathrm{d}}$ must be the velocity due to the head $h + h_\mathrm{a}$, and therefore

$$h + h_\mathrm{a} = \frac{v_\mathrm{o}^2}{c_\mathrm{d}^2 \times 2g}, \text{ and } h = \frac{v_\mathrm{o}^2}{c_\mathrm{d}^2 \times 2g} - \frac{v_\mathrm{a}^2}{2g}; \text{ hence}$$

$$\frac{h_\mathrm{a}}{h} = \frac{1}{\dfrac{v_\mathrm{o}^2}{v_\mathrm{a}^2 \times c_\mathrm{d}^2} - 1} = \frac{c_\mathrm{d}^2\, v_\mathrm{a}^2}{v_\mathrm{o}^2 + c_\mathrm{d}^2 v_\mathrm{a}^2} = \frac{c_\mathrm{d}^2\, \mathrm{A}^2}{\mathrm{C}^2 - c_\mathrm{d}^2\, \mathrm{A}^2}, \text{ for } \frac{v_\mathrm{o}^2}{v_\mathrm{a}^2} = \frac{\mathrm{C}^2}{\mathrm{A}^2}.$$

We have hence

(44.) $$\frac{h_\mathrm{a}}{h} = \frac{c_\mathrm{d}^2\, \mathrm{A}^2}{\mathrm{C}^2 - c_\mathrm{d}^2\, \mathrm{A}^2};$$

substituting this value in equations (41) and (42), there results

(45.) $$\mathrm{D} = \mathrm{A}\sqrt{2gh} \times c_\mathrm{d}\left\{1 + \frac{c_\mathrm{d}^2}{m^2 - c_\mathrm{d}^2}\right\}^{\frac{1}{2}},$$

in which $m = \dfrac{\mathrm{C}}{\mathrm{A}}$, for the discharge from an orifice at some depth, and for the discharge from a weir,

(46.) $$\mathrm{D} = \tfrac{2}{3}\mathrm{A}\sqrt{2gh_\mathrm{b}} \times c_\mathrm{d}\left\{\left(1 + \frac{c_\mathrm{d}^2}{m^2 - c_\mathrm{d}^2}\right)^{\frac{3}{2}} - \left(\frac{c_\mathrm{d}^2}{m^2 - c_\mathrm{d}^2}\right)^{\frac{3}{2}}\right\}.$$

The two last equations give the discharge when the ratio of the channel to the orifice $\dfrac{\mathrm{C}}{\mathrm{A}} = m$ is known, and also when the whole quantity of water passing through the orifice, that due to the velocity of approach as well as the pressure, suffers a contraction whose coefficient is $c_\mathrm{d}$. When $c_\mathrm{d} = 1$, equation (45) may be changed into

$$\mathrm{D} = \mathrm{A}\left\{\frac{2gh}{1 - \left(\dfrac{1}{m}\right)^2}\right\}^{\frac{1}{2}}.$$

This is the equation of Daniel Bernoulli, which is only a particular case of the one we have given.

If we put $n = \frac{c_d^2}{m^2 - c_d^2}$, the values of $\left\{1 + \frac{c_d^2}{m^2 - c_d^2}\right\}^{\frac{1}{2}}$, and of $\left\{1 + \frac{c_d^2}{m^2 - c_d^2}\right\}^{\frac{3}{2}} - \left\{\frac{c_d^2}{m^2 - c_d^2}\right\}^{\frac{3}{2}}$, can be easily had from TABLES III. and IV. We have, however, calculated TABLE V. for different ratios of the channel to the orifice, and for different values of the coefficient of discharge. This table gives at once the values of

$$c_d \left\{1 + \frac{c_d^2}{m^2 - c_d^2}\right\}^{\frac{1}{2}} \text{ and } c_d \left\{\left(1 + \frac{c_d^2}{m^2 - c_d^2}\right)^{\frac{3}{2}} - \left(\frac{c_d^2}{m^2 - c_d^2}\right)^{\frac{3}{2}}\right\}$$

as new coefficients, and the corresponding value of

$$\frac{h_a}{h} = \frac{c_d^2}{m^2 - c_d^2}.$$

It is equally applicable, therefore, to equations (41) and (42) as to equations (45) and (46). For instance, we find here at once the value of $628 \left\{(1{\cdot}047)^{\frac{3}{2}} - ({\cdot}047)^{\frac{3}{2}}\right\}$ in EXAMPLE II., p. 58, equal to ·666, as $\frac{h_a}{h} = {\cdot}047$, and the next value to it for the coefficient ·628, in the table, is ·046, opposite to which we find ·666, the new coefficient sought. The sectional area of the channel in this case, as appears from the first column, must be about three times that of the weir or notch.

TABLE V. is calculated from coefficients $c_d$ in still water, which vary from ·550 to 1. Those from ·606 to ·650, and the mean value ·628 are most suited to practice. When the channel is equal to the orifice, the supply must equal the discharge, and for open channels, with the mean coefficient ·628, we find accordingly from the table the new coefficient 1·002 for weirs, or 1 very nearly as it should be. We also find, in the same case, viz. when $A = c$, and $c_d = {\cdot}628$, that for short tubes, Fig. 13, the resulting new coefficient becomes ·807. This, as we shall afterwards see, agrees very closely with the experimental results. When the coefficients in still

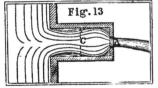

Fig. 13

water are less than ·628, or more correctly ·62725, the orifice, according to our formula, cannot equal the channel unless other resistances take place—as from friction in tubes longer than one and a half or two diameters, or in wide crested weirs; and for greater coefficients the junction of the short tube with the vessel must be rounded, Fig. 14, on one or more sides; and in weirs or notches the approaches must slope from the crest and ends to the bottom or sides, and the overfall be sudden. The converging form of the approaches must, however, increase the ve-

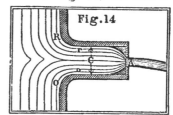
Fig. 14

locity of approach; and therefore $v_a$ is greater than $\frac{A}{C} \times v_o$ when c is measured between $r\,o$ and $R\,O$, Fig. 14, to find the discharge, or new coefficient of an orifice placed at $r\,o$.

As the coefficients in TABLE V. are calculated for orifices at the end of short cylindrical or prismatic tubes at right angles to the sides or bottom of a cistern, a correction is required when the junction is rounded off as at $R\,O\;r\,o$, Fig. 14. When the channel is equal to the orifice, the new coefficient in equation (45) becomes

$$c_d \left\{ 1 + \frac{c_d^2}{1 - c_d^2} \right\}^{\frac{1}{2}} = c_d \times \left\{ \frac{1}{1 - c_d^2} \right\}^{\frac{1}{2}}.$$

The velocity in the tube Fig. 14 is to that in the tube Fig. 13 as 1 to $c_d \left\{ \frac{1}{1 - c_d^2} \right\}^{\frac{1}{2}}$ nearly, or for the mean value $c_d = ·628$, as 1 to ·807. Now, as $\frac{c}{A}$ is assumed equal to $\frac{v_o}{v_a}$ in the cylindrical or prismatic tube, Fig. 13, $\frac{·807\,c}{A} = \frac{v_o}{v_a}$ in the tube Fig. 14 with the rounded junction, for $v_a$ becomes $\frac{v_a}{·807}$; hence, in order to find the discharge from orifices at the end of the short tube, Fig. 14, we have only to multiply the numbers representing the ratio $\frac{C}{A}$ in the first column, Table V., by ·807, or more generally by $c_d \left\{ \frac{1}{1-c_d^2} \right\}^{\frac{1}{2}}$, and find the coefficient opposite to the product. Thus if $c_d = ·628$, we find, when $\frac{C}{A} = 1$, $c_d \left\{ \frac{1}{1-c_d^2} \right\}^{\frac{1}{2}} = ·807$ in the table. If, again, we suppose $\frac{C}{A} = 3$, then $3 \times ·807 = 2·421$, the value of $\frac{v_o}{v_a}$ for the tube Fig. 14, and opposite this value of $\frac{C}{A}$, taken in column 1, we shall find ·651 for the new coefficient. For the cylindrical or prismatic tube, Fig. 13, the new coefficient would be ·642.

### DIFFERENT EFFECTS OF CENTRAL AND MEAN VELOCITIES.

There is, however, another circumstance to be taken into consideration, and which we shall have to refer to more particularly hereafter; it is this, that the central velocity directly facing the orifice is also the maximum velocity in the tube, and not the mean velocity. The ratio of these is as $1 : ·835$ nearly; hence, in the above example, where $\frac{C}{A} = 3$, we get $3 \times ·835 = 2·505$ for the value of $\frac{C}{A}$ in column 1, Table V., opposite to which we shall find ·649, the coefficient for an orifice of one-third of the section of the tube when cylindrical or prismatic, Fig. 13; and $3 \times ·835 \times ·807 = 2·02$ nearly, opposite to which we shall get ·661 for the coefficient when the orifice is at the end of the short tube, Fig. 14, with a rounded junction. We have, therefore, $\frac{C}{A} \times ·835$ equal to the new value of $\frac{C}{A}$ for finding the discharge from orifices at the end of cylindrical or prismatic tubes, and $\frac{C}{A} \times ·835 \times ·807$

$= \frac{C}{A} \times \cdot 67$ nearly for the new value of $\frac{C}{A}$ when finding the discharge from orifices at the end of a short tube with a rounded junction, nearly.

The ratio of the mean velocity in the tube to that facing the orifice cannot be less than ·835 to 1, and varies up to 1 to 1; the first ratio obtaining when the orifice is pretty small compared with the section of the tube, and the other when they are equal. If we suppose the curve D C, whose abscissæ (A $b$) represent the ratio of the orifice to the section of the tube, and whose ordinates ($b$ $c$) represent the ratio of the mean velocity in the tube to that facing the orifice, to be a parabola, we shall find the following values :—

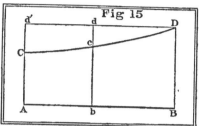

Fig 15

| Ratios $\frac{A}{C} = \frac{A\,b}{A\,B}$ | Values of $d\,c.$ | Values of $b\,c.$ |
|---|---|---|
| 0 | ·165 | ·835 |
| ·1 | ·163 | ·837 |
| ·2 | ·158 | ·842 |
| ·3 | ·150 | ·850 |
| ·4 | ·139 | ·861 |
| ·5 | ·124 | ·876 |
| ·6 | ·106 | ·894· |
| ·7 | ·084 | ·916 |
| ·8 | ·059 | ·941 |
| ·9 | ·031 | ·969 |
| 1·0 | ·000 | 1·000 |

These values of $b\,c$ are to be multiplied by the corresponding ratio $\frac{C}{A}$ in order to find a new value, opposite to which will be found, in the table, the coefficient for orifices at the ends of short prismatic or cylindrical tubes; and this new value again multiplied by ·807, or more generally by $c_d \left\{ \dfrac{1}{1 - c_d^2} \right\}^{\frac{1}{4}}$, will give another new value of $\dfrac{c}{A}$, opposite to which, in the table, will be found the coefficient for orifices at the ends of short tubes with rounded junctions.

EXAMPLE III. *What is the discharge from an orifice* A, *Fig.* 16, 2 *feet long by* 1 *foot deep, the value of* $\dfrac{c}{A}$ *being* 3, *and the depth of the centre of* A 1 *foot* 6 *inches below the surface?* We have
$$D_t = 2 \times 1 \times \dfrac{117 \cdot 945}{12}$$

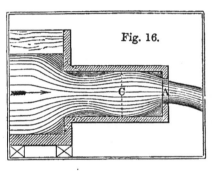

Fig. 16.

(TABLE II.) $= 2 \times 9\cdot 829 = 19\cdot 658$ cubic feet per second for the theoretical discharge. From the foregoing table the coefficient for the mean velocity, facing the orifice, is about ·86; hence $\dfrac{c}{A} \times ·86 = 3 \times ·86 = 2·58$. If we take the coefficient from TABLE I., we shall find it (opposite to 2, the ratio of the length of the orifice to its depth) to be ·617; and, for this coefficient, opposite to 2·58, in TABLE V., or the next number to it, we find the required coefficient ·636; hence the discharge is ·636 × 19·658 = 12·502 cubic feet per second. If we assume the coefficient in still water to be ·628, then we shall obtain the new coefficient ·647, and the discharge would be ·647 × 19·658 = 12·719 cubic feet. If the junction of the tube with the cistern be rounded, as shown by the dotted lines, we have to multiply 2·58 by ·807, which gives 2·08 for the new value of $\dfrac{c}{A}$, opposite which we shall find, in TABLE V., when the first coefficient is ·628, the new coefficient ·659; and the discharge in this case would be ·659 × 19·658 = 12·955 cubic feet per second.

It is not necessary to take out the coefficient of mean velocity facing the orifice to more than two places of decimals. For sluices in streams and mill races, Fig. 17, the mean coefficient ·628 in still water

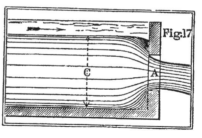

may be assumed, and thence the new coefficient suited to the ratio $\frac{c}{A}$ may be found, as in the first portion of EXAMPLE III.

EXAMPLE IV. *What is the discharge through the aperture* A, *equal 2 feet by 1 foot, when the channel is to the orifice as* 3·375 *to* 1, *and the depth of the centre is* 1·25 *foot below the surface, taken at about 3 feet above the orifice?* Here the coefficient of the approaching velocity is ·85 nearly, whence the new value of $\frac{c}{A}$ is 3·375 × ·85 = 2·87; and as $c_d = $ ·628, we shall get from TABLE V. the new coefficient ·644. Hence

$D = 2 \times 1 \times \frac{107·669}{12} \times $ ·644 (TABLE II.) $= 2 \times 8·972 \times $ ·644

$= 17·944 \times $ ·644 $= 11·556$ cubic feet per second.

Weisbach finds the discharge, by an empirical formula, to be 11·31 cubic feet. If the coefficient be sought in TABLE I., we shall find it ·617 nearly, from which, in TABLE V., we shall find the new coefficient to be ·632; hence 17·944 × ·632 = 11·341 cubic feet per second. If the coefficient ·6225 were used, we should find the new coefficient equals ·638, and the discharge 11·468 cubic feet. *Or thus:* The ratio of the head at the upper edge to the depth of the orifice is $\frac{9}{12} = $ ·75, and from TABLE IV. we find $(1·75)^{\frac{3}{2}} - (·75)^{\frac{3}{2}} = 1·6655$. Assuming the coefficient to be ·644, we find from TABLE VI. the discharge per minute over a weir 12 inches deep and 1 foot long to be $\frac{208·650 + 205·119}{2} = 206·884$ cubic feet nearly;

F

and as the length of the orifice is 2 feet, we have
$\frac{2 \times 206 \cdot 884 \times 1 \cdot 6655}{60} = 11 \cdot 482$ cubic feet per second, which is
the correct theoretical discharge for the coefficient ·644, and
less than the approximate result, 11·556 cubic feet above
found, by only a very small difference. The velocity of
approach in this example must be derived from the surface
inclination of the stream.

For notches or Poncelet weirs the approaching velocity is a
maximum at or near the surface. If the central velocity at the
surface facing the notch be 1, the mean velocity from side to
side will be ·914. We may therefore assume the variation of
the central to the mean velocity to be from 1 to ·914; and hence
the ratio of the mean velocity at the surface of the channel
to that facing the notch or weir cannot be less than ·914 to 1,
and varies up to 1 to 1; the first ratio obtaining when the
notch or weir occupies a very small portion of the side or
width of the channel, and the other when the weir extends
for the whole width. Following the same mode of calculation
as at p. 63, Fig. 15, we shall find as follows:—

| Ratio of the width of the notch to the width of the channel. | Values of $de$, Fig. 15. | Values of $bc$, Fig. 15. |
|---|---|---|
| 0 | ·086 | ·914 |
| ·1 | ·085 | ·915 |
| ·2 | ·083 | ·917 |
| ·3 | ·078 | ·922 |
| ·4 | ·072 | ·928 |
| ·5 | ·064 | ·936 |
| ·6 | ·055 | ·945 |
| ·7 | ·044 | ·956 |
| ·8 | ·031 | ·969 |
| ·9 | ·016 | ·984 |
| 1·0 | ·000 | 1·000 |

These values of $bc$ are to be used as before in order to find
the value of $\frac{C}{A}$, opposite to which in the tables, and under the
heading for weirs, will be found the new coefficient.

EXAMPLE V. *The length of a weir is* 10 *feet; the width of the approaching channel is* 20 *feet; the head, measured about* 6 *feet above the weir, is* 9 *inches; and the depth of the channel* 3 *feet: what is the discharge?* Assuming the circumstances of the overfall to be such that the coefficient of discharge for heads, measured from still water in a deep weir basin or reservoir, will be ·617, we find from TABLE VI. the discharge to be 128·642 × 10 = 1286·42 cubic feet per minute; but from the smallness of the channel the water approaches the weir with some velocity, and $\frac{C}{A} = \frac{20 \times 3}{10 \times \frac{3}{4}} = 8$. We have also the width of the channel equal to twice the width of the weir, and hence (small table, p. 30) 8 × ·936 = 7·488 for the new value of $\frac{C}{A}$. From TABLE V. we now find the new coefficient $\frac{·622 + ·624}{2} = ·623$, and hence the discharge is

$$\frac{1286·42 \times ·623}{·617} = 1298·93 \text{ cubic feet per minute.}$$

*Or thus:* As the theoretical discharge, TABLE VI., is 2084·96 cubic feet, we get 2084·96 × ·623 = 1298·93, the same as before. In this example, however, the mean velocity approaching the overfall bears to the mean velocity in the channel a greater ratio than 1 : ·936, as, though the head is pretty large in proportion to the depth of the channel, the ratio of the sections $\frac{A}{C} = \frac{1}{8}$ is small. We shall therefore be more correct by finding the multiplier from the small table, p. 63. By doing so the new value of $\frac{C}{A}$ is 8 × ·838 = 6·704. From this and the coefficient ·617 we shall find, as before from TABLE V., the new coefficient to be ·627; hence we get 2084·96 × ·627 = 1307·27 cubic feet per minute for the discharge.

The foregoing solution takes for granted that the velocity of approach is subject to contraction before arriving at the overfall or in passing through it; now, as this reduces the mean velocity of approach from 1 to ·784, when the coefficient for heads in still water is ·617, we have to multiply the value of $\frac{C}{A} = 6·704$, last found, by ·784, and we get $6·704 \times ·784 = 5·26$ for the value $\frac{C}{A}$ due to this correction, from which we find the corresponding coefficient in TABLE V. to be ·629, and hence the corrected discharge is $2084·96 \times ·629 = 1311·44$ cubic feet.

It is to be remarked that the value of $\frac{C}{A}$ in TABLE V. *is simply an approximate value for the ratio of the velocity in the channel facing the orifice to the velocity in the orifice itself;* and the corrections applied in the foregoing examples were for the purpose of finding this ratio more correctly than the simple expression $\frac{C}{A}$ gives it. The following auxiliary table will enable us to find the correction, and thence the new coefficient, with facility. Thus, if the channel be five times the orifice, and a loss in the approaching velocity takes

AUXILIARY TABLE TO TABLE V.

| Ratio of the orifice to the channel, or $\frac{A}{C}$ | Multipliers due to the difference of the central and mean velocity. | Multipliers for finding the new values of $\frac{C}{A}$ in TABLE V., when the water approaches without contraction or loss of velocity. | | | | | | |
|---|---|---|---|---|---|---|---|---|
| | | Coeffict. ·639 | Coeffict. ·628 | Coeffict. ·617 | Coeffict. ·606 | Coeffict. ·595 | Coeffict. ·584 | Coeffict. ·573 |
| 0 | ·835 | ·69 | ·67 | ·65 | ·64 | ·62 | ·60 | ·58 |
| ·1 | ·837 | ·70 | ·68 | ·66 | ·64 | ·62 | ·60 | ·59 |
| ·2 | ·842 | ·70 | ·68 | ·66 | ·64 | ·62 | ·61 | ·59 |
| ·3 | ·850 | ·71 | ·69 | ·67 | ·65 | ·63 | ·61 | ·59 |
| ·4 | ·861 | ·72 | ·69 | ·68 | ·66 | ·64 | ·62 | ·60 |
| ·5 | ·876 | ·73 | ·71 | ·69 | ·67 | ·65 | ·63 | ·61 |
| ·6 | ·894 | ·74 | ·72 | ·70 | ·68 | ·66 | ·64 | ·62 |
| ·7 | ·916 | ·76 | ·74 | ·72 | ·70 | ·68 | ·66 | ·64 |
| ·8 | ·941 | ·78 | ·76 | ·74 | ·72 | ·70 | ·68 | ·66 |
| ·9 | ·969 | ·81 | ·78 | ·76 | ·74 | ·72 | ·70 | ·68 |
| 1·0 | 1·000 | ·831 | ·807 | ·784 | ·762 | ·740 | ·719 | ·699 |

place equal to that in a short cylindrical tube, we get $5 \times \cdot 842 = 4\cdot 210$ for the new value of $\frac{C}{A}$, opposite to which, in TABLE V., will be found the coefficient sought. If the coefficient for still water be ·606, we shall find it to be ·612 for orifices and ·623 for weirs. But when the water approaches without loss of velocity, we find from the auxiliary table ·64 for the multiplier instead of ·842, and consequently the new value of $\frac{C}{A}$ becomes $5 \times \cdot 64 = 3\cdot 2$, from which we shall find ·617 to be the new coefficient for orifices and ·636 for weirs. The auxiliary table is calculated by multiplying the numbers in column 2 (see p. 63) by $c_d \left\{ \frac{1}{1-c_d^2} \right\}^{\frac{1}{2}}$, which will be found for the different values of $c_d$ in the table, viz. ·639, ·628, ·617, ·606, ·595, ·584, and ·573, to be ·831, ·807, ·784, ·762, ·740, ·719, ·699 respectively, as given in the top and bottom lines of figures.

In weirs at right angles to channels with parallel sides, the sectional area can never equal that of the channel unless it be measured at or above the point A, where the sinking of the overfall commences; and unless also the bed C D and surface A B have the same inclination. In all open channels, as mill

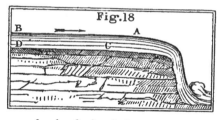

Fig. 18

races, streams, rivers, the supply is derived from a surface inclination A B, and this inclination regulates itself to the discharging power of the overfall. When the overfall and channel have the same width, and it is considerable, we have, as shall appear hereafter, $91 \sqrt{hs}$ for the mean velocity in the channel, where $h$ is the depth in feet and $s$ the rate of inclination of the surface A B. We have also $\frac{2}{3}\sqrt{2gh}$ for the theoretical velocity of discharge at the overfall, of equal depth with the channel, and, when both velocities are equal,

$$\tfrac{2}{3}\sqrt{2gh} = 5{\cdot}35\sqrt{h} = 91\sqrt{hs};$$

from which we find

$$s = \frac{1}{289},$$

the inclination of B A when the supply is equal to the theoretical discharge at the overfall. If the coefficient at the overfall were ·628, or, which is nearly the same thing, if a large and deep weir basin intervene between the weir and

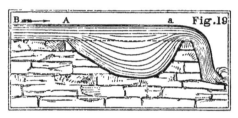

channel, Fig. 19, the velocity of approach would be destroyed, and we should have

$$5{\cdot}35 \times {\cdot}628 \sqrt{h} = 3{\cdot}36 \sqrt{h} = 91 \sqrt{hs};$$

and thence the inclination of A B

$$s = \frac{1}{734}$$

very nearly. When we come to discuss the surface inclination of rivers, we shall see that the conditions here assumed and the resulting surface inclinations would involve a considerable loss of head. If the quantity discharged under both circumstances be the same, and $h$ be the depth in the first case, Fig. 18, we shall then have the head in the latter case, Fig. 19, equal $\left(\frac{5{\cdot}35}{3{\cdot}36}\right)^{\frac{2}{3}} h = 1{\cdot}36\, h$ very nearly, from which and the surface inclination the extent of the backwater may be found with sufficient accuracy. When, in Fig. 19, the inclination of A B exceeds $\frac{1}{734}$, the head at $a$ must exceed the depth of the river above A. We must refer to pages further on for some remarks on the backwater curve.

## SECTION V.

### SUBMERGED ORIFICES AND WEIRS.—CONTRACTED RIVER CHANNELS.

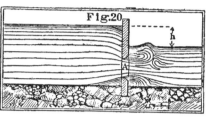

Fig.20

The available pressure at any point in the depth of the orifice A, Fig. 20, is equal to the difference of the pressures on each side. This difference is equal to the pressure due to the height $h$, between the water surfaces on each side of the orifice; in this case, the velocity is

(47.) $$v = c_d \sqrt{2gh};$$

and the discharge

(48.) $$\text{D} = ldc_d \sqrt{2gh};$$

in which, as before, $l$ is the length, and $d$ the depth of the rectangular orifice A.

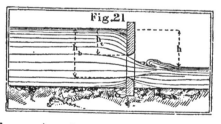

Fig.21

When the orifice is partly submerged, as in Fig. 21, we may put $h_b - h = d_2$ for the submerged depth, and $h - h_t = d_1$, the remaining portion of the depth; whence $d_1 + d_2 = d$ is the entire depth. The discharge through the submerged depth $d_2$ is $c_d l d_2 \sqrt{2gh}$, and the discharge though the upper portion $d_1$ is

$$\tfrac{2}{3} c_d l \sqrt{2g}(h^{\frac{3}{2}} - h_t^{\frac{3}{2}});$$

whence the whole discharge—assuming the coefficient of discharge $c_d$ is the same for the upper and lower depths—is

(49.) $$\text{D} = c_d l \sqrt{2g} \left\{ d_2 \sqrt{h} + \tfrac{2}{3}(h^{\frac{3}{2}} - h_t^{\frac{3}{2}}) \right\}.$$

We may, however, equation (31), assume that

$$\tfrac{2}{3} c_d l \sqrt{2g} (h^{\frac{3}{2}} - h_t^{\frac{3}{2}}) = c_d d_1 l \sqrt{2g\left(h - \tfrac{d_1}{2}\right)}$$

very nearly, and hence

(50.)  $D = c_d l d_2 \sqrt{2gh} + c_d l d_1 \sqrt{2g\left(h - \tfrac{d_1}{2}\right)}.$

As $h_t + \tfrac{d_1}{2} = h - \tfrac{d_1}{2}$, this equation may be changed into

(51.)  $D = c_d l d_2 \sqrt{2gh} + c_d l d_1 \sqrt{2g\left(h_t + \tfrac{d_1}{2}\right)}.$

In either of these forms the values of

$$c_d \sqrt{2gh},\ c_d \sqrt{2g\left(h - \tfrac{d_1}{2}\right)},\ \text{and}\ c_d \sqrt{2g\left(h_t + \tfrac{d_1}{2}\right)}$$

can be had from TABLE II., and the value of the discharge D thence easily found.

When the water approaches the orifice with a determinate velocity, the height $h_a$ due to that velocity can be found from TABLE II., and the discharge is then found by substituting $h + h_a$ and $h_t + h_a$ for $h$ and $h_t$ in the above equations.

In the submerged weir, Fig. 22, $h$ becomes equal to $d_1$, and $h_t = 0$; the discharge, equation (49), then becomes

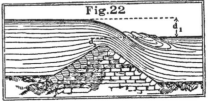

Fig. 22

(52.)  $D = c_d l d_2 \sqrt{2gd_1} + \tfrac{2}{3} c_d l d_1 \sqrt{2gd_1}$
$\qquad = c_d l \sqrt{2gd_1} \left\{ d_2 + \tfrac{2}{3} d_1 \right\}.$

When the water approaches with a velocity due to the height $h_a$, then $h$ becomes $h + h_a$, $h_t = h_a$, and equation (49) becomes

(53.)  $D = c_d l \sqrt{2g} \left\{ d_2 \sqrt{d_1 + h_a} + \tfrac{2}{3} (d_1 + h_a)^{\frac{3}{2}} - h_a^{\frac{3}{2}} \right\}.$

In the improvement of the navigation of rivers, it is sometimes necessary to construct weirs so as to raise the upper waters by a given depth $d_1$. The discharge D is in such cases previously known, or easily determined, and from the values of $d_1$ and D we can easily determine, equation (52), the value of

(54.) $$d_2 = \frac{D}{c_d l \sqrt{2g d_1}} - \frac{2}{3} d_1;$$

or, by taking the velocity of approach into account,

(55.) $$d_2 = \frac{D}{c_d l \sqrt{2g(d_1 + h_a)}} - \frac{2}{3} \frac{(d_1 + h_a)^{\frac{3}{2}} - h_a^{\frac{3}{2}}}{\sqrt{d_1 + h_a}}.$$

This value of $d_2$ must be the depth of the top of the weir below the original surface of the water, in order that this surface should be raised by a given depth $d_1$. When $h_a$ is small compared with $d_2$, we may take

$$\frac{2}{3}(d_1 + h_a) = \frac{2}{3} \times \frac{(d_1 + h_a)^{\frac{3}{2}} - (h_a)^{\frac{3}{2}}}{\sqrt{d_1 + h_a}} \text{ in equation (55).}$$

EXAMPLE VI.—*A river whose width at the surface is 70 feet, whose hydraulic mean depth is 4·4 feet, and whose cross sectional area is 325 feet, has a surface inclination of 1 foot per mile; to what depth below, or height above the surface must a weir at right angles to the channel be raised, so that the depth of water immediately above it shall be increased by 3½ feet?*

When the hydraulic mean depth is 4·4 feet, and the fall per mile 1 foot, we find from TABLE VIII. that the mean velocity of the river is 29·98 or 30 inches very nearly per second. The discharge is, therefore, $325 \times 2\frac{1}{2} = 812·5$ cubic feet per second, or 48750 cubic feet per minute. Hence, $\frac{48750}{70} = 696·4$ cubic feet must pass over each foot in length of the weir per minute. Assuming the coefficient $c_d = ·628$ in the first instance, we find from TABLE VI. the head passing over a weir corresponding to this discharge to be 27·4 inches; but as the head is to be increased by 3½ feet, or

42 inches, it is clear that the weir must be *perfect;* that is, have a clear overfall, and rise $42 - 27·4 = 14·6$ inches over the original water surface. In order that the weir may be submerged, or *imperfect,* the head could not be increased by more than 27·4 inches. Let us, therefore, assume in the example, that the increase shall be only 18 instead of 42 inches; the weir then becomes submerged, and we have, from equation (54),

$$d_2 = \frac{696·4}{·628 \sqrt{18'' \times 2g}} - \frac{2}{3} \times 18'' \text{ (as } l = 1 \text{ foot).}$$

The value of the first part of this expression is found from TABLE VI. or TABLE II. equal to

$$\frac{696·4}{\frac{12}{18} \times \frac{3}{2} \times 370·341} = \frac{696·4}{370·341} = 1·88 \text{ feet} = 22·56 \text{ inches};$$

hence, $22·56 - \frac{36}{3} = 10·56$ inches is the values of $d_2$; that is, the submerged weir must be built within 10·56 inches of the surface to raise the head 18 inches above the former level. If, however, the velocity of approach be taken into account, we shall find this velocity equals $\frac{812·5}{430} = 2$ feet per second very nearly; and the height, or value of $h_a$, due to this velocity, taken from TABLE II., is $\frac{3}{4} = ·75$ inches nearly; therefore, from equation (55),

$$d_2 = \frac{696·4}{·628 \sqrt{2g \times 18·75''}} - \frac{2}{3} \times \frac{(18·75)^{\frac{3}{2}} - (·75)^{\frac{3}{2}}}{\sqrt{18·75}}.$$

The value of $\frac{696·4}{·628 \sqrt{2g \times 18·75''}} = $ (from TABLE VI.)

$$\frac{696·4}{\frac{3}{2} \times \frac{12}{18·75} \times 393·75} = \frac{696·4}{378·8}* = 1·84 \text{ feet} = 22·08 \text{ inches};$$

\* This is found from TABLE II. more readily.

ORIFICES, WEIRS, PIPES, AND RIVERS. 75

also, $\dfrac{2}{3} \times \dfrac{(18\cdot 75)^{\frac{3}{2}} - (\cdot 75)^{\frac{3}{2}}}{\sqrt{18\cdot 75}} = \dfrac{2}{3} \times 18\cdot 75 - \dfrac{2}{3} \times \dfrac{(\cdot 75)^{\frac{3}{2}}}{\sqrt{18\cdot 75}}$

$= 12\cdot 5 - \dfrac{2}{3} \times \dfrac{\cdot 65}{4\cdot 33} = 12\cdot 5 - \cdot 1 = 12\cdot 4.$

Hence $d_2 = 22\cdot 08 - 12\cdot 4 = 9\cdot 68$ inches, or about 1 inch less than the value previously found from equation (54). The mean coefficient of discharge was here assumed to be ·628. Experiments on submerged weirs show that the value of $c_d$ varies up to ·8, but as this coefficient would reduce the value of $d_2$, or the depth of the top of the weir below the surface, it is safer (where a given depth above a weir must be obtained) to use the lesser and ordinary coefficients of perfect weirs, with a clear overfall, for finding the crest levels of submerged weirs, when it is necessary to construct them. If the coefficient ·8 were used in the previous calculation, we should have found

$$d_2 = \dfrac{\cdot 628 \times 22\cdot 08}{\cdot 8} - 12\cdot 4 = 17\cdot 33 - 12\cdot 4 = 4\cdot 93 \text{ inches,}$$

or not much more than half the previous value; but this would only increase the whole height of the weir by $9\cdot 68 - 4\cdot 93 = 4\cdot 75$ inches.

As $D = \dfrac{2}{3} c_d l \sqrt{2g} \{(d_1 + h_a)^{\frac{3}{2}} - h_a^{\frac{3}{2}}\}$ for a perfect weir with a free overfall, it is clear that when D is greater than

$$\dfrac{2}{3} c_d l \sqrt{2g} \{(d_1 + h_a)^{\frac{3}{2}} - h_a^{\frac{3}{2}}\},$$

the weir is imperfect or submerged.

CONTRACTED RIVER CHANNELS.

When the banks of a river, whose bed has a uniform inclination, approach each other, and contract the width of

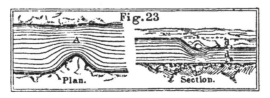
Fig. 23
Plan. Section.

the channel in any way, as in Fig. 23, the water will rise at the contracted portion, A, until the increased velocity of discharge compensates for the reduced cross section. If we put, as before, $d_1$ for the increase of depth at the contracted width, and $d_2$ for the previous depth of the channel, we shall find the water passing through the lower depth, $d_2$, equal to $c_d \, l d_2 \sqrt{2gd_1}$, in which $l$ is the width of the contracted channel at A, and the water overflowing through $d_1$ equal to $\frac{2}{3} c_d \, l \, d_1 \sqrt{2gd_1}$; and hence the whole discharge through A is

(56.) $$D = c_d \, l \sqrt{2gd_1} \left( d_2 + \frac{2}{3} d_1 \right).$$

When our object is to find the width $l$ of the contracted channel, so that the depth of water in the upper stretch shall be increased by a given depth $d_1$, we have

(57.) $$l = \frac{D}{c_d \sqrt{2gd_1} \left( d_2 + \frac{2}{3} d_1 \right)}.$$

When the velocity of approach is large, or the height $h_a$ due to it a large portion of $d_1$, it must not be neglected. In this case, as before, we find the discharge through the depth $d_2$ equal to $c_d \, l d_2 \sqrt{2g(d_1 + h_a)}$; and the discharge through the depth $d_1$ equal to $\frac{2}{3} c_d \, l \sqrt{2g} \{ (d_1 + h_a)^{\frac{3}{2}} - h_a^{\frac{3}{2}} \}$; and hence the whole discharge is

(58.) $$D = c_d \, l \sqrt{2g} \{ d_2 (d_1 + h_a)^{\frac{1}{2}} + \frac{2}{3} [ (d_1 + h_a)^{\frac{3}{2}} - h_a^{\frac{3}{2}} ] \};$$

from which we find

(59.) $$l = \frac{D}{c_d \sqrt{2g} \{ d_2 (d_1 + h_a)^{\frac{1}{2}} + \frac{2}{3} [ (d_1 + h_a)^{\frac{3}{2}} - h_a^{\frac{3}{2}} ] \}}.$$

If the projecting spur or jetty at A be itself submerged, these formulæ must be extended; the manner of doing so, however, presents no difficulty, as it is only necessary to find the discharges of the different sections according to the preceding formulæ and add them together; but the resulting formula so found is too complicated to be of any practical value.

### HEADS ABOVE BRIDGES.

Equations (56), (57), (58), and (59), are applicable to the cases of contraction in river channels caused by the construction of bridge-piers and abutments, when the width $l$ is put for the sum of the openings between them. The value of the coefficient $c_d$ will depend on the peculiar circumstances of each case; we have seen that it rises as high as ·8, in some cases of submerged weirs, and for cases of contracted channels it rises sometimes as high as ·956, particularly when they are analogous to those for the discharge through mouthpieces and short tubes. When the heads of the piers are square to the channel, the coefficient may be taken at about ·8; when the angles of the cut-waters or sterlings are obtuse, it may be taken at about ·9; and when curved and acute, at ·97. With this coefficient, a head of $1\frac{3}{4}$ inch will give a velocity of very nearly 36 inches, or 3 feet per second; but as a certain amount of loss takes place from the velocity of the tail-water being in general less than that through the arch, from obstructions in the passage, and from square-headed and very short piers, the coefficient may be so small in some cases as ·628, which would require a head of $4\frac{1}{4}$ inches to obtain the same velocity. This head is to the former as 17 to 7. The selection of the proper coefficient suited to any particular case is, therefore, a matter of the first importance in determining the effect of obstructions in river channels: we shall have to recur to this subject again, but it is necessary to observe here, that the form of the approaches, the length of the piers compared with the distance between them, or span, and the length and form of the obstruction compared with the width of the channel, must be duly considered before the coefficient suited to the particular case can be fixed upon. Indeed, the coefficients will always approximate towards those, given in the next section, for mouth-pieces, shoots, and short tubes similarly circumstanced. For some further remarks on contracted channels, see SECTION X.

## SECTION VI.

SHORT TUBES, MOUTH-PIECES, AND APPROACHES.—COEFFICIENTS OF DISCHARGE FOR SIMPLE AND COMPOUND SHORT TUBES.—SHOOTS.

The only orifices we have heretofore referred to were those in thin plates or planks, with a few incidental exceptions. It has been shown, page 20, Fig. 4, that a rounding off, next the water, of the mouth-piece increases the coefficient; and when the curving assumes the form of the *vena-contracta*, the coefficient increases to ·974, or nearly unity. The discharge from a short cylindrical tube A, Fig. 24, whose length

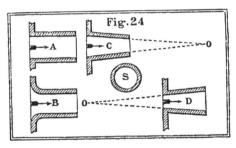

is from one and a half to three times the diameter, is found to be very nearly an arithmetical mean between the theoretical discharge and the discharge through a circular orifice in a thin plate of the same diameter as the tube, or ·814 nearly. If, however, the inner arris be rounded or chamfered off in any way, the coefficient will increase until, in the tube B, Fig. 24, with a properly-rounded junction, the coefficient becomes unity very nearly. In the conical short tubes C and D the coefficients are found to vary according to some function of the converging or diverging angles o, o, and according as we take the lesser or greater diameter to calculate from. When the length of the tube exceeds three times the diameter, the friction of the water against the sides should be taken into account. As for orifices in thin plates, so also for short tubes, the coefficients are found to vary according to the depth of the centre below the surface of the water, and to increase as the depths and diameter of the tube decrease. Poleni was the first who remarked that the discharge through a short tube was greater than that through a simple orifice of the same diameter, in the proportion of 133 to 100, or as ·617 to ·821.

### CYLINDRICAL SHORT TUBES, A, FIG. 24.

The experiments of Bossût, as reduced by Prony, give the following coefficients, at the corresponding depths, for a cylindrical tube A, Fig. 24, 1 inch in diameter and 2 inches

COEFFICIENTS FOR SHORT TUBES, FROM BOSSÛT.

| Heads, in feet. | Coefficients. | Heads, in feet. | Coefficients. | Heads, in feet. | Coefficients. |
|---|---|---|---|---|---|
| 1 | ·818 | 6 | ·806 | 11 | ·805 |
| 2 | ·807 | 7 | ·806 | 12 | ·804 |
| 3 | ·807 | 8 | ·805 | 13 | ·804 |
| 4 | ·807 | 9 | ·805 | 14 | ·804 |
| 5 | ·806 | 10 | ·805 | 15 | ·803 |

long. The depths are given in Paris feet in the original, but the coefficients remain the same practically for depths in English feet.

Venturi's experiments give a coefficient ·823 for a short tube A, $1\frac{1}{2}$ inch in diameter and $4\frac{1}{2}$ inches long, at a depth of 2 feet $8\frac{1}{2}$ inches, the coefficient through an orifice in a thin plate of the same diameter and at the same depth being ·622. We have calculated these coefficients from the original experiments. The measures in those were Paris feet and inches, from which the calculations were directly made; and as the difference *in the coefficient* for small changes of depth or dimensions is immaterial or vanishes, as may be seen by the foregoing small table, and as 1 Paris inch or foot is equal to 1·0658 English inches or feet, the former measures exceed the latter by only about $\frac{1}{15}$th. We may therefore assume that *the coefficient* for any orifice, at any depth, is the same, whether the dimensions be in Paris or English feet or inches. This remark will be found generally useful in the consideration of the older continental experiments, and will prevent unnecessary reductions from one standard to another where the coefficients only have to be considered.

The mean value derived from the experiments of Michelotti, at depths from 3 to 20 feet, and with short tubes a from $\frac{1}{2}$ inch to 3 inches in width, is $c_d = \cdot 814$. Buff's experiments* give the following results for a tube $\frac{3}{10}$ of an

BUFF'S COEFFICIENTS FOR SMALL SHORT TUBES.

| Head in inches. | Coefficient. | Head in inches. | Coefficient. | Head in inches. | Coefficient. |
|---|---|---|---|---|---|
| 1¼ | ·855 | 6 | ·840 | 23 | ·829 |
| 2½ | ·861 | 14 | ·840 | 32 | ·826 |

inch wide and $\frac{5}{10}$ of an inch long, nearly. The increase for smaller tubes and for lesser depths appears by comparing these results with the foregoing, and from the results in themselves, generally. Weisbach's experiments give a mean value for $c_d = \cdot 815$, and for depths of from 9 to 24 inches the coefficients ·843, ·832, ·821, ·810 respectively, for tubes $\frac{4}{10}$, $\frac{8}{10}$, $\frac{12}{10}$, and $\frac{16}{10}$ of an inch wide, the length of each tube being three times the diameter. D'Aubuisson and Castel's experiments with a tube ·61 inch diameter and 1·57 inch long, give ·829 for the coefficient at a depth of 10 feet.

We have calculated the coefficients in the following two short tables, from Rennie's experiments with glass orifices and tubes, Table 7, p. 435, *Philosophical Transactions* for 1831. The form of the orifices, or length of the shorter tubes is not stated, but it is probable, from the results, that the arrises of the ends were in some way rounded off; it is stated they were "enlarged." Indeed, the discharges from the short tube, or orifice of ¼ inch diameter exceed the theo-

* Annalen der Physik und Chemie von Poggendorff, 1839. Band 46, p. 243.

retical ones in the proportion of 1·261 to 1, and 1·346 to 1. These results could not have been derived from a simple cylindrical tube, but might have arisen from the arrises being more or less rounded at both ends, and the orifice partaking of the nature of a compound tube, which may be constructed, as we shall hereafter shew, so as to increase the theoretical discharge from 1 up to 1·553. The resulting coefficients

COEFFICIENTS FOR SHORT TUBES, THE ENDS ENLARGED.

| Head in feet. | ¼ inch diameter. | ½ inch diameter. | ¾ inch diamete . | 1 inch ameter. |
|---|---|---|---|---|
| 1 | 1·231 | ·831 | ·766 | ·912 |
| 2 | 1·261 | ·839 | ·820 | ·920 |
| 3 | 1·346 | ·838 | ·821 | ·880 |
| 4 | 1·261 | ·831 | ·829 | ·991 |

for the ½ and ¾ inch tubes, approach very closely to those obtained by other experimenters, but those for the inch tube are too high, unless the arris at the ends was also rounded. The coefficients derived from the experiments with a cylindrical glass tube 1 foot long, as here given, are very variable; like the others they are, however, valuable, as ex-

COEFFICIENTS DERIVED FROM EXPERIMENTS WITH A GLASS TUBE ONE FOOT LONG.

| Heads in feet. | ¼ inch diameter. | ½ inch diameter. | ¾ inch diameter. | 1 inch diameter. |
|---|---|---|---|---|
| 1 | ·892 | ·703 | ·691 | ·760 |
| 2 | ·914 | ·734 | ·718 | ·749 |
| 3 | ·931 | ·723 | ·709 | ·777 |
| 4 | ·914 | ·725 | ·677 | ·815 |

hibiting the uncertainty attending "experiments of this nature," and the necessity for minutely observing and recording every circumstance which tends to alter and modify them. Indeed, for small tubes, a very slight difference in the measurement of the diameter must alter the result a good deal, particularly when it is recollected that measurements are seldom taken more closely than the sixteenth of an inch,

G

unless in special cases. As the author, however, states, p. 433 of the work referred to, that the "diameters of the tubes at their extremities were carefully enlarged to prevent wire edges from diminishing the sections;" this circumstance alone must have modified the discharges, and would account for most of the differences.

The coefficient for rectangular short tubes differs in no way materially from those given for cylindrical ones, and may be taken on an average at ·814 or ·815.

SHORT TUBES WITH A ROUNDED MOUTH-PIECE, B, FIG. 24.

When the junction of a short tube with a vessel takes the form of the contracted vein, Figs. 3 and 4, page 20, the mean value of the coefficient $c_d = $ ·956, and the actual discharge is found to be from 93 to 98 per cent. of the theoretical discharge. Wiesbach, for a tube $1\frac{1}{2}$ inch long and $\frac{9}{10}$ inch diameter, rounded at the junction, found at 1 foot deep $c_d = $ ·958, at 5 feet deep $c_d = $ ·969, and at 10 feet deep $c_d = $ ·975. These experiments shew an increase of coefficient in this particular case for an increase of depth. Any other form of junction than that of the contracted vein, will reduce the discharge, and the coefficients will vary from ·807 to ·974, according to the change from the cylindrical to the properly curved form, beyond which the coefficient again decreases from ·974 to ·807. The coefficients derived from Venturi's experiments will be given hereafter.

SHORT CONICAL CONVERGENT TUBES, C, FIG. 24.

The experiments of D'Aubuisson and Castel lead to the following coefficients of discharge and velocity* from a conically convergent tube c at a depth of 10 feet. We have interpolated the original angles and coefficients so as to render the table more convenient to refer to for practical purposes than the original. The diameter of the tube at the smaller or discharging orifice in the experiments was ·61 inches, and

* Traité d'Hydraulique, Paris, p. 60.

ORIFICES, WEIRS, PIPES, AND RIVERS. 83

COEFFICIENTS FOR CONICAL CONVERGENT TUBES.

| Converging angle o. | Coefficient of discharge. | Coefficient of velocity. | Converging angle o. | Coefficient of discharge. | Coefficient of velocity. |
|---|---|---|---|---|---|
| 1° | ·858 | ·858 | 14° | ·943 | ·964 |
| 2° | ·873 | ·873 | 16° | ·937 | ·970 |
| 3° | ·908 | ·908 | 18° | ·931 | ·971 |
| 4° | ·910 | ·909 | 20° | ·922 | ·971 |
| 5° | ·920 | ·916 | 22° | ·917 | ·973 |
| 6° | ·925 | ·923 | 26° | ·904 | ·975 |
| 8° | ·931 | ·933 | 30° | ·895 | ·976 |
| 10° | ·937 | ·950 | 40° | ·869 | ·980 |
| 12° | ·942 | ·955 | 50° | ·844 | ·985 |

the length of the axis 1·57 inch; that is, the length was 2·6 times the smaller diameter of the tube. The coefficient was ·829 for the cylindrical tube, *i. e.* when the angle at o was nothing. The angle of convergence o determines, from the proportions, the length of the inner and longer diameter of the tube. The coefficients of discharge increase up to ·943 for an angle of 13½ or 14 degrees, after which they again decrease; but the coefficients of velocity increase as the angle of convergence o increases from ·829, when the angle is zero up to ·985 for an angle of 50 degrees.

When D is the discharge and A the area of the section, we have, as before shown, $D = c_d A \sqrt{2gh}$; but as, in conically convergent or divergent tubes the inner and outer areas (or, as they may be called, the receiving and discharging sections) vary, it is clear that (the discharge being the same, and also the theoretical velocity $\sqrt{2gh}$) the coefficient $c_d$ must vary inversely with the sectional area A, and that $c_d$ A must be constant. For the coefficients tabulated, the sectional area used is that at the smaller end of a convergent tube c.

For a short tube c, whose length is ·92 inch, lesser diameter 1·21 inch, and greater diameter 1·5 inch, we have found, from Venturi's experiments, that $c_d = ·607$ if the larger diameter be used in the calculation, and $c_d = ·934$

G 2

when the lesser diameter is made use of, the discharge taking place under a pressure of 2 feet 8½ inches.

The earlier experiments of Poleni, when reduced, furnish us with the following coefficients: A tube 7·67 inches long, 2·167 inches diameter at each end, $c_d = \cdot 854$; the like tube with the inner or receiving orifice increased to 2¾ inches, $c_d = \cdot 903$; increased to 3·5 inches, $c_d = \cdot 898$; increased to 5 inches, $c_d = \cdot 888$; and increased to 9·83 inches, $c_d = \cdot 864$. The depth or head was 21·33 inches, the discharging orifice 2·167 inches diameter, and the length 7·67 inches, in each case.

In the conically divergent tube D, Fig. 24, the coefficient of discharge is larger than for the same tube C, convergent, when the water fills both tubes, and the smaller areas, or those at the same distances from the centres o o, are made use of in the calculations. A tube whose angle of convergence o is 5° nearly, with a head of from 1 to 10 feet, whose axial length is 3½ inches, smaller diameter 1 inch, and larger diameter 1·3 inch, gives, when placed as at C, ·921 for the coefficient; but when placed as at D, the coefficient increases to ·948. In the first case the smaller area, used in both calculations, being the receiving, and in the other the discharging, orifice. The coefficient *of velocity* is, however, larger for the tube C than for the tube D, and the discharging jet of water has a greater amplitude in falling. The effects of conically diverging tubes will, however, be better perceived from the experiments on compound short tubes.

#### COMPOUND ADJUTAGES AND SHORT TUBES.

If the tube A, Fig. 24, be pierced all round with small holes at the distance of about half its diameter from the reservoir, the discharge will be immediately reduced in the proportion of ·814 to ·617. Venturi found the reduction for a tube 1½ inch diameter and 4½ inches long, at a depth of 2 feet 10½ inches, as 41 to 31, or as ·823 to ·622. As long as one hole remained open, the discharge continued at the same reduced rate; but when the last hole was stopped, the discharge again increased to the original quantity. If a small hole be pierced in a tube 4 diameters long, at the distance of 1½ or 2 diameters at farthest from the junction, the discharge will remain unaffected. This shows that the contraction in the cylindrical tube extends only a short distance from the junction, probably 1¼ or 1½ diameter, including the whole curvature of the contraction.

The contraction at the entrance into a tube from a reservoir accounts for the coefficients for a short tube A, Fig. 24, and the short tubes, diagrams 1 and 2, Fig. 25, being each the

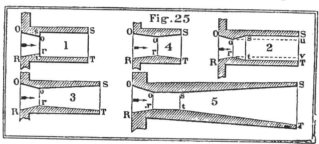

same decimal nearly, when OR : or :: 1 : ·8, or when or is not less than OR × ·79, and is at the distance of nearly $\frac{OR}{2}$ from OR. The form of the junction O o r R remaining as we have described it, the following coefficients will enable us to judge of the discharging powers of different short mouthpieces. They have been deduced and calculated principally from Venturi's experiments*.

These coefficients show very clearly that any calculations from the mere head and size of the orifice, without taking into consideration the form of the discharging tube and its connection with the reservoir, are very uncertain; and that the discharge can only be correctly obtained when all the circumstances of the case, including the form of the discharging orifice and its approaches, are duly considered.

* See Nicholson's translation of Venturi's Experimental Inquiries, published in the "Tracts on Hydraulics," London, 1836. The coefficients in the Table, next page, have been all calculated for the first time by us.

THE DISCHARGE OF WATER FROM

TABLE OF COEFFICIENTS FOR MOUTH-PIECES AND SHORT TUBES.

| Description of orifice, mouth-piece, or short tube. | Coefficients for the diameter $or$. | Coefficients for the diameter $or$. |
|---|---|---|
| 1. An orifice 1½ inch diameter in a thin plate | ·622 | ·974 |
| 2. A cylindrical tube 1½ inch diameter and 4½ inches long, A, Fig. 24 | ·823 | ·823 |
| 3. A cylindrical tube, B, Fig. 24, having the junction rounded, as in Fig. 4, page 20 | ·611 | ·956 |
| 4. A short conical convergent mouth-piece, C, Fig. 24, of the proportions of $o\,o\,r$ R, Fig. 25 | ·607 | ·934 |
| 5. The like tube divergent, with the smaller diameter at the junction with the reservoir; length 3½ inches, lesser diameter 1 inch, and greater diameter 1·3 inch | ·561 | ·948 |
| 6. The tube, $o\,o\,u\,v\,r$ R, diagram 2, Fig. 25, when $o\,R = 1½$ inch, $o\,r = 1·21$ inch, $u\,v = 1·21$ inch, and $o\,u = r\,v = 2$ inches, the cylindrical portion being shown by dotted lines | ·600 | ·923 |
| 7. The same tube when $o\,u = 11$ inches | ·567 | ·873 |
| „ The same tube when $o\,u = 23$ inches | ·531 | ·817 |
| 8. The tube, $o\,o\,s\,T\,t\,r$ R, diagram 2, Fig. 25, in which $o\,R = s\,t = s\,T = 1½$ inch, from $o$ to $s$ 1¾ inch, and $s\,s = 3$ inches, gives the same coefficient as the cylindrical tube, result No. 2 (see No. 18), viz. | ·823 | 1·266 |
| 9. The tube, diagram 1, Fig. 25, $o\,R = 1½$ inch | ·804 | 1·237 |
| 10. The same tube, having the spaces $o\,s\,o$ and $r\,t\,R$ between the mouth-piece $o\,o\,r$ R and the cylindrical tube $o\,s\,T\,R$ open to the influx of the water | ·785 | 1·209 |
| 11. The double conical tube, $o\,o\,s\,T\,r$ R, diagram 3, Fig. 25, when $o\,R = s\,T = 1½$ inch, $o\,r = 1·21$ inch, $o\,o = ·92$ inches, and $o\,s = 4·1$ inch | ·928 | 1·428 |
| 12. The like tube when, as in diagram 4, Fig. 25, $o\,o\,r$ R $= o\,s\,T\,r$, and $o\,o\,s = 1·84$ inch | ·823 | 1·266 |
| 13. The like tube when $s\,T = 1·46$ inch, and $o\,s = 2·17$ inches | ·823 | 1·266 |
| 14. The like tube when $s\,T = 3$ inches, and $o\,s = 9¾$ inches | ·911 | 1·400 |
| 15. The like tube when $o\,s = 6½$ inches, and $s\,T$ enlarged to 1·92 inch | 1·020 | 1·569 |
| 16. The like tube when $s\,T = 2¼$ inches, and $o\,s = 12½$ inches | 1·215 | 1·855 |
| 17. A tube, diagram 5, Fig. 25, when $o\,s = r\,t = 3$ inches, $o\,r = s\,t = 1·21$ inch, and the tube $o\,s\,T\,r$ the same as described in No. 11, viz. $s\,T = 1½$ inch, and $s\,s = 4·1$ inches | ·895 | 1·377 |
| 18. The tube, diagram 2, Fig. 25, when $s\,T$ is enlarged to 1·97 inch, and $s\,s$ to 7 inches, the other dimensions remaining as in No. 8 | ·945 | 1·454 |
| 19. When the junction of $o\,s\,r\,t$ with $s\,s\,T\,t$, diagram 2, Fig. 25, is improved, the other parts remaining as described in No. 8 | ·850 | 1·309 |
| Another experiment gives | ·847 | 1·303 |

# ORIFICES, WEIRS, PIPES, AND RIVERS.

When a tube similar to diagram 5, Fig. 25, has the junction o *o r* R rounded, as in Fig. 4, page 20, the outer extremity *s t* s T, such that $st = or$, $ss = 9st$, and the diameter s T $= 1\cdot 8$ times the diameter *s t*, with a *short* central cylindrical piece *o r s t* between, the coefficient of discharge corresponding to the diameter $or = rs$ will increase to $1\cdot 493$ or $1\cdot 555$; that is, the discharge is $\frac{1\cdot 493}{\cdot 622} = 2\cdot 4$, or $\frac{1\cdot 555}{\cdot 622} = 2\cdot 5$ times as much as through an orifice (whose diameter is *o r*) in a thin plate, and $\frac{1\cdot 555}{\cdot 822} = 1\cdot 9$ times as much as through a short cylindrical tube A, Fig. 24, whose diameter is also *o r*. Venturi was of opinion that this discharge continued even when the central cylindrical portion *o r s t* was of considerable length; but this was a mistake, as the maximum discharge is obtained when it is reduced so that o *o r* R and s *s t* T shall join, as in diagram 3, Fig. 25. We see from No. 15 of the foregoing coefficients that $\frac{1\cdot 569}{\cdot 622} = 2\cdot 52$ and $\frac{1\cdot 569}{\cdot 822} = 1\cdot 91$ are, perhaps, nearer to the maximum results obtainable in comparing the discharge from a compound tube o *o* s T *r* R, diagram 3, Fig. 25, with those through an orifice in a thin plate, and through a short cylindrical tube. When the form of the tube becomes curvilineal through-

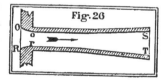

Fig. 26

out, as in Fig. 26, s T $= 1\cdot 8\, or$ and $os = 9\, or$, the coefficient suited to the diameter *o r* will be $1\cdot 57$ nearly, and the discharge will be $\frac{1\cdot 57}{\cdot 622} = 2\cdot 52$ times as much as through an orifice *o r* in a thin plate.

The whole of the preceding coefficients have been determined from circumstances in which the coefficient for an orifice in a thin plate was $\cdot 622$, and for a short cylindrical tube $\cdot 822$ or $\cdot 823$. When the circumstances of head and approaches in the reservoir are such as to increase or decrease those primary coefficients, the other coefficients for compound adjutages will have to be increased or decreased proportionately.

On examining the foregoing results, it appears sufficiently clear that the utmost effect produced by the formation of the compound mouth-piece o o s T r R, with the exception of No. 16, is simply a restoration of the loss effected by contraction in passing through the orifice o R in a thin plate, *and that the coefficient* 2·5 *applied to the contracted section at o r is simply equal to the theoretical discharge, or the coefficient unity, applied to the primary orifice* o R; for as, orifice o R : orifice *o r* :: 1 : ·64, very nearly, when o *o r* R takes the form of the *vena-contracta*, and the coefficient of discharge for an orifice *o r* in a thin plate is ·622, we get the theoretical discharge through the orifice o R to the actual discharge through an orifice *o r*, so is 1 to ·622 × ·64, so is 1 : ·39808 :: 1 : ·4 very nearly; and as ·4 × 2·5 = 1, it is clear that the form of the tube o *o* s T *r* R, when it produces the foregoing effect, simply restores the loss caused by contraction in the *vena-contracta*. Venturi's sixteenth experiment, from which we have derived the coefficients in No. 16, gives the coefficient 1·215 for the orifice o R. This indicates that a greater discharge than the theoretical, through the receiving orifice, may be obtained. It is, however, observable that Venturi, in his seventh proposition, does not rely on this result, and Eytelwein's experiments do not give a larger coefficient than 2·5 applied to the contracted orifice *o r*, which, we have above shown, is equal to the theoretical discharge through o R.

### SHOOTS.

When the sides and under edge of an orifice or notch increase in thickness, so as to be converted into a shoot or small channel, open at the top, the coefficients reduce very considerably, and to an extent beyond what the increased resistance from friction, particularly for small depths, indicates. Poncelet and Lesbros* found for orifices 8″ × 8″, that the addition of a horizontal shoot 21 inches long reduced the coefficient from ·604 to ·601, with a head of about 4 feet; but for a head of 4½ inches the coefficient fell from ·572 to ·483. For notches 8″ wide, with the addition of an horizontal shoot 9′ 10″ long, the coefficient fell from ·582 to ·479 for a head of 8″; and from ·622 to ·340 for a head of 1″. Castel also found for a notch 8″ wide, with the addition of a shoot 8″ long, inclined 4° 18′, the mean coefficient for heads from 2″ to 4½″, to be ·527 nearly.

* Traité d'Hydraulique, pp. 46 et 94.

## SECTION VII.

LATERAL CONTACT OF THE WATER AND TUBE.—ATMOSPHERIC PRESSURE.—HEAD MEASURED TO THE DISCHARGING ORIFICE.—COEFFICIENT OF RESISTANCE.—FORMULA FOR THE DISCHARGE FROM A SHORT TUBE.—DIAPHRAGMS.—OBLIQUE JUNCTIONS.—FORMULA FOR THE TIME OF THE SURFACE SINKING A GIVEN DEPTH.

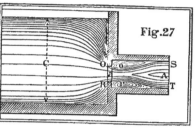

Fig. 27

The contracted vein $o\,r$ is about ·8 times the diameter $o\,R$; but it is found, notwithstanding, that water, in passing through a short tube of not less than $1\frac{1}{2}$ diameter in length, fills the whole of the discharging orifice $s\,T$. This is partly effected by the outflowing column of water carrying forward and exhausting a portion of the air between it and the tube, and by the external air then pressing on the column so as to enlarge its diameter and fill the whole tube. When once the water approaches closely to the tube, or is caused to approach, it is attracted and adheres with some force to it. The water between the tube and the *vena-contracta* is, however, rather in a state of eddy than of forward motion, as appears from the experiments, with the tube, diagram 2, Fig. 25, giving the same discharge as the simple cylindrical tube. If the entrance be even contracted by a diaphragm, as at $o\,R$, Fig. 27, the water will generally fill the tube if it be only sufficiently long. Short cylindrical tubes do not fill when the discharge takes place in an exhausted receiver; but even diverging tubes, D, Fig. 24, will be filled, in the air, when the angle of divergence o does not exceed 7 or 8 degrees, and the length be not very great nor very short.

*When a tube is fitted, with any inclination, to the bottom or side of a vessel, it is found that the discharge is that due to the head measured from the surface of the water to the lower or discharging extremity of the tube.* It must, however, be sufficiently long, and not too long, to get filled throughout. Guiglielmini was the first who referred this effect to atmospheric pressure, but the first simple explanation is that given by Dr. Mathew Young, in the *Transactions of the Royal Irish Academy,* vol. 7, p. 53. Venturi also, in his fourth proposition, gives a demonstration.

The values of the coefficients for short cylindrical tubes, which we have given above, have been derived from experiment. Coefficients which agree pretty closely with them, and which are derived from the coefficients for the discharge through an orifice in a thin plate, may, however, be calculated as follows: Let C be the area of the approaching section, Fig. 27, A the area of the discharging short tube, and $a$ the area of the orifice O R which admits the water from the vessel into the tube: also put, as before, $h$ for the head measured from the surface of the water to the centre of the tube, and diaphragm O R; $v$ for the velocity of discharge at S T; $v_a$ for the velocity of approach in the section C towards the diaphragm O R; and $c_c$ for the coefficient of contraction in passing from O R to $or$; then we have $C \times v_a = A v$, the contracted section $o r = c_c a$, and consequently the velocity at the contracted section $= \dfrac{A v}{a c_c} = \dfrac{C v_a}{a c_c}$. Now a head equal to

$$\frac{v^2 - v_a^2}{2g} = \frac{v^2 \left(1 - \dfrac{A^2}{C^2}\right)}{2g}$$

is necessary to change the velocity $v_a$ into $v$ if there be no loss in the passage; but as the water at the contracted section $or$, moving with a velocity $\dfrac{A v}{a c_c}$, strikes against the water between it and T S, moving, from the nature of the case, with a slower velocity, a certain loss of effect takes place from the impact. If this be sudden (as it is generally supposed) it is then shown by writers on mechanics *that a loss of head equal to that due to the difference of the velocities* $\dfrac{A v}{a c_c} - v$, *before and after the impact must take place.* This loss of head is therefore equal to

$$\frac{\left(\dfrac{A}{a\,c_c} - 1\right)^2 v^2}{2g},$$

whence we must have the whole head,

(60.) $$h = \frac{\left(1 - \dfrac{A^2}{C^2}\right) v^2 + \left(\dfrac{A}{a\,c_c} - 1\right)^2 v^2}{2g},$$

from which we find the velocity

(61.) $$v = \sqrt{2gh} \left\{ \frac{1}{1 - \dfrac{A^2}{C^2} + \left(\dfrac{A}{a\,c_c} - 1\right)^2} \right\}^{\frac{1}{2}}.$$

Now, as $v = \sqrt{2gh}$ would be the velocity of discharge were there no resistances, or loss sustained, it is evident that

$$\left\{ \frac{1}{1 - \dfrac{A^2}{C^2} + \left(\dfrac{A}{a\,c_c} - 1\right)^2} \right\}^{\frac{1}{2}}$$

is the coefficient of velocity. When the diameter of the diaphragm OR becomes equal to the diameter ST of the tube, $A = a$, and as the coefficient of velocity becomes equal to the coefficient of discharge when there is no contraction, we get in this case the coefficient of discharge

(62.) $$c_d = \left\{ \frac{1}{1 - \dfrac{A^2}{C^2} + \left(\dfrac{1}{c_c} - 1\right)^2} \right\}^{\frac{1}{2}},$$

and when the approaching section C is very large compared with the area A,

(63.) $$c_d = \left\{ \frac{1}{1 + \left(\dfrac{1}{c_c} - 1\right)^2} \right\}^{\frac{1}{2}}.$$

If $c_c = \cdot 64$, we shall find from the last equation $c_d = \cdot 872$; if $c_c = \cdot 601$, $c_d = \cdot 833$; if $c_c = \cdot 617$, $c_d = \cdot 847$; and if

$c_c = \cdot 621$, $c_d = \cdot 856$. These results are in excess of those derived from experiment with cylindrical short tubes, perfectly square at the ends and of uniform bore. As some loss, however, takes place in the eddy between $or$ and the tube, and from the friction at the sides, not taken into account in the above calculation, it will account for the differences of not more than from 4 to 6 per cent. between the calculation and experiment. If $c_c$ be assumed for calculation equal $\cdot 590$, then $c_d = \cdot 821$; and as this result agrees very closely with the experimental one, $c_c$ should be taken of this value in using the above formulæ for practical purposes.

COEFFICIENT OF RESISTANCE.— LOSS OF MECHANICAL POWER IN THE PASSAGE OF WATER THROUGH THIN PLATES AND PRISMATIC TUBES.

The coefficients of contraction, velocity, and discharge have been already defined. *The coefficient of resistance is the ratio of the head due to the resistance, to the theoretical head due to the discharging or final velocity.* If $v$ be this velocity, the theoretical head due to it is $\dfrac{v^2}{2g}$; and if $c_r$ be the coefficient of resistance, then the head due to the resistance itself is, from our definition, $c_r \times \dfrac{v^2}{2g}$. Now if $c_v$ be the coefficient of velocity, the theoretical velocity of discharge must be $\dfrac{v}{c_v}$, and the head due to it is equal $\dfrac{v^2}{c_v^2 \times 2g}$; but as the theoretical head due to $v$ is $\dfrac{v^2}{2g}$, we have

$$\frac{v^2}{c_v^2 \times 2g} - \frac{v^2}{2g} = \left(\frac{1}{c_v^2} - 1\right)\frac{v^2}{2g}$$

for the height due to the resistance; and, therefore, from our definition

(64.) $$c_r = \frac{1}{c_v^2} - 1;$$

from which we find

(65.) $$c_v = \left\{\frac{1}{c_r + 1}\right\}^{\frac{1}{2}}.$$

These equations enable us to calculate the coefficient of resistance from the coefficient of velocity, and *vice versâ*. If $c_v = 1$, $c_r = 0$, as it should be. The following short Table, calculated from equation (65), will be of use.

COEFFICIENTS OF VELOCITY AND RESISTANCE*.

| Coefficient of velocity. | Coefficient of resistance. | Coefficient of velocity. | Coefficient of resistance. | Coefficient of velocity. | Coefficient of resistance. |
|---|---|---|---|---|---|
| ·990 | ·020 | ·910 | ·208 | ·830 | ·452 |
| ·970 | ·063 | ·890 | ·263 | ·820 | ·488 |
| ·950 | ·109 | ·870 | ·320 | ·814 | ·508 |
| ·930 | ·156 | ·850 | ·383 | ·810 | ·525 |

The coefficient of velocity for an orifice in a thin plate, or for a mouth-piece, Fig. 4, is ·974; while that for a short prismatic tube, A, Fig. 24, is ·814 nearly. The coefficient of resistance in the former case is ·054, and in the latter ·508; there is, therefore, 9·4 times as great a loss of mechanical power in the passage through short prismatic tubes, as through orifices in thin plates or tubes with a rounded junction, as in Fig. 4, the quantities of water discharged and the discharging velocities being the same.

If the quantities discharged and the heads be the same in both cases, then we shall have

$$\frac{v_t^2}{2g \times ·814^2} = \frac{v_o^2}{2g \times ·974^2} \text{ equal the head;}$$

that is, $\dfrac{v_t^2}{·663 \times 2g} = \dfrac{v_o^2}{·949 \times 2g}$, or $·949\,v_t^2 = ·663\,v_o^2$;

whence we get $v_t^2 = ·698\,v_o^2$ and $v_o^2 = 1·431\,v_t^2$ for the relation of the discharging velocities, $v_o$ from an orifice, and $v_t$ from a short tube. The height due to the resistance is, therefore, $\left(\dfrac{1}{·814^2} - 1\right)\dfrac{v_t^2}{2g}$ for short prismatic tubes, and

---

* See the Tables of resistances, discharge, and contraction, pp. 96 and 98.

$$\left(\frac{1}{\cdot 974^2} - 1\right) \frac{1 \cdot 431 \, v_t^2}{2g}$$

for orifices in thin plates. These are to each other as ·508 to ·054 × 1·431, or as 5·08 to ·773; that is to say, *the loss of mechanical power arising from the resistance in passing through short tubes is* 15½ *times as great as when the water passes through thin plates or mouth-pieces, as in* Fig. 4; and the discharging mechanical power in plates, is to that in tubes as 1·431 to 1, or as 1 : ·698, the heads and quantities discharged being the same.

The whole loss of mechanical power in the passage is 5·4 per cent. for the plates, and about 51 per cent. for short tubes. If the loss compared with the whole head be sought, we get, when $v$ is the discharging velocity, $\frac{v}{\cdot 814}$ for the theoretical velocity due to the head in short tubes, and its square $\frac{v^2}{\cdot 814^2} = \frac{v^2}{\cdot 663}$ is as the whole head; therefore, the whole head is to the head due to the discharging velocity, as $\frac{v^2}{\cdot 663}$ to $v^2$, or as 1 to ·663, and as ·508 is the coefficient of resistance* for the discharging velocity, ·508 × ·663 = ·337 is the coefficient of resistance due to the whole head; this is equal to a loss of 34 per cent. nearly, or about one-third. In like manner, we find ·974² × ·054 = ·0512 for the coefficient when the discharge takes place through thin plates, or 5⅛ per cent. of the whole head.

### DIAPHRAGMS.

When a diaphragm, o R, Fig. 27, is placed at the entrance of a short tube, we have shown, page 91, that a loss of head equal $\dfrac{\left(\frac{A}{a c_c} - 1\right)^2 v^2}{2g}$ takes place when $v$ is the discharging velocity, whence the coefficient of resistance is

---

* Table, p. 93.

equal to $\left(\dfrac{\text{A}}{ac_c} - 1\right)^2$,* according to our definition. The coefficient of construction, $c_c$, as we have before shown, page 92, should be taken equal to ·590 in the application of formula (63); and, as it must also be taken equal to about ·621 when the area of the tube A is very large compared with the area $a$ of the orifice o r in the diaphragm, we may assume that when $\dfrac{a}{\text{A}}$ is equal to

$$0,\ \dfrac{1}{10},\ \dfrac{2}{10},\ \dfrac{3}{10},\ \dfrac{4}{10},\ \dfrac{5}{10},\ \dfrac{6}{10},\ \dfrac{7}{10},\ \dfrac{8}{10},\ \dfrac{9}{10}$$

and 1 successively, the coefficient $c_c$ must be taken equal to ·621, ·618, ·615, ·612, ·609, ·606, ·603, ·600, ·597, ·593 and ·590, in the same order. As the approaching section c may be considered exceedingly large, the value of the coefficient of discharge or velocity, as the tube o r s t is supposed full, in equation (61), becomes

(66.) $$c_d = \left\{\dfrac{1}{1 + \left(\dfrac{\text{A}}{ac_c} - 1\right)^2}\right\}^{\frac{1}{2}},$$

and the coefficient of resistance

(67.) $$c_r = \left(\dfrac{\text{A}}{ac_c} - 1\right)^2;$$

from which equations and the above values of $c_c$, corresponding to $\dfrac{a}{\text{A}}$, we have calculated the following values of the coefficients of discharge and resistance through the tube o r s t, Fig. 27.

---

* When the alteration in the velocity passing through a diaphragm is sudden, we must reject the hypothesis of D'Aubuisson, "Traité d'Hydraulique," p. 238, and adopt that of Navier, taking the loss of head to correspond to the square of the difference and not to the difference of the squares of the velocities *in* and *after* passing the orifice. The coefficient of contraction must, however, be varied to suit the ratio of the channels, as in this and the following pages.

COEFFICIENTS OF CONTRACTION, DISCHARGE, AND RESISTANCE FOR DIAPHRAGMS.

| Ratio $\frac{a}{A}$ | Coefficient $c_c$ | Coefficient $c_d$ | Coefficient $c_r$ | Ratio $\frac{a}{A}$ | Coefficient $c_c$ | Coefficient $c_d$ | Coefficient $c_r$ |
|---|---|---|---|---|---|---|---|
| 0·0 | ·621 | ·000 | infinite. | 0·6 | ·603 | ·493 | 3·115 |
| 0·1 | ·618 | ·066 | 231· | 0·7 | ·600 | ·587 | 1·907 |
| 0·2 | ·615 | ·139 | 50·8 | 0·8 | ·597 | ·675 | 1·198 |
| 0·3 | ·612 | ·219 | 19·8 | 0·9 | .593 | ·753 | ·762 |
| 0·4 | ·609 | ·307 | 9·6 | 1·0 | ·590 | ·821 | ·483 |
| 0·5 | ·606 | ·399 | 5·3 | ... | ... | ... | ... |

In this table $c_c$ is the coefficient of contraction, $c_d$ the coefficient of discharge, suited to the larger section of the pipe A, at ST; and $c_r$ the coefficient of resistance. The discharge is found from equation (61), as c is here very large compared with A, to be

(67A.) $$D = A \sqrt{2gh} \left\{ \frac{1}{1 + \left(\frac{A}{ac_c} - 1\right)^2} \right\}^{\frac{1}{2}}$$
$$= A \sqrt{2gh} \left\{ \frac{1}{1 + c_r} \right\}^{\frac{1}{2}} = c_d A \sqrt{2gh}.$$

The coefficient of resistance, $c_r$, is here equal $\left(\frac{A}{ac_c} - 1\right)^2$, and the coefficient of discharge $c_d = \frac{1}{(1 + c_r)^{\frac{1}{2}}}$.*

The tube must be so placed, that the water, after passing the diaphragm, shall fill it; for instance, between two cisterns, when the height $h$ must be measured between the water surfaces, or when the tube is sufficiently long to be filled; in this case, however, *the height must be determined from the discharging velocity*, as a portion of the head is required to overcome the friction, which we shall have immediately to refer to.

* For the loss sustained by contraction in the bore of a pipe by a diaphragm, see further on.

The table shows that the head due to the resistance is 5·3 times that due to the discharging velocity, when the area of the diaphragm is half the area of the tube; that is, the whole head required is 6·3 times that due to the velocity, and that the coefficient of discharge is reduced to ·399. In order to find the coefficients suited to the smaller area of the orifice in the diaphragm o R, when it is to be used in calculations of the discharge, we have only to divide the numbers corresponding to $\frac{a}{A}$ into those of $c_d$, opposite to them in the table. Thus, when $\frac{a}{A} = ·8$, we have the coefficient of discharge suited to the area $a$, equal $\frac{·675}{·8} = ·844$, and so of other values of the ratio $\frac{a}{A}$. The coefficients in the opposite page are for the larger orifice A.

### TUBES OBLIQUE AT THE JUNCTION.

When a tube is attached obliquely, as in Fig. 28, we have found that if the number of degrees in the angle T O S, formed by the direction of the tube O S, with the perpendicular O T, be re-

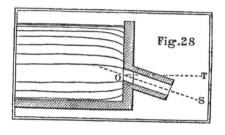

Fig.28

presented by $\varphi$, then $·814 - ·0016\varphi$ will give the coefficient of discharge corresponding to the obliquely attached short tube in the Figure. This formula is, however, empirical, but it is simple, and agrees pretty closely with experimental results. As the coefficient of resistance is equal $\frac{1}{c_v^2} - 1$, equation (64), we have here $c_r = \frac{1}{(·814 - ·0016\,\varphi)^2} - 1$; from these equations we have calculated the following Table:—

COEFFICIENTS OF DISCHARGE AND RESISTANCE FOR OBLIQUE JUNCTIONS.

| φ in degrees. | Coefficient of discharge. | Coefficient of resistance. | φ in degrees. | Coefficient of discharge. | Coefficient of resistance. |
| --- | --- | --- | --- | --- | --- |
| 0° | ·814 | ·508 | 35° | ·758 | ·740 |
| 5 | ·806 | ·539 | 40 | ·750 | ·778 |
| 10 | ·798 | ·569 | 45 | ·742 | ·816 |
| 15 | ·790 | ·602 | 50 | ·734 | ·856 |
| 20 | ·782 | ·635 | 55 | ·726 | ·897 |
| 25 | ·774 | ·669 | 60 | ·718 | ·940 |
| 30 | ·766 | ·704 | 65 | ·710 | ·984 |

The coefficient of resistance for a tube at right angles to the side, is to the like coefficient when it makes an angle of 45 degrees as ·508 to ·816, or as 1 to 1·6 nearly; and the loss of head is greater in the same proportion. If the short tube be more than three or four diameters in length, friction will have to be taken into account.

FORMULA FOR FINDING THE TIME THE SURFACE OF WATER IN A CISTERN TAKES TO SINK A GIVEN DEPTH.

In experiments for finding the value of the coefficients of discharge, one of the best methods is to observe the time the water discharged from the orifice takes to sink the surface in a prismatic cistern a given depth; the ratio of the observed to the theoretical time will then give the coefficient sought. A formula for finding the theoretical time is, therefore, of much practical value. In Fig. 29, the time of falling from $st$ to $sT$, in seconds, is

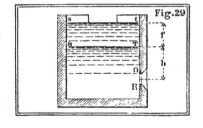
Fig. 29

(68.) $$t = \frac{A}{4\cdot 0125\, a} \{(h+f)^{\frac{1}{2}} - h^{\frac{1}{2}}\},$$

in which $a$ is the area of the orifice o R, and A the area of the prismatic vessel at $st$ or $ST$; this formula is for measures in feet. For measures in inches, we have

(69.) $$t = \frac{A}{13 \cdot 9 \, a} \{(h + f)^{\frac{1}{2}} - h^{\frac{1}{2}}\}.*$$

EXAMPLE VII.—*A cylindrical vessel 5·74 inches in diameter has an orifice ·2 inch in diameter at a depth of 16 inches below the surface, measured to the centre; it is found that the water sinks 4 inches in 51 seconds; what is the coefficient of discharge?*

The theoretical time, $t$, is found from equation (69), equal

$$\frac{5 \cdot 74^2 \times \cdot 7854}{13 \cdot 9 \times \cdot 2^2 \times \cdot 7854} \{16^{\frac{1}{2}} - 12^{\frac{1}{2}}\} = \frac{32 \cdot 9476}{\cdot 556} \{4 - 3 \cdot 4641\}$$

$$= \frac{17 \cdot 6566}{\cdot 556} \times \cdot 5359 = 31 \cdot 8 \text{ seconds; hence, } \frac{31 \cdot 8}{51} = \cdot 624 \text{ is}$$

the coefficient sought. When the orifice $OR$ and the horizontal section of the vessel are similar figures, $\frac{A}{a}$ is equal $\frac{ST^2}{OR^2}$; and therefore, for circular cisterns and orifices, it is unnecessary to introduce the multiplier ·7854.

## SECTION VIII.

FLOW OF WATER IN UNIFORM CHANNELS.—MEAN VELOCITY.
—MEAN RADII AND HYDRAULIC MEAN DEPTHS.—BORDER.
—TRAIN.—HYDRAULIC INCLINATION.—EFFECTS OF FRICTION.—FORMULÆ FOR CALCULATING THE MEAN VELOCITY.
—APPLICATION OF THE FORMULÆ AND TABLES TO THE SOLUTIONS OF THREE USEFUL PROBLEMS.

In rivers the velocity is a maximum along the central line of the surface, or, more correctly, over the deepest part of the channel; and it decreases thence to the sides and bottom: but when backwater arises from any obstruction, either a

* See NOTE A for formulæ for finding the time of filling *Lock-chambers.*

submerged weir, Fig. 22, or a contracted channel, Fig. 23, the velocity in the channel approaching the obstruction is a maximum at the depth of the backwater below the surface, and it decreases thence to the surface, sides, and bottom. When water flows in a pipe of any length, the velocity at the centre is greatest, and it decreases thence to the sides or circumference of the pipe. If the pipe be supposed divided into two portions in the direction of its length, the lower portion or channel will be analogous to a small river or stream, in which the velocity is greatest at the central line of the surface, and the upper portion will be simply the lower reversed. A pipe flowing full may, therefore, be looked upon as a double stream, and we shall soon see that the formulæ for the discharge from each kind are all but identical, though a pipe may discharge full at all inclinations, while the inclinations in rivers or streams, having uniform motion, never exceed a few feet per mile.

### MEAN VELOCITY.

It is found, by experiment, that the mean velocity is nearly independent of the depth or width of the channel, the central or maximum velocity being the same. From a number of experiments, Du Buât derived empirical formulæ equivalent to

$$v = \frac{v_b + v}{2} = v - v^{\frac{1}{2}} + \tfrac{1}{2}, \; v_b = (v^{\frac{1}{2}} - 1)^2, \text{ and } v = (v_b^{\frac{1}{2}} + 1)^2;$$

in these equations $v$ is the mean velocity, v the maximum velocity, and $v_b$ the velocity at the sides, or bottom, expressed in French inches. Tables calculated from these formulæ do not give correct results for measures in English inches, though they are those generally adopted. Disregarding the difference in the measures, which are as 1 to 1·0678, it will be found that, in the generality of channels, the mean velocity is not an arithmetical mean between the velocity at the central surface line and that at the bottom, though nearly so between the mean bottom and mean surface velocities. Dr. Young[*], modifying Du Buât's formula, assumes for

---
[*] Philosophical Transactions, 1808, p. 487.

English inches that $v + v^{\frac{1}{2}} = \mathrm{v}$, and hence $v = \mathrm{v} + \tfrac{1}{2} - (\mathrm{v} + \tfrac{1}{4})^{\frac{1}{2}}$. This gives results very nearly the same as the other formula for $v$, but something less, particularly for small surface velocities. For instance, Du Buât's formula gives ·5 inch for the mean velocity when the central surface velocity is 1 inch, whereas Dr. Young's makes it ·38 inch. For large velocities both formulæ agree very closely, disregarding the difference between the measures, which is only 7 per cent. They are best suited to very small channels or pipes, but unless at mean velocities of about 3 feet per second, they are wholly inapplicable to rivers.

Prony found, from Du Buât's experiments, that for measures in metres $v = \left(\dfrac{2 \cdot 37187 + \mathrm{v}}{3 \cdot 15312 + \mathrm{v}}\right)\mathrm{v}$. This, reduced for measures in English feet, becomes

(70.) $$v = \left(\dfrac{7 \cdot 783 + \mathrm{v}}{10 \cdot 345 + \mathrm{v}}\right)\mathrm{v};$$

and for measures in English inches,

(71.) $$v = \left(\dfrac{93 \cdot 39 + \mathrm{v}}{124 \cdot 14 + \mathrm{v}}\right)\mathrm{v}.$$

For medium velocities $v = \cdot 81\, \mathrm{v}$. The experiments from which these formulæ were derived were made with small channels. We have calculated the values of $v$ from that of v, equation (71), and given the results in columns 3, 6, and 9, in TABLE VII. Ximenes, Funk, and Brünnings' experiments in larger channels give the mean velocity at the centre of the depth equal $\cdot 914\, \mathrm{v}$, when the central surface velocity is v; but as the velocity also decreases in nearly the same ratio from the centre to the sides of the channel, we shall get the mean velocity in the whole section equal $\cdot 914 \times \cdot 914\, \mathrm{v} = \cdot 835\, \mathrm{v}$; and hence we have, for large channels,

(72.) $$v = \cdot 835\, \mathrm{v}.$$

We have also calculated the values of $v$ from this formula, and given the results in columns 2, 5, and 8 of TABLE VII. This table will be found to vary considerably from those calculated from Du Buât's formula in French inches, hitherto generally used in this country, and much more applicable for all practical purposes.

## MEAN RADIUS.—HYDRAULIC MEAN DEPTH.—BORDER.—COEFFICIENT OF FRICTION.

If, in the diagrams 1 and 2, Fig. 30, exhibiting the sections of cylindrical and rectangular tubes filled with flowing water, the areas be divided respectively by the perimeters A C B D A and A B D C A, the quotients are termed "*the mean radii*" of the tubes, diagrams 1 and 2, and the perimeters in contact with the flowing water are termed "*the borders.*" In the diagrams 3 and 4, the surface A B is not in contact with the channel, and the length of the bed and sides A C D B become "*the border.*" "*The mean radius*" is equal to the area A B D C A divided by the length of the border A C D B. "*The hydraulic mean depth*" is the same as "*the mean radius,*" this latter term being perhaps most applicable to pipes flowing full, as in diagrams 1 and 2; and the former to streams and rivers which have a surface line A B, diagrams 3 and 4. We shall, throughout the following equations, designate the value of the mean radius, hydraulic mean depth, or quotient, $\frac{\text{area A B D C A}}{\text{border B D C A}}$,* by the letter $r$, only remarking here that *for cylindrical pipes flowing full, or rivers with semicircular beds, it is always equal to half the radius, or one-fourth of the diameter.*

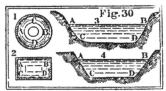
Fig. 30

Du Buât was the first to observe that the head due to the resistance of friction for water flowing in a uniform channel increased directly as the length of the channel $l$, directly as the border, and inversely as the area of the cross-flowing

---

* M. Girard has conceived it necessary to introduce the coefficient of correction 1·7 as a multiplier to the border for finding $r$, to allow for the increased resistance from aquatic plants; so that, according to his reduction,

$$r = \frac{\text{area}}{1\cdot 7 \text{ border}}.$$

See Rennie's First Report on Hydraulics as a Branch of Engineering; Third Report of the British Association, p. 167. Also equation (85), p. 112.

section*, very nearly; that is, as $\frac{l}{r}$. It also increases as the square of the velocity, nearly; therefore the head due to the resistance must be proportionate to $\frac{v^2 l}{2gr}$. If $c_f \times \frac{v^2 l}{2gr} = h_f$, then $c_f$ is the coefficient for the head due to the resistance of friction, as $h_f$ is the head necessary to overcome the friction, $c_f$ is therefore termed "*the coefficient of friction.*"

### HYDRAULIC INCLINATION.—TRAIN.

If $l$ be the length of a pipe or channel, and $h_f$ the height due to the resistance of friction of water flowing in it, then $\frac{h_f}{l}$ is *the hydraulic inclination*. In Fig. 31 the tubes A B, C D, of

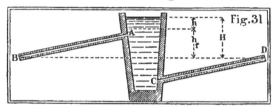

the same length $l$, and whose discharging extremities B and D are on the same horizontal plane B D, will have the same hydraulic inclination and the same discharge, no matter what the actual inclinations or the depth of the entrances at A and C may be, so they be of the same kind and bore; and as the velocities in A B and C D are the same, the height $h$ due to them must be the same when the circumstances of the orifices of entry A and C are alike. We have the whole head $H = h + h_f$ (see pp. 89 and 90). The hydraulic inclination is not therefore the whole head H, divided by the length $l$ of the pipe, as it is sometimes mistaken, but the height $h_f$ (found by subtracting the height $h$, due to the entrance at A or C, and the velocity in the pipe, from the whole height) divided by the length $l$. When the height $h$ is very small compared with the whole height H, as it is in very long tubes with moderate heads,

* Pitot had previously, in 1726, remarked that the diminution arising from friction in pipes is, *cæteris paribus*, inversely as the diameters.

$\frac{H}{l}$ may be substituted for $\frac{h_f}{l}$ without error; but for short pipes up to 100 feet in length the latter only should be used in applying Du Buât's and some other formulæ; otherwise the results will be too large, and only fit to be used approximately in order to determine the height $h$ from the velocity of discharge thus found. When the horizontal pipe C D, Fig. 32, is

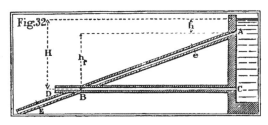

equal in every way to the inclined pipe A B, and the head at A is that due to the velocity in C D, the discharge from the pipe A B will be equal to that from C D; but a peculiar property belongs to the pipe A B in the position in which it is here placed; for *if we cut it short at any point e, or lengthen it to any extent, to E, the discharge will remain the same and equal to that through the horizontal pipe* C D. The velocity in A B at the angle of inclination A B C, when A C $= h_f$, and A B $=$ C D, is therefore such that it remains unaffected by the length A E or A $e$, to which it may be extended or cut short; and *at this inclination the water in the pipe* A B *is said to be "in train."* In like manner a river or stream is said to be *"in train"* when the inclination of its surface bears such a relation to the cross section that the mean velocity is neither decreased nor increased by the length of the channel; and we perceive from this *that the acceleration caused by the inclination is exactly counterbalanced by the resistances to the motion when the moving water in a pipe or river channel is in train.*

As $h = (1 + c_r) \frac{v^2}{2g}$ where $c_r$ is the coefficient of the height due to the resistance at the orifice of entry A or C, and $h_f = c_f \frac{v^2 l}{2 g r}$, we therefore get

(73.) $$\text{H} = (1 + c_r)\frac{v^2}{2g} + c_f \times \frac{lv^2}{2gr} = \left(1 + c_r + c_f\frac{l}{r}\right)\frac{v^2}{2g},$$

and hence we find the mean velocity of discharge

(74.) $$v = \left\{\frac{2g\text{H}}{1 + c_r + c_f\frac{l}{r}}\right\}^{\frac{1}{2}}.$$

When $h$ is small compared with $h_f$, or, which comes to the same thing, $1 + c_r$ small compared with $c_f \times \frac{l}{r}$,

(75.) $$\text{H} = c_f \times \frac{lv^2}{2gr},$$

and

(76.) $$v = \left\{\frac{2gr\frac{\text{H}}{l}}{c_f}\right\}^{\frac{1}{2}}.$$

If, in the last equation, we substitute $s$ for $\frac{\text{H}}{l}$, equal the sine of the angle of inclination A B C, we then have

(77.) $$v = \left\{\frac{2grs}{c_f}\right\}^{\frac{1}{2}}.$$

An average value of $c_f$ for all pipes with straight channels is ·0069914, from which we find equation (77) becomes, for measures in feet,

(78.) $$v = 96\sqrt{rs}.$$

As the mean value of the coefficient of resistance $c_r$ for the entrance into a tube is ·508, and as $2g = 64·403$, and $c_f = ·0069914$, equation (74), for measures in feet, becomes

$$v = \left\{\frac{64·403\,\text{H}}{1·508 + ·0069914\frac{l}{r}}\right\}^{\frac{1}{2}}, \text{ or}$$

(79.) $$v = \left\{\frac{\text{H}\,r}{·0234\,r + ·0001085\,l}\right\}^{\frac{1}{2}}.$$

This, multiplied by the section, gives the discharge.

## DU BUAT'S FORMULA.

The coefficient of friction $c_f$ is not, however, constant, as it varies with the velocity. That which we have just given answers for pipes when the velocity is from 18 to 24 inches per second. For pipes and rivers it is found to increase as the velocity decreases; that is, the loss of head is proportionately greater for small than for large velocities. Du Buât found the loss of head to be also greater for small than large channels, and applied a correction accordingly in his formula. This, expressed in French inches, is

(80.) $$v = \frac{297\,(r^{\frac{1}{2}} - 0\cdot 1)}{\left(\frac{1}{s}\right)^{\frac{1}{2}} - \text{hyp. log.}\left(\frac{1}{s} + 1\cdot 6\right)^{\frac{1}{2}}} - \cdot 3\,(r^{\frac{1}{2}} - 0\cdot 1),$$

maintaining the preceding notation, in which $s = \frac{h_t}{l}$. In this formula 0·1, in the numerator of the first term, is deducted as a correction due to the hydraulic mean depth, as it was found that $297\,(r^{\frac{1}{2}} - 0\cdot 1)$ agreed more exactly with experiment than $297\,r^{\frac{1}{2}}$ simply. The second term, hyp. log. $\left(\frac{1}{s} + 1\cdot 6\right)$, of the denominator is also deducted to compensate for the observed loss of head being greater for less velocities, and the last term $\cdot 3\,(r^{\frac{1}{2}} - 0\cdot 1)$ is a deduction for a general loss of velocity sustained from the unequal motions of the particles of water in the cross section as they move along the channel. These corrections are empirical; they were, however, determined separately, and after being tested by experiment, applied, as above, to the radical formula $v = 297\,\sqrt{rs}$.

Du Buât's formula was published in his *Principes d'Hydraulique*, in 1786. It is, as we have seen, partly empirical, but deduced by an ingenious train of reasoning and with considerable penetration from about 125 experiments, made with pipes from the 19th part of an inch to 18 inches in diameter, laid horizontally, inclined at various inclinations, and vertical; and also from experiments on open channels with sectional areas from 19 to 40,000 square inches, and inclinations of from 1 in 112 to 1 in 36,000. The lengths of the pipes experimented with varied from 1 to 3, and from 3 to 3600 feet.

In several experiments by which we have tested this formula, the resulting velocities found from it were from 1 to 5 per cent. too large for small pipes, and too small for straight rivers in nearly the same proportion. As the experiments from which it was derived were made with great care, those with pipes particularly so, this was to be expected. Experiments with pipes of moderate or short lengths should have the circumstances of the orifice of entry from the reservoir duly noted; for the close agreement of this formula with them must depend a great deal, in such pipes, on the coefficient due to the height $h$, which must be deducted from the whole head H before the hydraulic inclination, $\dfrac{h_t}{l} = s$, can be obtained; but for very long pipes and uniform channels this is not necessary.

Under all the circumstances, and after comparing the results obtained from this and other formulæ, we have preferred calculating tables for the values of $v$ from this formula reduced for measures in English inches, which is

$$v = \frac{306 \cdot 596 \, (r^{\frac{1}{2}} - \cdot 1032)}{\left(\dfrac{1}{s}\right)^{\frac{1}{2}} - \text{hyp. log.} \left(\dfrac{1}{s} + 1 \cdot 6\right)} - \cdot 2906 \, (r^{\frac{1}{2}} - \cdot 1032),$$

or more simply,

(81.) $$v = \frac{307 \, (r^{\frac{1}{2}} - \cdot 1)}{\left(\dfrac{1}{s}\right)^{\frac{1}{2}} - \text{hyp. log.} \left(\dfrac{1}{s} + 1 \cdot 6\right)} - \cdot 3 \, (r^{\frac{1}{2}} - \cdot 1).$$

This gives the value of $v$ a little larger than the original formula, but the difference is immaterial. For measures in English feet it becomes

(82.) $$v = \frac{88 \cdot 51 \, (r^{\frac{1}{2}} - \cdot 03)}{\left(\dfrac{1}{s}\right)^{\frac{1}{2}} - \text{hyp. log.} \left(\dfrac{1}{s} + 1 \cdot 6\right)} - \cdot 84 \, (r^{\frac{1}{2}} - \cdot 03).$$

The results of equation (81) are calculated for different

values of $s$ and $r$, and tabulated in·TABLE VIII., the first eight pages of which contain the velocities for values of $r$ varying from $\frac{1}{10}$th inch to 6 inches; or if pipes, diameters from $\frac{1}{4}$ inch to 2 feet, and of various inclinations from horizontal to vertical. The five last pages contain the velocities for values of $r$ from 6 inches to 12 feet, and with falls from 6 inches to 12 feet per mile.

EXAMPLE VIII.—*A pipe, $1\frac{1}{2}$ inch diameter and 100 feet long, has a constant head of 2 feet over the discharging extremity; what is the velocity of discharge per second?*

The mean radius $r = \frac{1\frac{1}{2}}{4} = \frac{3}{8}$ inches, and $\frac{100}{2} = 50 = \frac{1}{s}$, is the approximate hydraulic inclination. At page 2 of TABLE VIII., in the column under the mean radius $\frac{3}{8}$, and opposite to the inclination 1 in 50, we find 30·69 inches for the velocity sought. This, however, is but approximative, as the head due to the velocity should be subtracted from the whole head of 2 feet, before finding the true hydraulic inclination. This head depends on the coefficient of resistance at the entrance orifice, or the coefficient of discharge for a short tube. In all Du Buât's experiments this latter was taken at ·8125, but it will depend on the nature of the junction, as, if the tube runs into the cistern, it will become as small as ·715; and, if the junction be rounded into the form of the contracted vein, it will rise ·974, or 1 nearly. In this case, the coefficient of discharge may be assumed ·815\*, from which, in TABLE II., we find the head due to a velocity of 30·69 inches to be $1\frac{7}{8} = 1·87$ inch nearly, which is the value of $h$; and hence, $\text{H} - h = h_t = 24 - 1·87 = 22·13$ inches; and $\frac{l}{h_t} = \frac{100 \times 12}{22·13} = 54·2 = \frac{1}{s}$, the hydraulic inclination more correctly. With this new inclination and the mean radius $\frac{3}{8}$, we find the velocity by interpolating between the inclinations 1 in 50 and 1 in 60, given in the Table to be 30·69 — 1·34 = 29·35 inches per

\* See EXAMPLE 16, pp. 10. 12.

second. This operation may be repeated until $v$ is found to any degree of accuracy according to the formula; but it is, practically, unnecessary to do so. If we now wish to find the discharge per minute in cubic feet, we can easily do so from TABLE IX., in which for an inch and a half pipe we get

|  |  | Inches. |  |  | Cubic feet. |  |  |
|---|---|---|---|---|---|---|---|
| For a velocity of | | 20·00 | per second, | | 1·22718 | per minute. | |
| „ | „ | 9·00 | „ | „ | ·55223 | „ | „ |
| „ | „ | ·30 | „ | „ | ·01841 | „ | „ |
| „ | „ | ·04 | „ | „ | ·00245 | „ | „ |
| „ | „ | 29·34 | „ | „ | 1·80027 | „ | „ |

The discharge found experimentally by Mr. Provis, for a tube of the same length, bore, and head, was 1·745 cubic foot per minute.

If we suppose the coefficient of discharge due to the orifice of entry and stop-cock in Mr. Provis's 208 experiments[*] with $1\frac{1}{2}$ inch lead pipes of 20, 40, 60, 80, and 100 feet lengths, to be ·715, the results calculated by the tables will agree with the experimental results with very great accuracy, and it is very probable from the circumstances described, that the ordinary coefficient ·815 due to the entry was reduced by the circumstances of the stopcock and fixing to about ·715; but even with ·815 for the coefficient, the difference between calculation and experiment is not much, the calculation being then in excess in every experiment, the average being about 5 per cent., and not so much in the example we have given.

TABLE VIII. will give the velocity, and thence the discharge immediately, for long pipes, and TABLE X. enables us to calculate the cubic feet discharged per minute, with great facility. For rivers, the mean velocity, and thence the discharge, is also found with quickness. See also TABLES XI., XII., and XIII.

EXAMPLE IX.—*A watercourse is 7 feet wide at the bottom, the length of each sloping side is 6·8 feet, the width at the surface is 18 feet, the depth 4 feet, and the inclina-*

[*] Proceedings of the Institution of Civil Engineers, pp. 201, 210, vol. ii.

tion of the surface 4 inches in a mile; what is the quantity flowing down per minute?

Here $\dfrac{(18 + 7) \times \frac{4}{2}}{7 + 2 \times 6\cdot 8} = \dfrac{50}{20\cdot 6} = 2\cdot 4272$ feet $= 29\cdot 126$ inches

$= r$, is the hydraulic mean depth; and as the fall is 4 inches per mile, we find at the 11th page of TABLE VIII., the velocity $v = 12\cdot 03 - \cdot 16 = 11\cdot 87$ inches per second; the discharge in cubic feet per minute is, therefore,

$$50 \times \dfrac{11\cdot 87}{12} \times 60 = 2967\cdot 5.$$

Watt, in a canal of the fall and dimensions here given, found the mean velocity about $13\frac{1}{3}$ inches per second. This corresponds to a fall of 5 inches in the mile, according to the formula.

In one of the original experiments with which the formula was tested on the canal of Jard, the measurements accorded very nearly with those in this example, viz. $\dfrac{1}{s} = 15360$, and $r = 29\cdot 1$ French inches; the observed velocity at the surface was $15\cdot 74$, and the calculated mean velocity from the formula $11\cdot 61$ French inches*. TABLE VII. will give $12\cdot 29$ inches for the mean velocity, corresponding to a superficial velocity of $15\cdot 74$ inches. This shews that the formula also gives too small a value for $v$ in this case, by about $\frac{1}{17}$th of the result, it being about $\dfrac{1}{8\cdot 3}$ part in the other. The probable error in the formula *applied to straight clear rivers* of about 2 feet 6 inches hydraulic mean depth is nearly $\frac{1}{12}$th or 8 per cent. of the tabulated velocity, and this must be added for the more correct result; the watercourse being supposed nearly straight and free from aquatic plants.

Notwithstanding the differences above remarked on, we are of opinion that the results of this formula, which we

---

* These measures reduced to English inches, give $r = 31\cdot 014$, $v = 12\cdot 374$; and the surface velocity $16\cdot 775$ inches; reduced for mean velocity $13\cdot 101$ inches.

have calculated and tabulated, may be more safely relied on as applied to *general practical purposes* than most of those others which we shall proceed to lay before our readers. Rivers or watercourses are seldom straight or clear from weeds, and even if the sections, during any improvements, be made uniform, they will seldom continue so, as "*the regimen,*" or adaptation of the velocity to the tenacity of the banks, must vary with the soil and bends of the channel, and can seldom continue permanent for any length of time unless protected. From these causes a loss of velocity takes place, difficult, if not impossible, to estimate accurately, but which may be taken at from 10 to 15 per cent. of that in the clear unobstructed direct channel; but, be this as it may, *it is safer to calculate the drainage or mechanical results obtainable from a given fall and river channel, from formulæ which give lesser, than from those which give larger velocities.* This is a principle engineers cannot too much observe.

We have before remarked, that for both pipes and rivers the coefficient of resistance increases as the velocity decreases. This is as much as to say, in the simple formula for the velocity, $v = m\sqrt{rs}$, that $m$ must increase with $v$, and as some function of it. This is the case in TABLE VIII., throughout which the velocities increase faster than $\sqrt{r}$, the $\sqrt{s}$, or the $\sqrt{rs}$. In all formulæ with which we are acquainted but Du Buât's and Young's, *the velocity found is constant when* $\sqrt{rs}$ *or* $r \times s$ *is constant.* In Du Buât's formula for $r \times s$ constant, $v$ obtains maximum values between $r = \frac{3}{8}$ inch and $r = 1$ inch; the differences of the velocities for different values of $r$ above 1 inch, $r \times s$ being constant, are not much. We may always find the maximum value, or nearly so, by assuming $r = \frac{3}{4}$ inch, and finding the corresponding inclination from the formula $\frac{4rs}{3}$, which is equal to it. For example, if $r = 12$ inches, and $s = \frac{1}{10560}$, the velocity is found equal 9·52 inches; but when $rs$ is constant, the inclination $s$ corresponding to $r = \frac{3}{4}$ inch is $\frac{4 \times 12''}{3 \times 10560} = \frac{1}{660}$, from which we find from the table $v = 10.25$ inches for the maximum velocity, making a difference of fully 7 per cent.

When $r = .01$ of an inch, or a pipe is $\frac{1}{25}$th part of an inch in diameter, Du Buât's formula fails, but it gives correct results for pipes $\frac{1}{8}$th of an inch in diameter, and two of the experiments from which it was derived were made with pipes 12 inches long and only $\frac{1}{18}$th part of an inch in diameter.

COULOMB having shown that the resistance opposed to a disc revolving in water increases as the function $av + bv^2$ of the velocity $v$, we may assume that the height due to the resistance of friction in pipes and rivers is also of this form; and that

(83.) $$h_f = (av + bv^2)\frac{l}{r},$$

and, consequently,

(84.) $$rs = av + bv^2.$$

GIRARD first gave values to the coefficients $a$ and $b$. He assumed them equal, and each equal to $.0003104$ for measures in metres, and thence the velocity in canals,

(85.) $$v = (3221.016\, rs + .25)^{\frac{1}{2}} - .5;\text{*}$$

which reduced for measures in English feet becomes

(86.) $$v = (10567.8\, rs + 2.67)^{\frac{1}{2}} - 1.64.$$

The value of $a = b = .0003104$ was obtained by means of twelve experiments by Du Buât and Chezy. Of course the value is four times this in the original, as we use the mean radius in all the formulæ instead of the diameter. This formula is only suited for very small velocities in canals containing aquatic plants, and the velocity must be increased by .7 foot per second for clear channels.

PRONY found from thirty experiments on canals, that $a = .000044450$ and $b = .000309314$,† for measures in metres, from which we find

(87.) $$v = (3232.96\, rs + .00516)^{\frac{1}{2}} - .0719;$$

---

\* See Brewster's Encyclopedia, Article Hydrodynamics, p. 529.

† Recherches Physico-Mathematiques sur la Théorie des Eaux Courantes.

this reduced for measures in English feet is,

(88.) $\quad v = (10607\cdot02\,rs + \cdot0556)^{\frac{1}{2}} - \cdot236;$*

the velocities did not exceed 3 feet per second in the experiments from which this was derived.

For pipes, Prony found,† from fifty-one experiments made by Du Buât, Bossut, and Couplet, with pipes from 1 to 5 inches diameter, from 30 to 17,000 feet in length, and one pipe 19 inches diameter and nearly 4000 feet long, that $a = \cdot00001733$, and $b = \cdot0003483$, from which values

(89.) $\quad v = (2871\cdot09\,rs + \cdot0006192)^{\frac{1}{2}} - \cdot0249,$

for measures in metres, and for measures in English feet,

(90.) $\quad v = (9419\cdot75\,rs + \cdot00665)^{\frac{1}{2}} - \cdot0816.$

Prony also gives the following formula applicable to pipes and rivers. It is derived from fifty-one selected experiments with pipes, and thirty-one with open channels:

(91.) $\quad v = (3041\cdot47\,rs + \cdot0022065)^{\frac{1}{2}} - \cdot0469734,$‡

for measures in metres, which, reduced for measures in English feet, is

(92.) $\quad v = (9978\cdot76\,rs + \cdot02375)^{\frac{1}{2}} - \cdot15412.$

EYTELWEIN, following the method of investigation pursued previously by Prony, found from a large number of experiments, $a = \cdot0000242651$, and $b = \cdot000365543$ in rivers, for measures in metres; and, therefore,

---

* For canals containing aquatic plants, reeds, &c., we must substitute $\frac{r}{1\cdot7}$ for $r$. See note, p. 102.

† Recherches Physico-Mathématiques sur la Théorie du Mouvement des Eaux Courantes, 1804.

‡ Recherches Physico-Mathematiques sur la Théorie des Eaux Courantes. A reduction of this formula into English feet is given at page 6, Article Hydrodynamics, Encyclopedia Britannica; at page 164, Third Report, British Association, by Rennie; and at pages 427 and 533, Article Hydrodynamics, Brewster's Encyclopedia. This reduction, $v = -0\cdot1541 + (\cdot02375 + 32806\cdot6\,rs)^{\frac{1}{2}}$ is entirely incorrect; and, being the same in each of those works, appears to have been copied one from the other.

(93.) $$v = (2735{\cdot}66\,rs + {\cdot}001102)^{\frac{1}{2}} - {\cdot}0332.$$*

This reduced for measures in English feet, is

(94.) $$v = (8975{\cdot}43\,rs + {\cdot}0118858)^{\frac{1}{2}} - {\cdot}1089.$$

He also shows,† that $\frac{10}{11}$ths of a mean proportional between the fall in two English miles and the hydraulic mean depth, gives the mean velocity very nearly. This rule for measures in inches is equivalent to

(95.) $$v = 324\,\sqrt{rs};$$

and for measures in feet

(96.) $$v = 93{\cdot}4\,\sqrt{rs}.$$

For the velocity of water in pipes, he found,‡ from the fifty-one experiments of Du Buât, Bossut, and Couplet, that $a = {\cdot}0000223$, and $b = {\cdot}0002803$, from which we get for measures in metres,

(97.) $$v = (3567{\cdot}29\,rs + {\cdot}00157)^{\frac{1}{2}} - {\cdot}0397\,;$$

which reduced for measures in English feet becomes

(98.) $$v = (11703{\cdot}95\,rs + {\cdot}01698)^{\frac{1}{2}} - {\cdot}1303.$$

Another formula given by Eytelwein for pipes, which includes the head due to the velocity for the orifice of entry, is

(99.) $$v = 50\left(\frac{dh}{l + 50\,d}\right)^{\frac{1}{2}};$$

in which $h$ is the head, $l$ the length, and $d$ the diameter of the pipe, all expressed in English feet. It must be here mentioned, that much of the valuable information presented by Eytelwein is but a modification of what Du Buât had previously given, and to whom for much that is attributed to the former we are primarily indebted.

Dr. THOMAS YOUNG§ also derives his formula from the supposition, that the head due to the resistance of friction

---

* Mémoires de l'Academie de Berlin, 1814 et 1815. See Equation (110).
† Handbuch der Mechanik und der Hydraulik, Berlin, 1801.
‡ Mémoires de l'Academie des Sciences de Berlin, 1814 et 1815.
§ Philosophical Transactions for 1808.

ORIFICES, WEIRS, PIPES, AND RIVERS. 115

assumes the form of equation (83); calling the diameter of a pipe $d$, he takes

$$h_t = (av + bv^2)\frac{l}{d},$$

and the whole height $\text{H} = h_t + \dfrac{v^2}{586}$, expressed in inches.

He found from some experiments of his own, those collected by Du Buât, and some of Gerstner's, that

(100.) $a = \cdot 0000002 \left\{ \dfrac{900\,d^2}{d^2 + 1136} \right.$
$\left. + \dfrac{1}{d^{\frac{1}{2}}}\left( 1085 + \dfrac{13\cdot 21}{d} + \dfrac{1\cdot 0563}{d^2} \right) \right\};$

and

(101.) $b = \cdot 0000001 \left\{ 413 + \dfrac{75}{d} - \dfrac{1440}{d + 12\cdot 8} - \dfrac{180}{d + \cdot 355} \right\};$

then as $\dfrac{1}{586} = \cdot 00171$, we get

(102.) $v = \left\{ \dfrac{\text{H}\,d}{bl + \cdot 00171\,d} \right.$
$\left. + \left( \dfrac{al}{2bl + \cdot 00341\,d} \right)^2 \right\}^{\frac{1}{2}} - \dfrac{al}{2bl + \cdot 00341\,d}.$

When the length $l$ of the pipe is very great compared with the head due to the orifice of entrance and velocity, $\cdot 00171\,v^2$, we have

(103.) $v = \left\{ \dfrac{\text{H}\,d}{bl} + \dfrac{a^2}{4b^2} \right\}^{\frac{1}{2}} - \dfrac{a}{2b};$

or by substituting for $\dfrac{\text{H}}{l}$ its value $s$, equal the sine of the inclination,

(104.) $v = \left\{ \dfrac{sd}{b} + \dfrac{a^2}{4b^2} \right\}^{\frac{1}{2}} - \dfrac{a}{2b}.$

The values of $a$ and $b$ are for measures in inches. For most rivers, *in which d must be taken equal* $4r$, he finds for French inch measures, $v = \sqrt{20000\,ds}$; this reduced for English inches is

(105.) $$v = 292\sqrt{rs};$$

which again reduced for feet measures, becomes

(106.) $$v = 84\cdot 3\sqrt{rs}.$$

These latter values, for rivers, are even smaller than those found from Du Buât's formula; less than the observed velocities, and less than those found from any other formula, with the exception of Girard's. The values of the coefficients $a$ and $b$ vary in this formula with the value of $d = 4r$; they are expressed generally in equations (101) and (102), from which we have calculated the following Table for different values of $d$ and $r$.

An examination of this Table will show that $a$ obtains a minimum value when $d$ is between 10 and 11 inches; and $b$ when the diameter is between ½ and ¾ of an inch. Now, it appears from equation (102), that $v$ increases with $\sqrt{\dfrac{\text{H}\,d}{bl}}$ nearly, or, which is the same thing, as $b$ decreases, there must, *cæteris paribus,* be a maximum value of $v$ for a given value of $\dfrac{\text{H}\,d}{l}$, or $rs$, when $d$ is between ½ and ¾ inch; but as $\dfrac{a}{2b}$ has a minimum value when $d$ is nearly 12 inches, the maximum value of $v$ referred to will be found between values of $d$ from ¾ inch to 12 inches; in fact, when $d = 10$ inches nearly. We have already pointed out a similar peculiarity in Du Buât's general theorem, at page 111. *It will not be necessary to take out the values of* $\dfrac{a}{2b}$ *and* $\dfrac{a^2}{4b^2}$ *to more than one place of decimals.*

The values of $\dfrac{a}{2b}$ are also given in the following Table, and may be used in equation (104) for finding the discharge from long pipes. It is, however, necessary to remark, that this equation is sometimes misapplied in finding the velocity from short pipes, and those of moderate lengths. It is necessary

ORIFICES, WEIRS, PIPES, AND RIVERS.

TABLE OF THE VALUES OF $a$, $b$, $\frac{a}{2b}$, AND $\frac{a^2}{4b^3}$, IN EQUATION (104.)

| $d$ in inches. | $r$ in inches. | $a$. | $b$. | $\frac{a}{2b}$ | $\frac{a^2}{4b^3}$ | $d$ in inches. | $r$ in inches. | $a$. | $b$. | $\frac{a}{2b}$ | $\frac{a^2}{4b^3}$ |
|---|---|---|---|---|---|---|---|---|---|---|---|
| ·1 | ·025 | ·000837 | ·000066 | 6·341 | 40·207 | 6 | 1·5 | ·000094 | ·000032 | 1·468 | 2·155 |
| ·2 | ·05 | ·000536 | ·000035 | 7·657 | 58·629 | 7 | 1·75 | ·000090 | ·000033 | 1·363 | 1·857 |
| ·3 | ·075 | ·000406 | ·000028 | 7·250 | 52·562 | 8 | 2 | ·000087 | ·000033 | 1·318 | 1·737 |
| ·4 | ·100 | ·000356 | ·000025 | 7·120 | 50·694 | 9 | 2·25 | ·000084 | ·000034 | 1·235 | 1·526 |
| ·5 | ·125 | ·000316 | ·000025 | 6·320 | 39·942 | 10 | 2·50 | ·000083 | ·000034 | 1·220 | 1·489 |
| ·6 | ·150 | ·000287 | ·000024 | 5·979 | 35·750 | 11 | 2·75 | ·000083 | ·000034 | 1·220 | 1·489 |
| ·7 | ·175 | ·000264 | ·000024 | 5·500 | 30·250 | 12 | 3·00 | ·000084 | ·000035 | 1·200 | 1·440 |
| ·8 | ·200 | ·000247 | ·000025 | 4·940 | 24·403 | 15 | 3·75 | ·000086 | ·000035 | 1·228 | 1·509 |
| ·9 | ·225 | ·000232 | ·000025 | 4·640 | 21·529 | 20 | 5 | ·000096 | ·000036 | 1·333 | 1·777 |
| 1 | ·250 | ·000220 | ·000025 | 4·400 | 19·360 | 40 | 10 | ·000140 | ·000038 | 1·842 | 3·393 |
| 1·5 | ·375 | ·000179 | ·000027 | 3·315 | 10·988 | 50 | 12·5 | ·000154 | ·000039 | 1·974 | 3·898 |
| 2 | ·500 | ·000155 | ·000028 | 2·767 | 7·661 | 60 | 15 | ·000165 | ·000039 | 2·115 | 4·474 |
| 2·5 | ·625 | ·000139 | ·000028 | 2·482 | 6·161 | 80 | 20 | ·000177 | ·000040 | 2·212 | 4·895 |
| 3·0 | ·750 | ·000127 | ·000029 | 2·189 | 4·794 | 100 | 25 | ·000184 | ·000040 | 2·300 | 5·290 |
| 3·5 | ·875 | ·000119 | ·000030 | 1·983 | 3·933 | 300 | 75 | ·000190 | ·000041 | 2·317 | 5·369 |
| 4 | 1·000 | ·000111 | ·000031 | 1·790 | 3·205 | 500 | 125 | ·000189 | ·000041 | 2·305 | 5·312 |
| 5 | 1·250 | ·000101 | ·000031 | 1·629 | 2·653 | Infinite | Infinite | ·000180 | ·000041 | 2·195 | 4·818 |

to use equation (102), which takes into consideration the head due to the velocity and orifice of entry for such pipes.

SIR JOHN LESLIE states,* that the mean velocity of a river in miles per hour, is $\frac{15}{13}$ths of the mean proportional between the hydraulic mean depth and the fall per mile, in feet. This rule is equivalent for measures in feet, to

(107.)    $v = 100 \sqrt{rs}$;

and is most applicable to very large rivers and velocities.

D'AUBUISSON, from an examination of the results obtained by Prony and Eytelwein, assumes† for measures in metres that $a = \cdot0000189$, and $b = \cdot0003425$ for pipes substituting these in equation (84) and resolving the quadratic

(108.)    $v = (2919 \cdot 71 \, rs + \cdot 00074)^{\frac{1}{2}} - \cdot 027$;

which reduced for measures in English feet becomes

(109.)    $v = (9579 \, rs + \cdot 00813)^{\frac{1}{2}} - \cdot 0902$.

For rivers he assumes with Eytelwein,‡ $a = \cdot 000024123$, and $b = \cdot 0003655$, for measures in metres, and hence

(110.)    $v = (2735 \cdot 98 \, rs + \cdot 0011)^{\frac{1}{2}} - \cdot 033$;

which for measures in English feet is

(111.)    $v = (8976 \cdot 5 \, rs + \cdot 012)^{\frac{1}{2}} - \cdot 109$.

When the velocity exceeds two feet per second, he assumes, from the experiments of Couplet, $a = 0$, and $b = \cdot 00035875$; these values give

(112.)    $v = \sqrt{2787 \cdot 46 \, rs}$;

for measures in metres, and

(113.)    $v = 95 \cdot 6 \sqrt{rs} = \sqrt{9145 \, rs}$

for measures in English feet. Equations (110) and (111) are the same as (93) and (94), found from Eytelwein's values of $a$ and $b$, and it may be remarked that D'Aubuisson's equations for the velocity generally are but simple modifications of Prony's and Eytelwein's.

* Natural Philosophy, p. 423.
† Traité d'Hydraulique, p. 224.
‡ Traité d'Hydraulique, p. 133. See Equation (93).

The values which we have found to agree best with experiments on clear straight rivers are $a = \cdot0000035$, and $b = \cdot0001150$ for measures in English feet, from which we find

(114.) $\qquad v = (8695 \cdot 6\, rs + \cdot 00023)^{\frac{1}{2}} - \cdot 0152,$

which for an average velocity of $1\frac{1}{2}$ foot per second will give $v = 92\cdot3\sqrt{rs}$ nearly, and for large velocities $v = 93\cdot3\sqrt{rs}$; for smaller velocities than $1\frac{1}{2}$ foot per second, the coefficients of $\sqrt{rs}$ decrease pretty rapidly. This formula will be found to agree more accurately with observation and experiment than any other we know of. *It is useful to remark in the preceding formulæ, that the second term and decimals both under the vinculum may be entirely neglected without error.*

WEISBACH is perhaps the only writer who has modified the form of the equation $rs = av + bv^2$. In Dr. Young's formula, $a$ and $b$ vary with $r$, but Weisbach assumes that $h_f = \left(a + \dfrac{b}{v^{\frac{1}{2}}}\right) \dfrac{l}{a} \times \dfrac{v^2}{2g}$, and finds from the fifty-one experiments of Couplet, Bossut and Du Buât, before referred to, one experiment by Guemard, and eleven by himself, all with pipes varying from an inch to five and a half inches in diameter, and with velocities varying from $1\frac{1}{2}$ inch to 15 feet per second, that $a = \cdot01439$, and $b = \cdot 0094711$ for measures in metres; hence we have for the metrical standard

(115.) $\qquad h_f = \left(\cdot01439 + \dfrac{\cdot 0094711}{v^{\frac{1}{2}}}\right) \dfrac{l}{d} \times \dfrac{v^2}{2g}.$

This reduced for the mean radius $r$ is

(116.) $\qquad h_f = \left(\cdot003597 + \dfrac{\cdot 0023678}{v^{\frac{1}{2}}}\right) \dfrac{l}{r} \times \dfrac{v^2}{2g};$

from which we find for measures in English feet

(117.) $\qquad h_f = \left(\cdot003597 + \dfrac{\cdot 0042887}{v^{\frac{1}{2}}}\right) \dfrac{l}{r} \times \dfrac{v^2}{2g},$

and thence

(118.) $$r s = \left(·003597 + \frac{·0042887}{v^{\frac{1}{2}}}\right) \frac{v^2}{2g};$$

and by substituting for $2g$, its value $64·403$,

(119.) $$r s = \left(·00005585 + \frac{·00006659}{v^{\frac{1}{2}}}\right) v^2.$$

In equation (117), $\left(·003597 + \frac{·0042887}{v^{\frac{1}{2}}}\right) = c_f$ is the coefficient of the head due to friction. The equation does not admit of a direct solution, but the coefficient should be first determined for different values of the velocity $v$ and tabulated, after which the true value of $v$ can be determined by finding an approximate value and thence taking out the corresponding coefficient from the table, which does not vary to any considerable extent for small changes of velocity. In the following small table we have calculated the coefficients of friction, and also those of $v^2$, in equation (119), for different values of the velocity $v$.

TABLE OF THE COEFFICIENTS OF FRICTION IN PIPES.

| Velocity in feet. | $c_f$ | $\dfrac{c_f}{64·4}$ | Velocity in feet. | $c_f$ | $\dfrac{c_f}{64·4}$ |
|---|---|---|---|---|---|
| ·0 | Infinity. | Infinity. | 1·75 | ·006839 | ·0001062 |
| ·1 | ·017159 | ·0002664 | 2 | ·006629 | ·0001029 |
| ·2 | ·013186 | ·0002047 | 2·5 | ·006309 | ·0000979 |
| ·3 | ·011427 | ·0001774 | 3 | ·006073 | ·0000943 |
| ·4 | ·010378 | ·0001611 | 3·5 | ·005890 | ·0000914 |
| ·5 | ·009662 | ·0001500 | 4 | ·005741 | ·0000891 |
| ·6 | ·009133 | ·0001418 | 5 | ·005514 | ·0000856 |
| ·7 | ·008723 | ·0001354 | 6 | ·005348 | ·0000830 |
| ·8 | ·008391 | ·0001303 | 7 | ·005218 | ·0000810 |
| ·9 | ·008117 | ·0001260 | 8 | ·005113 | ·0000794 |
| 1·0 | ·007886 | ·0001224 | 10 | ·004953 | ·0000769 |
| 1·25 | ·007433 | ·0001154 | 16 | ·004669 | ·0000725 |
| 1·5 | ·007098 | ·0001102 | 20 | ·004556 | ·0000707 |

ORIFICES, WEIRS, PIPES, AND RIVERS. 121

If the reciprocal of $\frac{c_f}{64\cdot 4}$, here found, be substituted in the equation $v = \sqrt{\frac{64\cdot 4}{c_f} rs}$, we shall have the value of $v$. According to this table the coefficient of friction for a velocity of seven inches is more than twice that for a velocity of twenty feet. On comparing these coefficients and those for pipes in the preceding formulæ, with those for rivers of the same hydraulic depth, we perceive that the loss from friction is greatest in the latter, as might have been anticipated.

COEFFICIENTS DUE TO THE ORIFICE OF ENTRY.—PROBLEMS.

Unless where otherwise expressed, the head due to the velocity and orifice of entry is not considered in the preceding equations. In equation (74), where it is taken into calculation generally, $v = \left\{ \dfrac{2g\mathrm{H}}{1 + c_r + c_f \times \dfrac{l}{r}} \right\}$ in which $1 + c_r$ is equal to $\left(\dfrac{1}{c_v}\right)^2$, $c_r$ being the coefficient of resistance due to the orifice of entry, and $c_v$ the coefficient of velocity or discharge from a short tube. If the tube project into the reservoir, and be of small thickness, $c_v$ will be equal ·715 nearly, and therefore $c_r = ·956$; if the tube be square at the junction, the mean value of $c_v$ will be ·814, and therefore $c_r = ·508$; and if the junction be rounded in the form of the contracted vein, $c_v$ is equal to unity very nearly, and $c_r = 0$. For other forms of junction the coefficients of discharge and resistance will vary between these limits, and particular attention must be paid to their values in finding the discharge from shorter tubes and those of moderate lengths, but in very long tubes $1 + c_r$ becomes very small compared with $c_f \times \dfrac{l}{r}$, and may be neglected without practical error. These remarks are necessary to prevent the misapplication of the tables and formulæ, as the height due to the velocity and orifice of entry is an important element in short tubes.

We have considered it unnecessary to give any formulæ for finding the discharge itself, because, the mean velocity once determined, the calculation of the discharge from the area of the section is one of simple mensuration; and the introduction of this element into the three problems to which this portion of hydraulic engineering applies itself, renders the equations of solution complex, though easily derived; and presents them with an appearance of difficulty and want of simplicity which excludes them, nearly altogether, from practical application. The three problems are as follows:—

I. *Given the fall, length, and diameter of a pipe or hydraulic mean depth of any channel, to find the discharge.*

Here all that is necessary is to find the mean velocity of discharge, which, multiplied by the area of the section (equal $d^2 \times \cdot 7854$ in a cylindrical pipe), gives the discharge sought. TABLE VIII. gives the velocity at once for long channels. TABLE IX. gives the discharge in cubic feet per minute for different diameters of pipes, and velocities in inches per second, when found from TABLE VIII. See also TABLES XI. and XII.

II. *Given the discharge, and cross section of a channel, to find the fall or hydraulic inclination.*

If the cross section be circular, as in most pipes, the hydraulic mean depth is one-fourth of the diameter; in other channels it is found by dividing the water and channel line of the section, wetted perimeter, or border, into the area. The velocity is found by dividing the area into the discharge, and reducing it to inches per second; then in TABLE VIII., under the hydraulic mean depth, find the velocity, corresponding to which the fall per mile will be found in the first column, and the hydraulic inclination in the second.

III. *Given the discharge, length, and fall, to find the diameter of a pipe, or hydraulic mean depth and dimensions of a channel.*

This is the most useful problem of the three. Assume any mean radius $r_a$ and find the discharge $D_a$ by Problem I. We shall then have for cylindrical pipes

$$r_a^{\frac{5}{2}} : r^{\frac{5}{2}} :: \text{D}_a : \text{D} :: 1 : \frac{\text{D}}{\text{D}_a};$$

and as $r_a$, D, and $\text{D}_a$ are known, $r^{\frac{5}{2}}$ becomes also known, and thence $r$. TABLE XIII. will enable us to find $r$ with great facility. Thus, if we had assumed $r_a = 1$ and found $\text{D}_a = 15$, D being 33, we then have

$$1 : r^{\frac{5}{2}} :: 1 : \frac{33}{15} :: 1 : 2\cdot 2, \text{ therefore } r^{\frac{5}{2}} = 2\cdot 2;$$

and thence by TABLE XIII., $r = 1\cdot 37$, the mean radius required, four times which is the diameter of the pipe. For other channels, the quantity thus found must be the hydraulic mean depth; and all channels, however varied in the cross section, will have the same velocity of discharge when the fall, length, and hydraulic mean depth are constant.

In order to find the dimensions of any polygonal channel whatever, which will give a discharge equal to D, we may assume any channel similar to that proposed, one of whose known sides is $s_a$, and find the corresponding discharge, $\text{D}_a$, by Problem I., or from TABLES XI. and XII.; then, if we call the like side of the required channel, s, we shall have $s = s_a \left( \frac{\text{D}}{\text{D}_a} \right)^{\frac{2}{5}}$, and thence the numerical value from TABLE XIII.

As it frequently happens that a deposit in, and encrustation of, a pipe takes place from the water passing through it, which diminishes the section considerably from time to time, it is always prudent, when calculating the necessary diameter, to take the largest coefficient of friction, $c_f$, or to increase its mean value by perhaps one half, when calculating the diameter from the formulæ. Some engineers, as D'Aubuisson, increase the quantity of water by one half to find the diameter; but much must depend on the peculiar circumstances of each case, as sometimes less may be sufficient, or more necessary. The discharge increases in similar figures, as $r^{\frac{5}{2}}$ or as $d^{\frac{5}{2}}$, and the corresponding increase in the diameter for any given or allowed increase in the discharge can be easily found by means of TABLE XIII., as shown above. If we increase the dimensions by one-sixth, the discharge will be increased by one-half nearly.

For shorter pipes, we have to take into consideration the head due to the velocity and orifice of entry. Taking the mean coefficient of velocity or discharge, we find, from TABLE II., the head due to the velocity, if it be known; this subtracted from the whole head, H, leaves the head, $h_f$, due to the hydraulic inclination, which is that we must make use of in the table. If the velocity be not given, we can find it approximately; the head found for this velocity, due to the orifice of entry, when deducted, as before, will give a close value of $h_f$, from which the velocity may be determined with greater accuracy, and so on to any degree of approximation. In general, one approximation to $h_f$ will be sufficient, unless the pipes be very short, in which case it is best to use equation (74). Example VIII., page 108, and the explanation of the use of the tables, SECTION I., may be usefully referred to.

TABLES XI., XII., and XIII. enable us to solve with considerable facility all questions connected with discharge, dimensions of channel, and the ordinary surface inclinations of rivers. The discharge corresponding to any intermediate channels or falls to those given in TABLES XI. or XII., will be found with abundant accuracy, by inspection and simple interpolation; and in the same manner the channels from the discharges. Rivers have seldom greater falls than those given in TABLE XII., but in such an event we have only to divide the fall by 4, then twice the corresponding discharge will be that required. TABLE XIII. gives the comparative discharging powers of all similar channels, whether pipes or rivers, and the comparative dimensions from the discharges. We perceive from it, that an increase of one-third in the dimensions doubles, and a decrease of one-fourth reduces the discharge to one-half. By means of this table, we can determine by a simple proportion, the dimensions of any given form of channel when the discharge is known. See EXAMPLE 17, page 12.

The mean widths in Tables XI. and XII. are calculated for rectangular channels, and those having side slopes of $1\frac{1}{3}$ to 1. Both these tables are, however, practically, equally applicable to any side slopes from 0 to 1 up to 2 to 1, or even higher, when the mean widths are taken and not those at top or bottom. A semihexagon of all trapezoidal channels of equal area has the greatest discharging power, and the semisquare and all rectangles exactly the same as channels of equal areas and depths with side slopes of $1\frac{1}{3}$ to 1. The maximum discharge is obtained between these for the semihexagon with side slopes, of nearly $\frac{1}{2}$ to 1, but for equal areas and depths the discharge decreases afterwards as the slope flattens. The question of "how much?" is here, however, a very important one; for, as we have already pointed out in equations (28) and (31), the differences for any practical purposes may be immaterial. This is particularly so in the case of channels with different side slopes, if, instead of the top or bottom, we make use of the mean width to calculate from. We then have only to subtract the ratio of the slope multiplied by the depth to find the bottom, and add it to find the top. If the mean width be 50 feet, the depth 5 feet, and the side slopes 2 to 1, we get $50 - (2 \times 5) = 40$ for the bottom, and $50 + (2 \times 5) = 60$ for the top width.

Side slopes of 2 to 1 present a greater difference from the mean slope of $1\frac{1}{3}$ to 1, than any others in general practice when new cuts are to be made. A triangular channel having slopes of 2 to 1, and bottom equal to zero, differs more in its discharging power from the half square, equal to it in depth and area, than if the bottom in each was equally increased, yet even here it is easy to show that this maximum difference is only $5\frac{1}{2}$ per cent. If the bottom be increased so as to equal the depth, it is only $4\frac{1}{2}$ per cent.; when equal to twice the depth, 3·8 per cent.; and when equal to four times the depth, to 2 per cent.; while the differences in the dimensions taken in the same order are only 2·2, 1·8, 1·5, and 0·8 per cent. For greater bottoms in proportion to the depth the differences become of no comparative value. It therefore appears pretty evident, that Tables XI. and XII. will be found equally applicable to all side slopes from 0 to 1 up to 2 to 1, by taking the mean widths. When new cuts are to be made, we see no reason whatever in starting from bottom rather than mean widths, to calculate the other dimensions; indeed, the necessary extra tables and calculations involved ought entirely to preclude us from doing so. Besides, the formulæ for finding the discharge vary in themselves, and for different velocities the coefficient of friction also varies.* Added to which the inequalities in every river channel, caused by bends and unequal regimen, precludes altogether any regularity in the working slopes and bottom, though the mean width would continue pretty uniform under all circumstances.

The quantities in TABLE XII. are calculated, from the velocities found from TABLE VIII., to correspond to a channel 70 feet wide and of different depths, the equivalents to which are given in TABLE XI. In order to apply these tables generally to all open channels, the latter are to be reduced to rectangular ones of the same depth and mean width, or the reverse, as already pointed out. If the dimensions of the given channel be not within the limits of TABLE XI., divide the dimensions of the larger channels by 4, and multiply the corresponding discharge found in TABLE XII. by 32; for smaller channels, multiply the dimensions by 4 and divide by 32. In like manner, if the discharge be given and exceed any to be found in TABLE XIII., divide by 32, and multiply the dimensions of the suitable equivalent channel found in TABLE XI. by 4. If we wish to find equivalent channels of less widths than 10 feet for small discharges, multiply the discharge by 32 and divide the dimensions of the corresponding equivalent by 4. Many other multipliers

* The coefficients of friction in rivers for velocities from 3 inches to 3 feet per second, varies from about ·0082 to ·0075; yet, strange to say, most tables are calculated from one coefficient alone; or, rather, from a formula equivalent to $94·17\,(rs)^{\frac{1}{2}}$, which gives larger results than either, see p. 14. Dimensions of channels calculated by means of this formula are too small. In pipes the variation of the coefficients is shown in the small table, p. 120.

and divisors as well as 4 and 32 may be found from TABLE XIII., such as 3 and 15·6, 6 and 88·2, 7 and 130, 9 and 243, 10 and 316, 12 and 499, &c. The differences indicated at pages 111 and 112, must be expected in the application of these rules, which will give, however, dimensions for new channels which can be depended on for doing duty.

It will be seen from TABLE XIII. that a very small increase in the dimensions increases the discharging power very considerably. TABLE XII. also shows that a small increase in the depth alone adds very much to the discharge. If we express in this latter case *a small increase* in the depth, $d$, by $\dfrac{d}{n}$, then it is easy to prove that the corresponding increase in the velocity, $v$, will be $\dfrac{v}{2n}$; and that in the discharge D, $\dfrac{3\,\text{D}}{2n}$, if the surface inclination continue unchanged; but as it is always observable in rivers that the surface inclinations increase with floods, the differences in practice will be found greater than these expressions make it. As in large rivers the surface inclinations are very small, four times the amount will add very little to the sectional areas, yet this increase will double fully the discharge, and we thence perceive how rivers may be absorbed into others without any great increase of the depths.

## SECTION IX.

### BEST FORMS OF THE CHANNEL.—REGIMEN.

We have seen above, that the determination of the hydraulic mean depth does not necessarily determine the section of the channel. If the form be a circle, the diameter is four times the mean radius; but, though this form be almost always adopted for pipes, the beds of rivers take almost every curvilineal and trapezoidal-shape. Other things being the same, that form of a river channel, in which the area divided by the border is a maximum, is the best. This is a semicircle having the diameter for the surface line, and in the same manner, half the regular figures, an octagon, hexagon, and square, in Fig. 33, are better forms for the channel, the areas being constant, than any others of the same number of sides. Of all rectangular

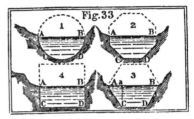

channels, Diagram 4, in which A B C D is half a square, is the best cross section; and in Diagram 3, A C D B, half a hexagon, is the best trapezoidal form of cross section. When the width of the bottom, C D, Diagram 3, is given, and the slope $\frac{Aa}{ca} = n$, then, in order that the discharge may be the greatest possible, we must have

$$ca = \left\{\frac{c}{2(n^2+1)^{\frac{1}{2}} - n}\right\}^{\frac{1}{2}}, \text{ and}$$

$$cD = \frac{c}{ca} - n \times ca$$

$$= \{[2(n^2+1)^{\frac{1}{2}} - n] \times c\}^{\frac{1}{2}} - n\left\{\frac{c}{2( +1)^{\frac{1}{2}} - n}\right\}^{\frac{1}{2}},$$

in which c is the given area of the channel. As, however, we have never known a river in which the slope of the natural banks continued uniform, even though made so for any improvements, we consider it almost unnecessary to give tables for different values of $n$. If, notwithstanding, we put $\varphi$ for the inclination of the slope A C, equal angle C A a, we shall find, as cot. $\varphi = n$, and $\sqrt{n^2 + 1} = \frac{1}{\sin. \varphi}$, that the foregoing equations become

(120.) $\quad ca = \left\{\frac{c \sin. \varphi}{2 - \cos. \varphi}\right\}^{\frac{1}{2}} = \frac{cD}{2\{(n^2+1)^{\frac{1}{2}} - n\}};$

ORIFICES, WEIRS, PIPES, AND RIVERS.   129

and

(121.) $$CD = \frac{C}{ca} - ca \times \cot. \varphi,$$

which will give the best dimensions for the channel when the angle of the slope for the banks is known.

When the discharge from a channel of a given area, with given side slopes, is a maximum, it is easy to show that THE HYDRAULIC MEAN DEPTH MUST BE HALF OF THE CENTRAL OR GREATEST DEPTH. This simple principle enables us to construct the best form of channel with great facility. *Describe any circle on the drawing-board; draw the diameter and produce it on both sides, outside the circle; draw a tangent to the lower circumference parallel to this diameter, and draw the side slopes at the given inclinations, touching the circumference also on each side and terminating on the parallel lines: the trapezoid thus formed will be the best form of channel, and the width at the surface will be equal to the sum of the two side slopes.* It is easy to perceive that this construction may be, simply, extended for finding the best form of a channel having any polygonal border whatever of more sides than three and of given inclinations.

Commencing with the best form of channel, which in practice will have the mean width, about double the depth, an equal discharging section of double the width of the first will have the contents one-eleventh greater, and the depth less in the proportion of 1 to 1·85. A channel of double the mean width of the second must have the sectional area further increased by about one-fifth, and a further decrease in the depth from 1·67 to 1 nearly. The greater expense of the excavation at greater depths will, in general, more than counterbalance these differences in the contents of the channel. When the banks rise above the flood line, and are unequal in their section, the wider channel involves further upper extra cutting, but there is greater capacity to discharge extra and extraordinary flooding, the banks are less liable to slip or give way, the slopes may be less, and the velocity being also less, the regimen will, in general, be better preserved.

K

TABLE OF THE RELATIVE DIMENSIONS OF MAXIMUM DISCHARGING CHANNELS.

| Angle of slope. | Engineering slope. | Depth in terms of the area. | Bottom in terms of the area. | Top in terms of the area. | Hydraulic mean depth in terms of the area. | Ratio of bottom to depth. | Ratio of top to depth. | Area in terms of the depth $d$. |
|---|---|---|---|---|---|---|---|---|
| 90° 0′ | 0 to 1 | $\cdot 707\sqrt{a}$ | $1\cdot 414\sqrt{a}$ | $1\cdot 414\sqrt{a}$ | $\cdot 354\sqrt{a}$ | 2· to 1 | 2· to 1 | $2d^2$ |
| 63 26 | ½ to 1 | $\cdot 759\sqrt{a}$ | $\cdot 938\sqrt{a}$ | $1\cdot 697\sqrt{a}$ | $\cdot 379\sqrt{a}$ | 1·236 to 1 | 2·236 to 1 | $1\cdot 736 d^2$ |
| 48 34¼ | ⅜ to 1 | $\cdot 748\sqrt{a}$ | $\cdot 675\sqrt{a}$ | $1\cdot 996\sqrt{a}$ | $\cdot 374\sqrt{a}$ | ·902 to 1 | 2·667 to 1 | $1\cdot 784 d^2$ |
| 45 0 | 1 to 1 | $\cdot 740\sqrt{a}$ | $\cdot 613\sqrt{a}$ | $2\cdot 093\sqrt{a}$ | $\cdot 370\sqrt{a}$ | ·828 to 1 | 2·828 to 1 | $1\cdot 828 d^2$ |
| 36 52 | 1⅓ to 1 | $\cdot 707\sqrt{a}$ | $\cdot 471\sqrt{a}$ | $2\cdot 357\sqrt{a}$ | $\cdot 354\sqrt{a}$ | ·667 to 1 | 3·333 to 1 | $2d^2$ |
| 33 41½ | 1½ to 1 | $\cdot 689\sqrt{a}$ | $\cdot 417\sqrt{a}$ | $2\cdot 484\sqrt{a}$ | $\cdot 345\sqrt{a}$ | ·605 to 1 | 3·605 to 1 | $2\cdot 105 d^2$ |
| 30 58 | 1⅔ to 1 | $\cdot 671\sqrt{a}$ | $\cdot 372\sqrt{a}$ | $2\cdot 608\sqrt{a}$ | $\cdot 336\sqrt{a}$ | ·554 to 1 | 3·888 to 1 | $2\cdot 221 d^2$ |
| 26 34 | 2 to 1 | $\cdot 636\sqrt{a}$ | $\cdot 300\sqrt{a}$ | $2\cdot 844\sqrt{a}$ | $\cdot 318\sqrt{a}$ | ·472 to 1 | 4·472 to 1 | $2\cdot 472 d^2$ |
| Semicircle | Curved | $\cdot 798\sqrt{a}$ | ·000 | $1\cdot 596\sqrt{a}$ | $\cdot 399\sqrt{a}$ | ·000 to 1 | 2 to 1 | $1\cdot 571 d^2$ |
| Circle | Curved | $1\cdot 128\sqrt{a}$ | ·000 | ·000 | $\cdot 282\sqrt{a}$ | ·000 to 1 | 0 to $d$ | $\cdot 785 d^2$ |

NOTE.—Half the top will give the length of the side slope. The top itself is the sum of the side slopes, and the border the sum of the top and bottom. The sum of the side slopes in any channel whatever is equal to the number corresponding to the slope in the eighth column multiplied by the depth; and the sum of the triangular side areas, corresponding to the slopes, is equal to the square of the depth multiplied by the numbers corresponding to the ratio in column 2. The bottom multiplied by the depth added to this gives the area of the section. The top of the best form of channel should be at the level of the extraordinary flood line, and not at that of the mean annual surface.

When the sectional area is given, the above table shows that the semicircle is the best discharging channel, and the complete circle the worst; the latter is so, however, only compared with the *open* channels given in the table, it being the very best form for an *enclosed* channel flowing full. *The best form of channel is particularly suited for new cuts in flat, marsh, callow, and fen lands, in which it is also often advisable to cut them with a level bed, up from the discharging point,* in order to increase the hydraulic mean depth, and consequently the velocity and discharge.

As the quantity of water coming down a river channel in a season varies very considerably,—we have observed it in one case to vary from one to thirty, and occasionally in the same channel from one to seventy-five,—the proportion of the water section to the channel itself must also vary, and those relations of the depth, sides, and width to each other, above referred to, cease to hold good and be the best under such circumstances. If the object be to construct a mill race, temporary drain for unwatering a river, or other small channel, in which the depth remains nearly constant, channels of the form of a half hexagon, diagram 3, Fig. 33, will be, perhaps, the best, if the tenacity of the banks permit the slope; but rivers, in which the quantity of water varies considerably, require wider channels in proportion to the depth; and also, that the velocity be so proportioned to the tenacity of the soil, or as it is termed "*the regimen,*" that the banks and bed shall not vary from time to time to any injurious extent, and that any deposits made during their summer state, and during light freshes, shall be carried off periodically by floods. Another circumstance, also, modifies the effects of the water on the banks. It is this, that at curves, and turns, the current acts with greatest effect against the bank, concave to the direction in which it is moving; deepening the channel there; undermining also the bank, as at A, Fig. 34; and raising the

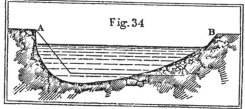

Fig. 34

bed towards the opposite side B. The reflexion of the current to the opposite bank from A acts also in a similar manner, lower down, upon it; and this natural operation proceeds, until the number of turns, increased length of channel, and loss of head from reflexion and unequal depths, brings the currents into regimen with the bed and banks. At all bends it is, therefore, prudent to widen the channel on the convex side B, Fig. 34, in order to reduce the velocity of approach, and if the bed be here also sunk below its natural inclination, as we see it in most rivers at bends, the velocity will be farther reduced, and the permanence of the bed be better established.

The circumstances to be considered in deciding on the dimensions and fall of a new river course, after the depth to which the surface of the water is to be brought has been decided on, are the following:—

The mean velocity must not be too slow, or aquatic plants will grow, and deposits take place, reducing the sectional area until a new and smaller channel is formed within the first with just sufficient velocity to keep itself clear. This velocity should not in general be less than from ten to fourteen inches per second. The velocity in a canal or river is increased very considerably by cutting or removing reeds and aquatic plants growing on the sides or bottom*.

The mean velocity must not be too quick, and should be so determined as to suit the tenacity and resistance of the channel, otherwise the bed and banks will change continually, unless artificially protected; it should not exceed

* M. Girard a fait observer, avec raison, que les plantes aquatiques, qui croissent toujours sur le fond et sur les berges des canaux, augmentent considérablement le perimètre mouillé, et par suite la résistance; il a rapellé que Dubuat, ayant mesuré la vitesse de l'eau dans le canal du Jard, avant et après la coupe des roseaux dont il était garni, avait trouvé un resultat bien moindre avant qu'après. En conséquence, il a presque doublé la pente donnée par le calcul . . ."—Traité d'Hydraulique, p. 135. When the fall does not exceed a few inches per mile, the velocity, as determined from the inclination, is very uncertain, and for this reason it is always prudent to increase the depths and sectional areas of channels in flat lands, as far as the regimen will permit. In such cases the section of the channel should approximate towards the best form. See p. 130.

25 feet per minute in soft alluvial deposits.
40 ,, ,, clayey beds.
60 ,, ,, sandy and silty beds.
120 ,, ,, gravelly.
180 ,, ,, strong gravelly shingle.
240 ,, ,, shingly.
300 ,, ,, shingly and rocky.
400 and upwards in rocky and shingly.

The fall per mile should decrease as the hydraulic mean depth increases, and both be so proportioned that floods may have sufficient power to carry off the deposits, if any, periodically. The proportion of the width to the depth of the channel should not be derived, for new cuts or river courses, from any formula, but taken from such portions of the old channel as approximate in depth and in the inclination of the surface to that proposed. When the depth is nearly half the width, the formula shows, *cæteris paribus*, that the discharge will be a maximum; but as (altogether apart from the question of expense) the quantity of water discharged daily, at different seasons, may vary from one to seventy, or more, and "*the regimen*" has to be maintained, the best proportion between the width and depth of a new cut should be obtained, as we have stated, from some selected portion of the old channel, whose general circumstances and surface inclination approximate to those of the one proposed; and the side slopes of the banks must be such as are best suited to the soil. The resistance of the banks to the current being in general less than that of the beds, which get covered with gravel, and the necessary provision required for floods, appears to be the principal reason why rivers are in general so very much wider than about twice the depth, the relation which gives the minimum of friction.

For enclosed channels, the circular form of sewer flowing full, will have the largest scouring power, at a given hydraulic inclination. For the area of the sections being the same, the velocity in the circular channel will be a maximum. When the supply is intermittent, and the channel too large, the egg-shaped form with the smaller end for the bottom,— or the sides vertical with an inverted ridge tile or v bottom for drains,—will have a hydrostatic flushing power to remove soil and obstructions, which a cylindrical channel, only partly full, does not possess; because a given quantity of water rises higher against the same obstruction, or obstacle, to the flow. It must be confessed, however, that for small drains and house-sewage, this gain is immaterial, and is at best but effected by a sacrifice of space, material, and friction in the upper part of drains, from 4 to 12 or 15 inches in diameter. Besides this, the mere hydrostatic pressure is only intermittent, and during an ordinary, or heavy, fall of rain, the hydrodynamic power is always more efficient in scouring properly-proportioned cylindrical drains; and the workmanship in the form and joints is less imperfect than for more compound forms, as those with egg-shaped and inverted tile bottoms. The moulds and joints of cylindrical stone-ware drains, exceeding 15 inches in diameter, are seldom, however, perfect; and the expense will exceed that of brick, stone, or other sufficient drains in most localities.

As to the increased discharging power which it is asserted by some, stone-ware cylindrical drains possess over other ordinary drains, no doubt it is true for small sizes, because the form, jointing, and surface are in general more smooth and circular; and *for sewage matter*\*, the friction and adherence to the sides and bottom is less; any advantage from these causes becomes, however, immaterial for the larger sizes, as these can be constructed of brick abundantly perfect to any form, and sufficiently smooth for all practical purposes, for in the larger properly-proportioned sizes the same amount of surface roughness opposed to the sewage matter is, comparatively, of no effect. The judicious inclination and form of the bottom, and properly curved junctions, are the principal points to be attended to. Smaller drains tile-bottomed, with brick or stone sides, and flat-covered, have one great advantage over circular pipes†. They can be opened up, for exa-

\* Weisbach found the coefficient of resistance 1·75 times as great for small wooden as for metallic pipes. All *permeable* pipes present greater resistance than *impermeable* ones; hence the principal advantage derived from glazing.

† Half-socket joints at bottom would remedy this imperfection in small pipes, and they could be better laid and cemented. A semicircular flange laid on at top would effectually protect the joint on the upper side.

mination and repairs at any time with facility, and at the smallest expense, but greater certainty must be attached to the working of small stone-ware drains than to equal sized small brick or stone drains, and they will be found, in general, also the cheaper. This, however, will depend on the locality.

In order that different sections should have the same discharging power, the inclination of the surface being the same, the areas must be inversely as the square roots of the hydraulic mean depths. The channel $adc\,\textsc{b}$, Fig. 35, will have the same discharge as the channel $\textsc{adcb}$ if they be to each other as $\left\{\dfrac{\textsc{adcb}}{\textsc{ad}+\textsc{dc}+\textsc{cb}}\right\}^{\frac{1}{2}}$ to $\left\{\dfrac{adc\,\textsc{b}}{ad+dc+c\,\textsc{b}}\right\}^{\frac{1}{2}}$,

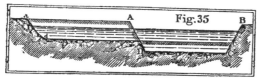

and hence the square root of the cube of the channel area, divided by the border, must be constant. With a fall of one or more feet to a mile, two channels, one 70 feet wide and 1 foot deep, and the other 20 feet wide and $2\frac{1}{2}$ feet deep, will have the same discharge. If we put $w$ for the width and $d$ for the depth of any rectangular channel, then $\left\{\dfrac{w^3 d^3}{w+2d}\right\}^{\frac{1}{2}} = m$; we therefore have the cubic equation

(122.) $$d^3 - \frac{2m^2}{w^3}d = \frac{m^2}{w^2}$$

for finding the depth $d$ of any other rectangular channel whose width is $w$, of the same discharging power. We have calculated the depths $d$ for different widths of channel from this equation, assuming a width of 70 feet and different depths to find $m$ from. This table will be found sufficiently accurate for all practical purposes, by taking a mean width, when the banks are sloped.

If the hydraulic inclinations vary, then the $\sqrt{rs}$ must be inversely as the areas of the channels when $\sqrt{rs} \times$ channel or the discharge is constant; and if the area of the channel and discharge be each constant, $r$ must vary inversely as $s$; and $r\,s$ be also constant. For instance, a channel which has a fall of four feet per mile, and a hydraulic mean depth of one foot, will have the same discharge as another channel of equal area, having a hydraulic mean depth of four feet and a fall per mile of only one foot.

TABLE OF EQUAL DISCHARGING RECTANGULAR CHANNELS *.

| Values of $m$. | The widths in feet are given in the top horizontal line, and the corresponding depths in feet in the other horizontal lines. | | | | | | | | |
|---|---|---|---|---|---|---|---|---|---|
| | 70 | 60 | 50 | 40 | 35 | 30 | 25 | 20 | 15 | 10 |
| 8·7 | 0·25 | ·27 | ·30 | ·35 | ·40 | ·45 | ·52 | ·58 | ·71 | ·98 |
| 24·6 | 0·5 | ·55 | ·62 | ·73 | ·80 | ·89 | 1·02 | 1·19 | 1·48 | 2·04 |
| 45·0 | ·75 | ·82 | ·94 | 1·10 | 1·20 | 1·35 | 1·56 | 1·82 | 2·28 | 3·22 |
| 69·0 | 1·0 | 1·10 | 1·26 | 1·48 | 1·62 | 1·81 | 2·10 | 2·46 | 3·11 | 4·50 |
| 96·9 | 1·25 | 1·39 | 1·58 | 1·86 | 2·04 | 2·28 | 2·65 | 3·12 | 3·98 | 5·89 |
| 126 | 1·5 | 1·67 | 1·90 | 2·24 | 2·46 | 2·75 | 3·20 | 3·80 | 4·88 | 7·31 |
| 158 | 1·75 | 1·95 | 2·22 | 2·62 | 2·88 | 3·23 | 3·75 | 4·50 | 5·80 | 8·86 |
| 193 | 2·0 | 2·23 | 2·54 | 3·00 | 3·31 | 3·72 | 4·32 | 5·22 | 6·78 | 10·50 |
| 267 | 2·5 | 2·79 | 3·18 | 3·76 | 4·16 | 4·70 | 5·50 | 6·68 | 8·84 | 14·00 |
| 349 | 3·0 | 3·35 | 3·84 | 4·54 | 5·04 | 5·72 | 6·69 | 8·22 | 11·03 | 17·68 |
| 437 | 3·5 | 3·91 | 4·50 | 5·33 | 5·95 | 6·75 | 7·93 | 9·82 | 13·32 | 21·68 |
| 531 | 4·0 | 4·48 | 5·14 | 6·13 | 6·85 | 7·81 | 9·21 | 11·48 | 15·75 | 26·00 |
| 629 | 4·5 | 5·05 | 5·79 | 6·95 | 7·75 | 8·90 | 10·50 | 13·19 | 18·22 | 30·36 |
| 732 | 5·0 | 5·62 | 6·45 | 7·75 | 8·66 | 10·00 | 11·79 | 14·96 | 20·80 | 35·00 |
| 839 | 5·5 | 6·18 | 7·12 | 8·57 | 9·62 | 11·10 | 13·24 | 16·77 | 23·47 | 39·81 |
| 951 | 6·0 | 6·75 | 7·80 | 9·40 | 10·60 | 12·22 | 14·65 | 18·65 | 26·25 | 44·86 |

## SECTION X.

EFFECTS OF ENLARGEMENTS AND CONTRACTIONS.—BACK-WATER WEIR CASE.—LONG AND SHORT WEIRS.

WHEN the flowing section in pipes or rivers expands or contracts suddenly, a loss of head always ensues; this is pro-

* This table is enlarged in TABLE XI., and is as equally applicable to all other measures, inches, yards, fathoms, &c., as to feet.

bably expended in forming eddies at the sides, or in giving the water its new section. A side current, moving slowly *upwards*, may be frequently observed in the wide parts of rivers, when the channel is unequal, though the downward current, at the centre, be pretty rapid; and though we may assume generally that the velocities are inversely as the sections, when the channels are uniform, we cannot properly do so where the motions are so uncertain as those referred to. When a pipe is contracted by a diaphragm at the orifice of entry, Fig. 27, we have seen (equation 60), that the loss of head is,

Fig. 36

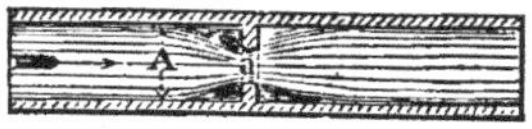

(123.) $$h = \frac{\left(1 - \frac{A^2}{C^2}\right)v^2 + \left(\frac{A}{a\,c_d} - 1\right)^2 v^2}{2g}.$$

When the diaphragm is placed in a uniform pipe, Fig. 36, then $A = C$, and we get the loss of head

(124.) $$h = \frac{\left(\frac{A}{a\,c_d} - 1\right)^2 v^2}{2g},$$

and the coefficient of resistance

(125.) $$c_r = \left(\frac{A}{a\,c_d} - 1\right)^2,$$

as in equation (67). The coefficient of discharge $c_d$ is here equal to the coefficient of contraction $c_c$, or very nearly. Now we have shown in equation (45), and the remarks following it, that the value of the coefficient of discharge, $c_d$, varies according to the ratio of the sections, $\frac{A}{a}$, and in TABLE V. we have calculated the new coefficients for different values of the ratios, and different values of the primary coefficient $c_d$. If we assume $c_d$, when A is very large compared with $a$, to be ·628, we then find by attending to the remarks at pp. 61 and 62, that the different values of $c_d$ corresponding to ·807 × $\frac{A}{a}$, taken from TABLE V., are those in columns Nos. 2 and 5 of

TABLE OF COEFFICIENTS FOR CONTRACTION, BY A DIAPHRAGM IN A PIPE.

| $\frac{a}{A}$ | $c_d$ | $c_r$ | $\frac{a}{A}$ | $c_d$ | $c_r$ |
|---|---|---|---|---|---|
| ·0 | ·628 | infinite | ·6 | ·713 | 1·790 |
| ·1 | ·630 | 221·2 | ·7 | ·753 | ·807 |
| ·2 | ·636 | 47·1 | ·8 | ·807 | ·301 |
| ·3 | ·647 | 17·2 | ·85 | ·845 | ·154 |
| ·4 | ·661 | 7·7 | ·9 | ·890 | ·062 |
| ·5 | ·683 | 3·7 | 1 | 1·000 | ·000 |

the above small table, the values of the coefficient of resistance, in columns 3 and 6, being calculated from equation (125) for the respective new values of the coefficient of discharge thus found. The table shows that when the aperture in a diaphragm is $\frac{2}{10}$ths of the section of the pipe, that 47 times the head due to the velocity is lost thereby. If the aperture in the diaphragm be rounded, the loss will not be so great, as the primary coefficient $c_d$ will then be greater than that due to an orifice in a thin plate : see coefficients, p. 96.

When there are a number of diaphragms in a tube, the loss of head for each must be found separately, unless the ratio $\frac{a}{A}$ be constant, and all added together for the total loss. If the diaphragms, however, approach each other, so that the water issuing from one of the orifices $a$, Fig. 36, shall pass into the next before it again takes the velocity due to the diameter of the pipe, the loss will not be so great as when the distance is sufficient to allow this change to take place. This view is fully borne out by the experiments of Eytelwein with tubes 1·03 inch in diameter, having apertures in the diaphragms of ·51 inch in diameter.

Venturi's twenty-fourth experiment, with tubes varying from ·75 inch to ·934 inch in diameter at the junction with the cistern, so as to take the form of the contracted vein, and expanding and contracting along the length from ·75 to 2 inches and from

2 inches to ·75 inch alternately, shows the great loss of head sustained by successive enlargements and contractions of a channel, even when the junction of the parts is gradual. Calling the coefficient for the short tube, with a junction of nearly the form of the contracted vein, 1, then the following coefficients are derivable from the experiment:—

Short tube with rounded junction . . . . . . . 1·
One enlargement . . . . . . . . . . . . . ·741
Three enlargements . . . . . . . . . . . . ·569
Five enlargements . . . . . . . . . . . . ·454
Simple tube with a rounded junction of the same
 length, 36 inches, as the tube with the five en-
 larged parts . . . . . . . . . . . . . . ·736

The head, in the experiment, was $32\frac{1}{2}$ inches. Venturi states that no observable differences occurred in the times of discharge when the enlarged portions were lengthened from $3\frac{1}{6}$ to $6\frac{1}{3}$ inches.

With reference to this experiment, so often quoted, it is necessary to remark that the diameters of the enlarged portions were 2 inches each, while the lengths varied only from $3\frac{1}{6}$ to $6\frac{1}{3}$ inches, and consequently were at most only $3\frac{1}{6}$th times the diameter. Now with such a large ratio of the width to the length of the enlarged portions, $a$ A B $b$, Fig. 37, it is pretty clear that a good deal of the head is lost by the impact of the moving water on the shoulders A and B.

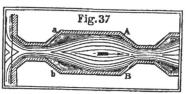

Fig. 37

That this is so is evident from the fact, stated by the experimenter, of the time of discharge remaining the same when $a$ A, in five different enlargements, was increased from $3\frac{1}{6}$ to $6\frac{1}{3}$ inches; though this must have lengthened the whole tube from 36 to 50 inches\*, thereby increasing the loss from friction proportionately, but which happened to be compensated for by the reduction in the resistances from

---

\* The dimensions throughout this experiment are given as in the original, viz. in French inches.

impact at A and B, and in the eddies, by doubling the lengths from $a$ to A.

If, however, the length from $a$ to A be very large compared with the diameter, and the junctions at $a$, A, B, and $b$, be well grafted, less loss will arise from the enlargement than if the smaller diameter continued all along uniform. The explanation is clear, as the resistance from friction is inversely as the square roots of the mean radii; and the length being the same, the loss must be less with a large than a small diameter.

These remarks, *mutatis mutandis*, apply equally to rivers as to pipes. We have already, pp. 72 and 75, pointed out the effects of submerged weirs and contracted river channels, and given formulæ for calculating them.

### BACKWATER FROM CONTRACTIONS IN RIVERS.

A river may be contracted in width or depth by jetties or weirs; and when the quantity to be discharged is known, we have given, in formulæ (9), (55), and (57), equations from which the increase of head may be determined. The effect of a weir, jetty, or contracted channel of any kind, is to increase the depth of water above; and this is sometimes necessary for navigation purposes or to obtain a head for mill power. When a weir is to rise over the surface, we can easily find, from the discharge and length, the discharge per minute over each foot of length, with which, and the coefficient due to the ratio of the sections, on and above the weir, found from TABLE V., we can find the head from TABLE VI. For submerged weirs and contracted widths of channel, the head can be best calculated, by approximation, from the equations above referred to.

The head once determined, the extent of the backwater is a question of some importance. If F C O D, Fig. 38, be the original surface of a river, and $a$ A B F the raised surface by backwater from the weir at $a$, then extent $a$ F of this backwater, in a regular channel, will be from 1·5 to 1·9 times $a$ c drawn parallel to the horizon to meet the original surface

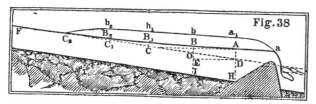

Fig. 38

in c. This rule will be found useful for practical purposes; but in order to determine more accurately the rise for a given length, $B_1 B_2$ or $B_1 B$, of the channel, it is necessary to commence at the weir and calculate the heights from A to B, B to $B_1$, and from $B_1$ to $B_2$ separately, the distance from A to $B_2$ being supposed divided into some convenient number of equal parts, so that the lengths A B, B $B_1$, &c., may be considered free from curvature. Now, as the head A D is known, or may be calculated by some of the preceding formulæ, the section of the channel at the head of the weir also becomes known, and thence the mean velocity in it, by means of the discharge over the weir. Putting A for the area of the channel at A H, $d$ for its depth A H, and $v$ for the mean velocity; also $A_1$ for the area of the channel at B I, $d_1$ for its depth, and $v_1$ for its mean velocity; $b_m$ the mean border between the sections at A H and B I; $r_m$ the mean hydraulic depth; $\dfrac{v + v_1}{2}$ the mean velocity; $A D = h$; $B O = h_1$; the sine of angle $O D E = s$; and the length $A B = D O$ nearly $= l$; we get $A \times v = A_1 \times v_1$ and $r_m = \dfrac{A + A_1}{2 b_m}$; but as, in passing from B to A, the velocity changes from $v_1$ to $v$, there is a loss of head equal $\dfrac{v_1^2 - v^2}{2g}$, and if $c_f$ be the coefficient of friction, there is a loss of head from this cause equal $c_f \times \dfrac{l}{r_m} \times \dfrac{\left(\dfrac{v_1 + v}{2}\right)^2}{2g}$; hence the whole change of head in passing from B to A is equal to $c_f \times \dfrac{l}{r_m} \times \dfrac{(v_1 + v)^2}{8g} - \dfrac{v_1^2 - v^2}{2g}$. But this change of head is equal to $BE - AD = BO + OE - AD = h_1 + ls - h$, whence we get

(126.) $h_1 - h = d_1 - d = c_f \times \dfrac{l}{r_m} \times \dfrac{(v_1 + v)^2}{8g} - \dfrac{v_1^2 - v^2}{2g} - ls;$

or as $v_1 = \dfrac{A v}{A_1}$, and $r_m = \dfrac{A + A_1}{2 b_m}$, we get, by a few reductions and change of signs,

(127.) $h - h_1 = \left( s - c_f \times b_m \times \dfrac{A + A_1}{2 A_1^2} \times \dfrac{v^2}{2g} \right) l + \dfrac{A^2 - A_1^2}{A_1^2} \times \dfrac{v^2}{2g};$

and therefore we get

(128.) $l = \dfrac{h - h_1 - \dfrac{A^2 - A_1^2}{A_1^2} \times \dfrac{v^2}{2g}}{s - c_f \times \dfrac{b_m \times (A + A_1)}{2 A_1^2} \times \dfrac{v^2}{2g}},$

from which we can calculate the length $l$ corresponding to any assumed change of level between A and B. Then, by a simple proposition we can find the change of level for any smaller length. To find the change of level directly from a given length does not admit of a direct solution, for the value of $h - h_1$ in equation (127) involves $A_1$, which depends again on $h - h_1$, and further reduction leads to an equation of a higher order; but the length corresponding to a given rise, $h_1$, is found directly by equation (128).

When the width of the channel, $w$, is constant, and the section equal to $w \times d$ nearly, the above equations admit of a further reduction for $A_1 = d_1 w$ and $A = dw$; by substituting these values in equation (127) it becomes, after a few reductions,

(129.) $h - h_1 = d - d_1$
$= \left( s - c_f \times b_m \times \dfrac{d \times d_1}{2 d_1^2 w} \times \dfrac{v^2}{2g} \right) l + \dfrac{d^2 - d_1^2}{d_1^2} \times \dfrac{v^2}{2g};$

or, as it may be further reduced,

(130.) $h - h_1 = \dfrac{s - c_f \times \dfrac{b_m}{d_1 w} \times \dfrac{d + d_1}{2 d_1} \times \dfrac{v^2}{2g}}{1 - \dfrac{d + d_1}{d_1^2} \times \dfrac{v^2}{2g}} \times l.$

# ORIFICES, WEIRS, PIPES, AND RIVERS. 143

Now, we may take in this equation for all practical purposes,
$$\frac{d + d_1}{2 d_1} \times \frac{b_m}{d_1 w} = \frac{b}{d w},$$
approximately, $b$ being the border of the section at $AH$; and also, $\dfrac{d + d_1}{d_1^2} = \dfrac{2}{d}$, approximately; therefore, we shall have

(131.) $$h - h_1 = \frac{s - c_f \times \dfrac{b}{d w} \times \dfrac{v^2}{2g}}{1 - \dfrac{2}{d} \times \dfrac{v^2}{2g}} \times l;$$

and

(132.) $$l = \frac{(h - h_1) \times \left(1 - \dfrac{2}{d} \times \dfrac{v^2}{2g}\right)}{s - c_f \times \dfrac{b}{d w} \times \dfrac{v^2}{2g}}.$$

Now, as $\dfrac{b}{d w} = \dfrac{1}{r}$, $2g = 64\cdot 4$, and the mean value of the coefficient of friction for small velocities $c_f = \cdot 0078$, we shall get

(133.) $$h_1 = h - \frac{64\cdot 4\, ds - \cdot 0078 \dfrac{d}{r} v^2}{64\cdot 4\, d - 2 v^2} \times l;$$

and

(134.) $$l = \frac{(h - h_1) \times (64\cdot 4\, d - 2 v^2)}{64\cdot 4\, ds - \cdot 0078 \dfrac{d}{r} v^2},$$

very nearly. Having by means of these equations found $AB$ from $BO$ or $BE$, and $BO$ from $AB$, we can in the same manner proceed up the channel and calculate $B_1 C$, $B_2 C_1$, &c., until the points $B$, $B_1$, $B_2$ in the curve of the backwater shall have been determined, and until the last nearly coincides with the original surface of the river. When $h_1 = 0$, we shall have

$$h = \frac{64\cdot 4\, ds - \cdot 0078 \dfrac{d}{r} v^2}{64\cdot 4\, d - 2 v^2} \times l.$$

If we examine equation (134) it appears that when $64\cdot 4\, d = 2v^2$, $l$ must be equal to zero; or when $\dfrac{d}{2} = \dfrac{v^2}{64\cdot 4}$, equal the height due to the velocity $v$. When $l$ is infinite, $64\cdot 4\, d$ must exceed $2v^2$, and $64\cdot 4\, ds$ equal to $\cdot 0078 \dfrac{d}{r} v^2$;

$$\text{or,}\ \frac{64\cdot 4\, rs}{\cdot 0078} = v^2,\ \text{and}\ v = 90\cdot 9\ \sqrt{rs}.$$

This is the velocity due to friction in a channel of the depth $d$, hydraulic mean depth $r$, and inclination $s$; and, as in wide rivers $r = d$ nearly, $v = 90\cdot 9\ \sqrt{ds}$, but when the numerator was zero we had from it $v = \sqrt{32\cdot 2\, d}$; equating these values of $v$, we get $s = \cdot 0039 = \dfrac{1}{256}$ nearly: see p. 70. Now, the larger the fraction $s$ is, the larger will the velocity $v$ become; and the larger $v$ becomes, the more nearly, in all practical cases, will the terms

$$64\cdot 4\, d - 2\, v^2\ \text{and}\ 64\cdot 4\, ds - \cdot 0078 \frac{d}{r} v^2,$$

in the numerator and denominator of equation (134), approach zero; when $64\cdot 4\, d - 2\, v^2$ becomes zero first, $l = 0$; when $64\cdot 4\, ds - \cdot 0078 \dfrac{d}{r} v^2$ becomes zero first, $l$ equal infinity; and when they both become zero at the same time, $l = h - h_1$, and $s = \dfrac{1}{256}$, see p. 70; if $s$ be larger than this fraction, the numerator in equation (134) will generally become zero before the denominator, or negative, in which cases $l$ will also be zero, or negative; and the backwater will take the form $\text{F}\,c_2\,b_2\,b_1\,b\,a_1\,a$, Fig. 38, with a hollow at $c_2$. Bidone first observed a hollow, as $\text{F}\,c_2\,b_2$, when the inclination $s$ was $\dfrac{1}{30}$. When the inclination of a river channel changes from greater to less, the velocity is obstructed, and a hollow similar to $\text{F}\,c_2\,b_2$ sometimes occurs; when the difference of velocity is considerable, the upper water at $b_2$ falls backwards towards $c_2$ and $\text{F}$, and forms a *bore*, a splendid instance of which is the *pororoca*, on the Amazon, which takes place where the inclination of the surface changes from 6 inches to ⅓th of an inch per mile, and the velocity from about 22 feet to 4½ feet per second.

## WEIR CASE, LONG AND SHORT WEIRS.

When a channel is of very unequal widths, above a weir, we have found the following simple method of calculating the backwater sufficiently accurate, and the results to agree very closely with observation. *Having ascertained the surface fall due to friction in the channel at a uniform mean section, add to this fall the height which the whole quantity of water flowing down would rise on a weir having its crest on the same level as the lower weir, and of the same length as the width of the channel in the contracted pass. The sum will be the head of water at some distance above such pass very nearly.* A weir was recently constructed on the river Blackwater, at the bounds of the counties Armagh and Tyrone, half a mile below certain mills, which, it was asserted, were injuriously affected by backwater thrown into the wheel-pits. The crest of the weir, 220 feet long, was 2 feet 6 inches below the pit; the river channel between varied from 50 and 57 feet to 123 feet in width, from 1 foot to 14 feet deep; and the fall of the surface, with 3 inches of water passing over the weir and the sluices down, was nearly 4 inches in the length of half a mile. Having seen the river in this state in summer, the writer had to calculate the backwater produced by different depths passing over the weir in autumn and winter, which in some cases of extraordinary floods were known to rise to 3 feet. The width of the channel about 60 feet above the weir averaged 120 feet. The width, 2050 feet above the weir and 550 feet below the mills, was narrowed by a slip in an adjacent canal bank, to 45 feet at the level of the top of the weir, the average width at this place as the water rose being 55 feet. The channel above and below the slip widened to 80 and 123 feet. Between the mills and the weir there were, therefore, two passes; one at the slip, averaging 55 feet wide; another above the weir, about 120 feet wide. Assuming as above, that the water rises to the heights due

L

to weirs 55 and 120 feet long, at these passes, we get, by an easy calculation, or by means of Table X., the heads in columns two and four of the following table, corresponding to the assumed ones on the weir, given in the first column.

TABLE OF CALCULATED AND OBSERVED HEIGHTS ABOVE M'KEAN'S WEIR ON THE RIVER BLACKWATER.

| Heights at M'Kean's weir 220 feet long, in inches. | Heights 60 feet above the weir channel 120 feet wide. | | Heights 2050 feet above the weir channel 55 feet wide; average. | |
|---|---|---|---|---|
| | Calculated inches. | Observed inches. | Calculated inches. | Observed inches. |
| 1½ | 2¼ | 2½ | 4¼ | 5½ |
| 2 | ... | ... | ... | ... |
| 3 | 4½ | ... | 7½ | 7 |
| 4 | 6 | ... | 10 | 9 |
| 5 | 7½ | ... | 12½ | 11¼ |
| 6 | 9 | 9 | 15 | 16¼ |
| 7 | 10½ | 10½ | 17½ | 18¾ |
| 8 | 12 | ... | 20 | 20½ |
| 9 | 13½ | 12½ | 22½ | 20¼ |
| 10 | 15 | ... | 24¼ | 20 |
| 11 | 16½ | ... | 27¾ | 24 |
| 12 | 18 | 17 | 30¼ | 31 |
| 13 | 19½ | 18½ | 32¾ | 33 |
| 15 | 22½ | 21 | 37¾ | 40 |
| 18 | 27 | 25 | 45¼ | 46 |
| 21 | 31½ | 29½ | 53 | 54 |
| 24 | 36 | 34 | 60½ | 62 |

As the length of the river was short, and the hydraulic mean depth pretty large, the fall due to friction for 60 feet above the weir was very small, and therefore no allowance was made for it; even the distance to the slip was comparatively short, being less than half a mile, and as the water approached it with considerable velocity, this was conceived, as the observations afterwards showed, to be a sufficient compensation for the loss of head below by friction. The observations were made by a separate party, over whom the writer had no control, and it is necessary to remark, that with the same head of water on the weir, they often differed more from each other than from the calculation. This, probably, arose from the different directions of the wind and the water rising during one observation, and falling during another.

The true principle for determining the head at $g$, Fig. 39, apart from that due to friction, is that pointed out at pages 75 and 76; but when the passes are very near each other, or the depth $d_2$, Fig. 23, is small, the effect of the discharge through $d_2$ is inconsiderable in reducing the head, as the contraction and loss of *vis-viva* are then large, and the head $d_1$ becomes that due to a weir of the width of the contracted channel at A, nearly. The reduction in the extent of the backwater, by lowering the head on a longer weir, is found by taking the difference of the amplitudes due to the heads at $g$, Fig. 39, in both cases, as determined from equations (56,) (128), *et seq.* This will seldom exceed a mile up the river, as the surface inclination is found to be considerably greater than that due to mere friction and velocity, and hence the general failure of drainage works designed on the assumption that the lowering of the head below, by means of long weirs, extends its effects all the way up a channel. We must nearly treble the length of a weir before the head passing over can be reduced by one-half, TABLE X., even supposing the circumstances of approach to be the same; surely several engineering appliances for shorter weirs, during periods of flood, would be found more effective and far less expensive than this alternative, with its extra sinking and weir basin.

The advocates for the necessity of weirs longer than the width of the channel, for drainage purposes, must show that the reduction of the head and extent of backwater above $g$, Fig 39, is not small, and that the effects extend the whole way up the channel, or at least as far as the district to be benefited. Practice has heretofore shown, that long weirs have failed (unless after the introduction of sluices or other appliances) in producing the expected arterial drainage results, notwithstanding the increased leakage from increased length, which must accompany their construction.

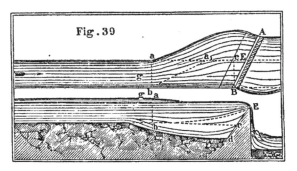

Fig. 39

The deepening in the weir basin $ab$ BEA is mostly of use in reducing the surface inclination between $ab$ and AB by increasing the hydraulic mean depth; but, thereby, the velocity of approach is lessened, and therefore the head at E increased. When the length of a weir basin $a$ E exceeds that point where these two opposite effects balance each other, there will be a gain by the difference of the surface inclinations in favour of the long weir; but unless $a$ E exceeds half a mile, this difference cannot amount to more than 3 or 4 inches, unless the river be very small indeed: and *if the channel be sunk for the long weirs* B A *or* $b\,a_1$, *it should also be sunk to at least the same depth and extent for the short weirs* B $e$, $b\,a$, *otherwise there is no fair comparison of their separate merits.* The effect of the widening between $ab$ and AB, the depth being the same, is also to reduce the surface inclination from $a$ to E; but, as before, unless $a$ E be of considerable length, this gain will also be small. Now AB, at best, is but a weir the direct width of the new channel at AB, and if the length $a$ E be considerable, we have an entirely new river channel with a direct weir at the lower end, and the saving of head effected arises entirely from the larger channel, with a *direct* transverse weir at the lower end.

By referring to TABLE VIII., it will be found that for a hydraulic mean depth of 5 feet a fall of $7\frac{1}{2}$ inches per mile will give a velocity of 2 feet per second: if we double the depth, a fall of 4 inches will give the same velocity; and for a depth of only 2 feet 6 inches, a fall of $12\frac{1}{2}$ inches is necessary. This is a velocity much larger than we have ever observed in a weir basin, yet we easily perceive that the difference in the inclinations for a short distance, E$a$ of a few hundred feet, must be small. If one section be double the other, the hydraulic mean depth remaining constant, the velocity must be one-half, and the fall per mile one-fourth, nearly. This would leave $7\frac{1}{4} - 2 = 5\frac{1}{4}$ inches per mile, or 1 inch per 1000 feet nearly, as the gain with a hydraulic mean depth of 5 feet for a double water channel. For greater depths the gain would be less, and the contrary for lesser depths.

Is the saving of head and amplitude of backwater we have calculated worth the increased cost of long weirs and the consequent necessity and expense of sinking and widening the channels for such long distances? We think not; indeed, *the sinking in the basin immediately at the weir is absolutely injurious, by destroying the velocity of approach,* and increasing the contraction. The gradual approach of the bottom towards the crest, shown by the upper dotted line $b$ E in the section, Fig. 39, and a sudden overfall, will be found more effective in reducing the head, unless so far as leakage takes place, than any depth of sinking for only 60 or 80 feet above long weirs.

In most instances, the extra head will be only perceived by an increased surface inclination, which may extend for a mile or so up the channel, according to the sinking and widening.

It is a general rule that, for shorter weirs, the coefficients of discharge decrease; this arises from the greater amount of lateral contraction, and is more marked in notches or Poncelet weirs, than for weirs extending from side to side of the channel; but for weirs exceeding 10 feet in length the decrease in the coefficients from this cause is immaterial, unless the head passing over bear a large ratio to the length; and we even see from the coefficients, page 45, derived from Mr. Blackwell's experiments, that with 10 inches head passing over a 2-inch plank, the coefficient for a length of 3 feet is ·614; for a length of 6 feet ·539; and for a length of 10 feet ·534; showing a decrease as the weir lengthens, but which may, in the particular cases, be accounted for. We have before referred to other circumstances which modify the coefficients, yet we may assume generally, without any error of practical value, that the coefficients are the same for different weirs extending from side to side of a river. If, then, we put $w$ and $w_1$ for the lengths of two such weirs, we shall have the relation of the heads $d$ and $d_1$ for the same quantity of water passing over, as in the following proportion:—

$$d : d_1 :: w_1^{\frac{3}{2}} : w^{\frac{3}{2}};$$

and therefore

(135.) $$d_1 = \left(\frac{w}{w_1}\right)^{\frac{3}{2}} \times d.$$

By means of this equation we have calculated TABLE X., the ratio $\frac{w}{w_1}$ being given in columns 1, 3, 5 and 7, and the value of $\left(\frac{w}{w_1}\right)^{\frac{3}{2}}$, or the coefficient by which $d$ is to be multiplied, to find $d_1$ in columns 2, 4, 6, and 8. It appears also, that *if we take the heads passing over any weir in a river in an arithmetical progression, the heads then passing over any other weir in the same river must also be in arithmetical progression, unless the quantity flowing down varies from erogation or supply, such as drawing off by mill races, &c.*

## SECTION XI.

BENDS AND CURVES.— BRANCH PIPES. — DIFFERENT LOSSES OF HEAD.—GENERAL EQUATION FOR FINDING THE VELOCITY.—HYDROSTATIC AND HYDRAULIC PRESSURE.—PIËZOMETER.—CATCHMENT BASINS.—RAIN FALL PER ANNUM.— WATER POWER.

The resistance or loss of head due to bends and curves has now to be considered. If we fix a bent pipe, F B C D E G, Fig. 40, between two cisterns, so as to be capable of revolving round in collars at F and G, we shall find the time the water takes to sink a given distance from *f* to F in the upper cistern the same, whether the tube

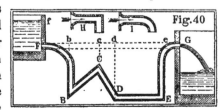

occupy the position shown in the figure or the horizontal position shown by the dotted line F $b\,c\,d\,e$ G. This shows that the resistances due to friction and to bends are independent of the pressure. If the tube were straight, the discharge would depend on the length, diameter, and difference of level between $f$ and G, and may be determined from the mean velocity of discharge, found from TABLE VIII. or equation (79.) Here, however, we have to take into consideration the loss sustained at the bends and curves, and our illustration shows that it is unaffected by the pressure.

The experiments of Bossut, Du Buât, and others, show that the loss of head from bends and curves—like that from friction—increases as the square of the velocity; but when the curves have large radii, and the bends are very obtuse, the loss is very small. With a head of nearly 3 feet, Venturi's twenty-third experiment, when reduced, gives—for a short straight tube 15 inches long, and 1¼ inch in diameter, having the junction of the form of the contracted vein, very nearly ·873 for the coefficient of discharge. When of the same length and diameter, but bent as in Diagram 1, Fig. 40, the coefficient is reduced to ·785; and when bent at a right angle as at H, Fig. 40, the coefficient is further reduced to ·560. In these respective cases we have therefore,

1.  $v = ·873 \sqrt{2gh}$, and $h = 1·312 \times \dfrac{v^2}{2g}$;

2.  $v = ·785 \sqrt{2gh}$, and $h = 1·623 \times \dfrac{v^2}{2g}$;

3.  $v = ·560 \sqrt{2gh}$, and $h = 3·188 \times \dfrac{v^2}{2g}$;

showing that the loss of head in the tube H, Fig. 40, from the bend, is $1·876 \times \dfrac{v^2}{2g}$, or nearly double the theoretical head due to the velocity in the tube. The loss of head by the circular bend is only $·311 \dfrac{v^2}{2g}$, or not quite one-sixth of the other.

Du Buât deduced, from about twenty-five experiments, that the head due to the resistance in any bent tube A B C D E F G H, Diagram 1, Fig. 41, depends on the number

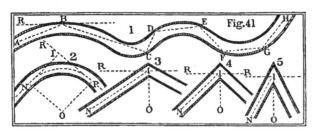

of deflections between the entrance at A and the departure at H; that it increases at each reflection as the square of the sine of the deflected angle, A B R for instance, and as the square of the velocity; and that if $\varphi$, $\varphi_1$, $\varphi_2$, $\varphi_3$, &c., be the number of degrees in the angles of deflection at B, C, D, E, &c., then for measures in French inches the height $h_b$, due to the resistances from curves, is

(136.) $$h_b = \frac{v^2 (\sin.^2 \varphi + \sin.^2 \varphi_1 + \sin.^2 \varphi_2 + \sin.^2 \varphi_3 + \&c.)}{3000}$$

which for measures in English inches becomes

(137.) $$h_b = \frac{v^2 (\sin.^2 \varphi + \sin.^2 \varphi_1 + \sin.^2 \varphi_2 + \sin.^2 \varphi_3 + \&c.)}{3197}$$

and for measures in English feet,

(138.) $$h_b = \frac{v^2 (\sin.^2 \varphi + \sin.^2 \varphi_1 + \sin.^2 \varphi_2 + \sin.^2 \varphi_3 + \&c.)}{266 \cdot 4}$$

or, as it may be more generally expressed for all measures,

(139.) $$h_b = (\sin.^2 \varphi + \sin.^2 \varphi_1 + \sin.^2 \varphi_2 + \sin.^2 \varphi_3 + \&c.) \frac{v^2}{8 \cdot 27 g},$$

in which $\frac{v^2}{8 \cdot 27 g} = \frac{1}{266 \cdot 4} = \cdot 00375$ in feet.

The angle of deflection, in the experiments from which equation (136) was derived, did not exceed 36°. We have already shown the loss of head from the circular bend in diagram 1, Fig. 40, where the angle of deflection is nearly 45°, to be $311 \frac{v^2}{2g} = \cdot00483 \, v^2$, but as the sin. 45° = ·707 sin.² 45° = ·5, we get $\cdot00483 \, v^2 = \cdot00966 \, v^2 \times$ sin.² 45°, or more than two and a half times as much as Du Buât's formula would give; and if we compare it with Rennie's experiments[*], with a pipe 15 feet long, ½ inch diameter, bent into fifteen curves, each 3¼ inches radius, we should find the formula gives a loss of head not much more than one-half of that which may be derived from the observed change, ·419 to ·370 cubic feet per minute in the discharge. See p. 156.

Dr. Young[†] first perceived the necessity of taking into consideration the length of the curve and the radius of curvature. In the twenty-five experiments made by Du Buât, he rejected ten in framing his formula, and the remaining fifteen agreed with it very closely. Dr. Young finds

(140.) $$h_b = \frac{\cdot0000045 \, \varphi \, \rho^{\frac{1}{2}} \times v^2}{\rho}$$

where $\varphi$ is the number of degrees in the curve N P, diagram 2, Fig. 41, equal the angle N O P; $\rho =$ O N the radius of curvature of the axis; $h_b$ the head due to the resistance of the curve, and $v$ the velocity, all expressed in French inches. This formula reduced for measures in English inches is

(141.) $$h_b = \frac{\cdot0000044 \, \varphi \, \rho^{\frac{1}{2}} \times v^2}{\rho};$$

and for measures in English feet,

(142.) $$h_b = \frac{\cdot000006 \, \varphi \, \rho^{\frac{1}{2}} \times v^2}{\rho}.$$

Equation (140) agrees to $\frac{1}{25}$th of the whole with twenty of Du Buât's experiments, his own formula agreeing so closely with only fifteen of them. The resistance must evidently increase with the number of bends or curves; but when they come close upon, and are grafted into, each other, as in

---

[*] Philosophical Transactions for 1831, p. 438.
[†] Philosophical Transactions for 1808, pp. 173–175.

diagram 1, Fig. 41, and in the tube F B C D E G, Fig. 40, the motion in one bend or curve immediately affects those in the adjacent bends or curves, and this law does not hold.

Neither Du Buât nor Young took any notice of the relation that must exist between the resistance and the ratio of the radius of curvature to the radius of the pipe. Weisbach does, and combining Du Buât's experiments with some of his own, finds for circular tubes,

(143.) $$h_b = \frac{\varphi}{180} \times \left\{ \cdot 131 + 1 \cdot 847 \left(\frac{d}{2\rho}\right)^{\frac{7}{2}} \right\} \times \frac{v^2}{2g};$$

and for quadrangular tubes,

(144.) $$h_b = \frac{\varphi}{180} \times \left\{ \cdot 124 + 3 \cdot 104 \left(\frac{d}{2\rho}\right)^{\frac{7}{2}} \right\} \times \frac{v^2}{2g};$$

in which $\varphi$ is equal the angle N O P = N I R, diagram 2, Fig. 41; $d$ the mean diameter of the tube, and $\rho$ the radius N O of the axis. When $\frac{d}{2\rho}$ exceeds ·2, the value of $\cdot 131 + 1 \cdot 847 \left(\frac{d}{2\rho}\right)^{\frac{7}{2}}$ exceeds $\cdot 124 + 3 \cdot 104 \left(\frac{d}{2\rho}\right)^{\frac{7}{2}}$, and the resistance due to the quadrangular tube exceeds that due to the circular one. We have arranged and calculated the following table of the numerical values of these two expressions for the more easy application of equations (143) and (144). This table will be found of considerable use in calculating the values of equations (143) and (144), as the second and fifth columns contain the values of $\cdot 131 + 1 \cdot 847 \left(\frac{d}{2\rho}\right)^{\frac{7}{2}}$, and the third and sixth columns the values of $\cdot 124 + 3 \cdot 104 \left(\frac{d}{2\rho}\right)^{\frac{7}{2}}$, corresponding to different values of $\frac{d}{2\rho}$; and it is carried to twice the extent of those given by Weisbach.

For bent tubes, diagrams 3, 4, and 5, Fig. 41, the loss of head is considerably greater than for rounded tubes. If, as before, we put the angle N I R = $\varphi$, I R being at right angles to I O the line bisecting the angle or bend, we shall find, by

ORIFICES, WEIRS, PIPES, AND RIVERS. 155

TABLE OF THE VALUES OF THE EXPRESSIONS

$$\cdot 131 + 1\cdot 847 \left(\frac{d}{2\varrho}\right)^{\frac{7}{4}} \text{ and } \cdot 124 + 3\cdot 104 \left(\frac{d}{2\varrho}\right)^{\frac{7}{4}}.$$

| $\frac{d}{2\varrho}$. | Circular tubes. | Quadrangular tubes. | $\frac{d}{2\varrho}$. | Circular tubes. | Quadrangular tubes. |
|---|---|---|---|---|---|
| ·1 | ·131 | ·124 | ·6 | ·440 | ·643 |
| ·15 | ·133 | ·128 | ·65 | ·540 | ·811 |
| ·2 | ·138 | ·135 | ·7 | ·661 | 1·015 |
| ·25 | ·145 | ·148 | ·75 | ·806 | 1·258 |
| ·3 | ·158 | ·170 | ·8 | ·977 | 1·545 |
| ·35 | ·178 | ·203 | ·85 | 1·177 | 1·881 |
| ·4 | ·206 | ·250 | ·9 | 1·408 | 2·271 |
| ·45 | ·244 | ·314 | ·95 | 1·674 | 2·718 |
| ·5* | ·294 | ·398 | 1·00 | 1·978 | 3·228 |

decomposing the motion, that the head $\frac{v^2}{2g}$ becomes $\frac{v^2}{2g} \times \cos.^2 2\varphi$ from the change of direction; and that a loss of head

(145.) $\quad h_b = (1 - \cos.^2 2\varphi)\frac{v^2}{2g} = \sin.^2 2\varphi \frac{v^2}{2g}$

must take place. When the angle is a right angle, as in diagram 4, $\cos. 2\varphi = 0$, and $h_b = \frac{v^2}{2g}$; that is to say, the loss of head is exactly equal to the theoretical head. When the angle or bend is acute, as in diagram 5, the loss of head is $(1 + \cos.^2 2\varphi)\frac{v^2}{2g}$, for then $\cos. 2\varphi$ becomes negative. Weisbach does not find the loss of head in a right angular bend greater than $\cdot 984 \frac{v^2}{2g}$; while Venturi's twenty-third experiment, made with extreme care, p. 151, shows the loss to be $1\cdot 876 \frac{v^2}{2g}$. When the pipes are long, however, the value of

* The values corresponding to $\frac{d}{2\varrho} = \cdot 55$ are ·359 and ·507 for circular and quadrangular tubes.

$\frac{v^2}{2g}$ is in general small, and this correction does not affect the final results in any material degree.

Rennie's experiments*, with a pipe 15 feet long, ½ inch in diameter, and with 4 feet head, give the discharge per second

|   |   | Cubic feet. |
|---|---|---|
| 1. | Straight . . . . . . . . . | ·00699 |
| 2. | Fifteen semicircular bends . . . | ·00617 |
| 3. | One bend, a right angle 8½ inches from the end of the pipe . . . | ·00556 |
| 4. | Twenty-four right angles . . . | ·00253 |

From these data we find consecutively, the theoretical discharge being ·021885 cubic feet per second, and the theoretical head $H = \frac{v^2}{2g}$, that

1. $v = ·319 \sqrt{2gH}$, and therefore $H = 9·82 \times \frac{v^2}{2g}$;

2. $v = ·282 \sqrt{2gH}$, „ „ $H = 12·58 \times \frac{v^2}{2g}$;

3. $v = ·254 \sqrt{2gH}$, „ „ $H = 15·50 \times \frac{v^2}{2g}$;

4. $v = ·116 \sqrt{2gH}$, „ „ $H = 74·34 \times \frac{v^2}{2g}$;

The loss of head, therefore, by the introduction of 15 semicircular bends, is $2·76 \frac{v^2}{2g}$; by the introduction of one right angle, $5·68 \frac{v^2}{2g}$; and by the introduction of 24 right angles, $64·52 \frac{v^2}{2g}$, or about 12 times the loss due to one right angle. This shows that the resistance does not increase as the number of bends, as we before remarked, p. 138, when they are close to each other. The loss of head from one right angle, $5·68 \frac{v^2}{2g}$, is more than double the loss from 15 semicircular bends, or

* Philosophical Transactions for 1831, p. 438.

$2\cdot76\dfrac{v^2}{2g}$. The loss of head for a right angular bend, determined from Venturi's experiment, is $1\cdot876\dfrac{v^2}{2g}$; formula (145) makes it $\dfrac{v^2}{2g}$; and Weisbach's empirical formula, ($\cdot9457\sin.\varphi + 2\cdot047\sin.^4\varphi)\dfrac{v^2}{2g}$, makes it only $\cdot984\dfrac{v^2}{2g}$. The formulæ now in use give, therefore, results considerably under the truth. It appears to us, that perhaps the *quantity* of water moving directly towards the bend must also be taken into consideration, and certainly the loss of mechanical effect from contraction, and eddies at the bend, as well as the loss arising from change of direction.

### BRANCH PIPES.

When a pipe is joined to another, the quantity of water flowing below the junction B, diagram 1, Fig. 42, must be

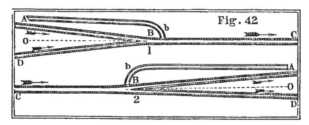

equal to the sum of the quantities flowing in the upper branches in the case of supply; and when the branch pipe draws off a portion of the water, as in diagram 2, the quantity flowing above the junction must be equal to the quantities flowing in the lower branches. Both cases differ only in the motion being *from* or *to* the branches, which, in pipes, are generally grafted at right angles to the main, for practical convenience, as shown at $bb$, and then carried on in any given direction. The loss of head arising from change of direction, equation (145), is $\sin.^2 2\varphi\dfrac{v^2}{2g}$, in which $2\varphi$ = angle A B O; but as in general $2\varphi$ is a right angle for branches to mains, this source of loss becomes then simply

$\frac{v^2}{2g}$. In addition to this, a loss of head is sustained at the junction, from a certain amount of force required to unite or separate the water in the new channel. In the case of drawing off, diagram 2, this loss was estimated by D'Aubuisson, from experiments by Génieys, to be about twice the theoretical head due to the velocity in the branch, or $\frac{2v^2}{2g}$, so that the whole loss of head arising from the junction is $\frac{v^2}{2g} + \frac{2v^2}{2g} = \frac{3v^2}{2g}$, or three times the theoretical head due to the velocity. In the case of supply, the loss is probably nearly the same. The actual loss is, however, very uncertain; but, as was before observed when discussing the loss of head occasioned by bends, two or three times $\frac{v^2}{2g}$ is in general so comparatively small, that its omission does not materially effect the final results. A loss also arises from contraction, &c. See pp. 97, 98.

The calculations for mains and branches become often very troublesome, but they may always be simplified by rejecting at first any minor corrections for contraction at orifice of entry, bends, junctions, or curves. If, in diagram 2, Fig. 42, we put $h$ for the head at B, or height of the surface of the reservoir over it; $h_a$ for the fall from B to A; $h_d$ for the fall from B to D; $l$ equal the length of pipe from B to the reservoir; $l_a$ equal the length B A; $l_d$ equal the length B D; $r$ equal the mean radius of the pipe B C; $r_a$ the mean radius of the pipe B A; $r_d$ the mean radius of B D; $v$ the mean velocity in B C; $v_a$ the velocity in B A; and $v_d$ the velocity in B D, we then find, by means of equation (73), the fall from the reservoir to A equal to

(146.) $\quad h + h_a = \left(c_r + c_f \times \frac{l}{r}\right)\frac{v^2}{2g} + \left(1 + c_f \times \frac{l_a}{r_a}\right)\frac{v_a^2}{2g};$

the fall from the reservoir to D equal to

(147.) $\quad h + h_d = \left(c_r + c_f \times \frac{l}{r}\right)\frac{v^2}{2g} + \left(1 + c_f \times \frac{l_d}{r_d}\right)\frac{v_d^2}{2g};$

and, as the quantity of water passing from C to B is equal to the sum of the quantities passing from B to A and from B to D,

(148.) $\quad v r^2 = v_a r_a^2 + v_d r_d^2.$

By means of these three equations we can find any three of the quantities $h$, $h_a$, $h_d$, $r$, $r_a$, $r_d$, $b$, $b_a$, $b_d$, the others being given. Equations (146) and (147) may be simplified by neglecting $c_r$, the coefficient due to the orifice of entry from the reservoir, and 1, the coefficient of velocity. They will then become

(148.) $\quad h + h_a = c_f \times \left( \dfrac{l}{r} \times \dfrac{v^2}{2g} + \dfrac{l_a}{r_a} \times \dfrac{v_a^2}{2g} \right).$

and

(149.) $\quad h + h_d = c_f \times \left( \dfrac{l}{r} \times \dfrac{v^2}{2g} + \dfrac{l_d}{r_d} \times \dfrac{v_d^2}{2g} \right).$

The mean value of $c_f$ for a velocity of 4 feet per second is ·005741, and of $\dfrac{c_f}{2g}$, ·0000891. The values for any other velocities may be had from the table of coefficients of friction given at p. 120. When $l$, $h$, and $r$ are given, the velocity $v$ can be had from the equation, $v = \left( \dfrac{2g}{c_f} \times \dfrac{rh}{l} \right)^{\frac{1}{2}}$, or more immediately from TABLE VIII.

## GENERAL EQUATION FOR MEAN VELOCITY.

We are now enabled to give a general equation for finding the whole head H, and the mean velocity $v$, in any channel; and to extend equations (73) and (74)* so as to comprehend the corrections due to bends, curves, &c. Designating, as before, the height due to the resistance at the orifice of entry by $h_r$, and the corresponding coefficient by $c_r$;

$h_f$ the head due to friction, and $c_f$ the coefficient of friction;
$h_b$ the head due to bends, and $c_b$ the coefficient of bends;
$h_c$ the head due to curves, and $c_c$ the coefficient of curves;
$h_e$ the head due to erogation, and $c_e$ the coefficient of erogation; and
$h_x$ the head due to other resistances, and $c_x$ their mean coefficient; then we get

* See Note B.

(150.) $\mathrm{H} = h_r + h_f + h_b + h_c + h_e + h_x + \dfrac{v^2}{2g}$;

that is to say, by substituting for $h_r$, $h_f$, &c., their values as previously found,

$\mathrm{H} = (1 + c_r)\dfrac{v^2}{2g} + c_f \dfrac{l}{r} \times \dfrac{v^2}{2g} + c_b \times \dfrac{v^2}{2g}$
$\qquad + c_c \times \dfrac{v^2}{2g} + c_e \times \dfrac{v^2}{2g} + c_x \times \dfrac{v^2}{2g}$;

or, more briefly,

(151.) $\mathrm{H} = \left(1 + c_r + c_f \times \dfrac{l}{r} + c_b + c_c + c_e + c_x\right)\dfrac{v^2}{2g}$;

from which we find

(152.) $v = \left\{ \dfrac{2g\mathrm{H}}{1 + c_r + c_f \times \dfrac{l}{r} + c_b + c_c + c_e + c_x} \right\}^{\frac{1}{2}}$.

It is to be observed here, that for very long uniform channels, the value of the mean velocity will be found in general equal to $\left\{\dfrac{2gr\mathrm{H}}{c_f l}\right\}^{\frac{1}{2}}$, as the other resistances and the head due to the velocity are all trifling compared with the friction, and may be rejected without error; but, as we before observed, it is advisable in practice, when determining the diameter of pipes, p. 123, to increase the value of $c_f$, TABLE, p. 120, or to increase the diameter found from the formula by one-sixth, which will increase the discharging power by one-half. (See TABLE XIII.)

### HYDROSTATIC AND HYDRAULIC PRESSURE.—PIËZOMETER.

When water is at rest in any vessel or channel, the pressure on a unit of surface is proportionate to the head at its centre[*], measured to the surface, and is expressed in lbs. for

---

[*] This is only correct when the surface is small in depth compared with the head. If H be the depth of a rectangular surface in feet, and also the head of water measured to the lower horizontal edge, then the pressure in lbs. is expressed by $31\frac{1}{4} \mathrm{H}^2$; and the centre of pressure is at $\frac{2}{3}$rds of the depth.

ORIFICES, WEIRS, PIPES, AND RIVERS.     161

measures in feet, by 62½ H S, in which H is the head, and S the surface exposed to the pressure, both in feet measures. This is the *hydrostatic pressure*. In the pipe A B C D F E, Fig. 43, the pressures at the points B, C, D, F, and E, on the sides of the tube will be respectively as the heads B $b$, C $c$, D $d$, F $f$, and E $e$, if all motion in the tube be prevented by stopping

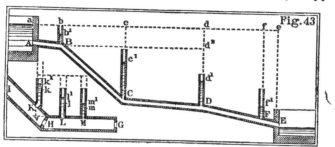

the discharging orifice at E. In this case the pressure is a maximum and hydrostatic; but if the discharging orifice at E be partially or entirely open, a portion of each pressure at B, C, D, F, &c., is absorbed in overcoming the different resistances of friction, bends, &c., between it and the orifice of entry at A, and also by the velocity in the tube, and the difference is the *hydraulic pressure*.

Bernouilli first showed that *the head due to the pressure at any point, in any tube, is equal to the effective head at that point, minus the head due to the velocity.* When the resistances in a tube are nothing, the effective head becomes the hydrostatic head, and by representing the former by $h_{ef}$ we shall have, adopting the notation in equation (150),

$$h_{ef} = H - (h_r + h_f + h_c + \&c.),$$

and consequently the head due to the hydraulic pressure equal

(153.)    $h_p = h_{ef} - \dfrac{v^2}{2g} = H - (h_r + h_f + h_b + \&c.) - \dfrac{v^2}{2g}.$

If small tubes be inserted, as shown in Fig. 43, at the points B, C, D, and F, the heights B $b^1$, C $c^1$, D $d^1$, F $f^1$, which the water rises to in these, will be represented by the corresponding values of $h_p$ in the preceding equation; and the difference between the heights C $c^1$, F $f^1$, at C and F, for instance, added

M

to the fall from C to F, will evidently express the head due to all the resistances between C and F. When H $=$ E$e$; and the orifice at E is open, we have, from equation (150)

$$\text{H} = h_r + h_f + h_b + h_c + \&c. + \frac{v^2}{2g}, \text{ and therefore } h_p = 0,$$

that is, the pressure at the discharging orifice is nothing.

The vertical tubes at B, C, D, F, when properly graduated, are termed *piëzometers* or *pressure gauges;* they not only show the actual pressure at the points where placed, but also the difference between any two, D $d^1$ — B $b^1$, for instance, added to the difference of head between B and D, or D $d^2$ will give D $d^1$ — B $b^1$ + D $d^2$ for the head or pressure due to the resistances between B and D. This instrument affords, perhaps, the very best means of determining the loss of head due to bends, curves, diaphragms, &c. The loss of head due to friction, bend, diaphragms, &c., between K and L, Fig. 43, is equal to K $k$ — L $l$ + K $v$. If M be the same distance from L as K is, L $l$ — M $m$ will be the height due to the friction (L and M being on the same level); therefore K $k$ — L $l$ + K $v$ — L $l$ + M $m$ $=$ K $k$ + K $v$ + M $m$ — 2 L $l$ is the head due to the diaphragm and bend both together. If the diaphragm be absent, we get the head due to the bend, and if the bend be absent, the head due to the diaphragm in like manner.

When the discharging orifice, as at E, is quite open, we have seen that the pressure there is zero; but when, as at G, it is only partly open, this is no longer the case, and the hydraulic pressure increases from zero to hydrostatic pressure, as the orifice decreases from the full section to one indefinitely small compared with it. A piëzometer, placed a short distance inside G, will give this pressure; and the difference between it and the whole head will be the head due to the resistances and velocity in the pipe; from which, and also the length and diameter, the discharge may be calculated as before shown. Again, by means of the head M $m^1$, and that due to the velocity of approach, we can also find the discharge through the diaphragm G; see equation (45) and the remarks following it. This result must be equal to the other, and we may in this way test the formulæ anew or correct them by the practical results.

The velocity of discharge of the tube A C D E, may be calculated by means of any piëzometric height $c\,c^1$; for, by putting the whole fall from $c^1$ to E equal to $H_{c^1}$, we get, disregarding bends, $v = \left\{ \dfrac{2\,g\,r\,H_{c^1}}{c_f\,l_{c^1}} \right\}^{\frac{1}{2}}$, in which $l_{c^1} = $ C E. This is evident from equation (152), as we have supposed that no part of the head is absorbed in generating velocity, or in overcoming the resistance of bends. If the bend at D were taken into consideration, then $v = \left\{ \dfrac{2\,g\,H_{c^1}}{c_f \times \dfrac{l_{c^1}}{r} + c_b} \right\}^{\frac{1}{2}}.$

### CATCHMENT BASINS.—FALL OF RAIN.

A catchment basin is a district which drains itself into a river and its tributaries. It is bounded, generally, by the summits of the neighbouring hills, ridges, or high lands; and may vary in extent from a few square miles to many thousands; that of the Shannon is over 4500 square miles. The average quantity of water which discharges itself into a river will, *cæteris paribus*, depend on the extent of its catchment basin, and the whole quantity of rain discharged on the area of the catchment basin, including lakes and rivers. The quantity of rain which falls annually varies with the district and the year; and it also varies at different parts of the same district. The average quantity in Ireland may be taken at about 34 inches deep, that which falls in Dublin being 27 inches[*], and that in Cork 41 inches nearly. The quantity

---

[*] The average yearly fall in Dublin for seven years, ending with 1849, was 26·407 inches; and the maximum fall in any month took place in April, 1846, being 5·082 inches. The average fall in inches per month for seven years, ending with 1849, was as follows:—October, 3·060; August, 2·936; January, 2·544; April, 2·503; November, 2·300; July, 2·116; June, 2·005; December, 1·938; September, 1·860; May, 1·814; March, 1·739; February, 1·534.—*Proceedings of the Royal Irish Academy*, vol. v. p. 18.

varies a good deal with the altitude of the district. In parts of Westmoreland it rises sometimes to 140 inches; in London, an average of 20 years' observations gives a fall of nearly 25 inches. In the district surrounding the Bann reservoirs in the County Down, the average fall has been found so high as 72 inches. Indeed, it is requisite to obtain the fall from observation for any particular district, when it is necessary to apply the results to scientific purposes.

The proportion between the quantity which falls, and that which passes from a catchment basin into its river, also varies very considerably. When the sides of a catchment basin are steep, and the water passes off rapidly into the adjacent river or tributaries, there is less loss by evaporation and percolation than when they are nearly flat. The soil, subsoil, and stratification have also considerable effect on the proportion. Reservoirs being generally constructed adjacent to steep side falls, give a much larger proportion of the quantity fallen than can be obtained from rivers in flatter districts; besides, the quantity of rain which falls on the high summits, near reservoirs, almost always exceeds the average fall, considerably. As 640 acres is equal to one square mile, and one acre is equal to 43,560 square feet, a fall of one inch of rain is equal to 3630 cubic feet per acre, and to $3630 \times 640 = 2323200$ cubic feet per square mile: the proportion of this fall, for each acre, or square mile of the catchment basin, which enters the river, must depend entirely on the district and local circumstances, the full or maximum quantity being retained on lakes.

It is too often taken for granted that the discharge from a catchment basin takes place, into the conveying channels, in nearly the same time that a given quantity of rain falls. Perhaps the largest registry on record in Great Britain is a fall of 2 inches in an hour. The maximum fall in any hour of any year seldom exceeds half of this amount, and then perhaps only once in several years. The quantity which falls will not be discharged into the channels in the same time. The quantity discharged, and time, will depend a good deal

QUANTITY PER ACRE FOR A GIVEN DEPTH OF FALL.

| Fall in inches per hour. | Cubic feet per acre. | Fall in inches per hour. | Cubic feet per acre. | Fall in inches per hour. | Cubit feet per acre. | Fall in inches per hour. | Cubic feet per acre. |
|---|---|---|---|---|---|---|---|
| 2 | 7260 | $\frac{1}{2}$ | 1815 | $\frac{1}{8}$ | 454 | $\frac{1}{20}$ | 181 |
| $1\frac{3}{4}$ | 6352 | $\frac{3}{8}$ | 1361 | $\frac{1}{9}$ | 403 | $\frac{1}{30}$ | 121 |
| $1\frac{1}{2}$ | 5445 | $\frac{1}{4}$ | 907 | $\frac{1}{10}$ | 363 | $\frac{1}{40}$ | 91 |
| $1\frac{1}{4}$ | 4537 | $\frac{1}{5}$ | 726 | $\frac{1}{12}$ | 302 | $\frac{1}{50}$ | 73 |
| 1 | 3630 | $\frac{1}{6}$ | 605 | $\frac{1}{14}$ | 259 | $\frac{1}{60}$ | 61 |
| $\frac{3}{4}$ | 2723 | $\frac{1}{7}$ | 519 | $\frac{1}{16}$ | 227 | $\frac{1}{70}$ | 52 |

on the season and district. The arterial channel receives the supply at different places and from different distances, and the water in passing into and from it does not encounter the same amount of resistance as if it all passed first into the upper end. Less sectional area is, therefore, necessary than if the whole discharge had to pass through the whole length of the channel. The relation of the quantity of rainfall to the portion which flows into the main channel, as well as the time which it takes to arrive at it, and the places of arrival, must be known before the proper size of a new channel can be determined, particularly sewers in urban districts. A pipe sufficient to discharge the water from 200 acres need not be 20 times the discharging power of one exactly suited to 10 acres of the same district, for the discharge from the outlying 190 acres will not arrive at the main in the same time as that from the adjacent 10 acres.

In the Holborn and Finsbury divisions Mr. Roe calculated that an 18-inch cylindrical pipe, laid at an inclination of 1 in 80, is sufficient for 20 acres of house-sewage, while a 5-inch pipe laid at an inclination of 1 in 20 is necessary for 1 acre, and a 3-inch pipe, laid also at 1 in 20, for $\frac{1}{4}$ acre. In each of these cases, however, the discharge must depend on the head and length of the pipe as well as the inclination at which it is laid. Assuming the inclination of those pipes to correspond with the hydraulic inclination, we have calculated their discharging powers to be respectively 807, 72, and 20 cubic feet per minute, the areas to be drained being 20, 1, and $\frac{1}{4}$ acres. *In all calculations of this kind it is necessary, for accuracy, to ascertain not only the maximum rainfall per hour, but also the proportions discharged per hour, according to the season and district, into the main channel, and also the junctions or places of arrival.* In urban districts 1500, 2100, and sometimes 3600 cubic feet per hour per acre, have to be discharged after extraordinary rainfalls. These may be taken as maximum results.

In urban districts, however, a much larger quantity of water is conveyed more rapidly, *cæteris paribus*, to the mains, than in suburban districts and catchment basins generally, in which the maximum discharge per acre per hour, even in the steeper and higher districts, seldom exceeds 700 cubic feet, and varies from about 20 cubic feet for the larger and flatter districts upwards. This arises from the impervious nature of the surfaces it falls upon in towns, and the lesser waste in passing to the drains, as well as a portion of the supply being often artificial. From 70 to 90 cubic feet* per acre per hour, is generally taken for the maximum discharge from the average number of catchment basins; this is equal to a supply of from $\frac{1}{50}$th to $\frac{1}{40}$th of an inch in depth from the whole area. Thorough drainage increases the supply and discharge. *Every catchment basin has, however, its own peculiar data, and a knowledge of these is necessary before we can draw any correct conclusions for new waterworks in connection with it.* It may be remarked, however, that any conclusions drawn from experiments on the supply of tributaries, particularly in high districts, are wholly inapplicable to the main channel into which they flow. The flow into tributaries and mountain streams, or rivers, is always more rapid than into main channels and rivers in flat districts, and the supply from

---

* Some interesting observations on rainfall and flood discharges are given in the Transactions of the Institution of Civil Engineers, Ireland, for 1851, pp. 19–33 and pp. 44–52.

springs often forms a large portion of the water flowing in them.

The effects of evaporation are very variable; sometimes 58 or 60 per cent. of the annual fall is carried off in this way from ordinary flat tillage soils, and other estimates are much higher; much, however, depends on the soil, subsoil, inclination, stratification, and season. The evaporation from water surfaces exceeds the annual fall in these countries by about one-third; and that from flat marsh and callow lands exceeds the evaporation from ordinary tillage, porous, and high lands. When the flat lands along the banks of rivers extend considerably on both sides, an extra fall is necessary into the main channel, along the normal drains, otherwise such lands must suffer from excessive eveporation as well as floods. Evaporation also varies with the climate.

### HORSE POWER.

Taking the value of a horse's power at 33,000 lbs.* raised 1 foot high in one minute, the theoretical horse power of an overfall, or a given quantity of water, is expressed by the fall in feet, multiplied by the discharge in cubic feet per minute, multiplied by $62\frac{1}{2}$ and divided by 33,000. The effective power depends on the nature, circumstances, and construction of the machine to which the theoretical power is applied. For an overshot wheel, the ratio of the power to the effect may be taken as 3 : 2, and, therefore, the effective horse power will be 49,500 lbs. weight of water falling 1 foot in one minute†. The maximum effect of overshot wheels is found to vary :— Smeaton found it ·76 times the theoretical effect; D'Aubuisson also ·76 times; and Weisbach found ·78 times the theoretical effect in the wheel of a stamp mill, at Frieberg, which was 23 feet high, 3 feet wide, and contained 48 buckets. To find the effective horse power we

---

\* 22,000 lbs. is much nearer the average power of horses as their work for twelve hours is ordinarily performed through the country.

† Small wheels are not so effective as larger ones under similar circumstances.

must divide these decimals, viz. ·76 and ·78, into 33,000, which will give 43,300 and 43,600 lbs nearly, falling 1 foot for a horse power.

For breast wheels, the ratio of the theoretical to the effective power must vary very considerably; the mean value may be taken at 1 : ·5; and, therefore, the value of a horse power would be 66,000 lbs. falling 1 foot in a minute. M. Morin found the efficiency to vary from 1 : ·7 to 1 : ·52; and Egen, with a wheel 23 feet in diameter, $4\frac{1}{3}$ feet wide, having 69 ventilated buckets exceedingly well constructed, found the ratio at best but as 1 : ·52, and under ordinary circumstances as 1 : ·48, the mean being as 1 : ·5. Very wide wheels give a larger effect, sometimes as high as ·73 in one, but a good deal depends on the manner of bringing on the water and the construction of the buckets.

For undershot wheels, the mean effect may be taken at ·33 in one, or 100,000 lbs. falling 1 foot high in one minute, for the effective horse power. Here, however, the height must be (in general) determined from the velocity. A maximum effect of ·5 in 1 is sometimes obtained, and a minimum effect of ·16 in 1.

Turbines may be said to give a useful effect of $\frac{2}{3}$ in 1, or 49,500 lbs. falling 1 foot in a minute for the effective horse power, the same as for an overshot wheel. A maximum effect of ·821 in 1 has been obtained*, and the efficiency varies from ·57 to ·80, or less.

Poncelet wheels give a useful effect of from ·5 to ·6 in 1; floating wheels ·38 in 1; impact wheels from ·16 to ·4; and Barker's mill from ·15 to ·35 in 1.

Corn mills will grind about a bushel of corn per horse power per hour, but a good deal depends on the state of the stones, and on the grain. The value of the work done in an hour being once known, the value of a horse power may be determined accordingly.

---

\* Proceedings Royal Irish Academy, vol. v. p. 214.

## NOTE A.

### DISCHARGE FROM ONE VESSEL OR CHAMBER INTO ANOTHER. —LOCK-CHAMBERS.

WE have given, at pages 98 and 99, formulæ for the time water in a prismatic vessel takes to fall a given depth, when

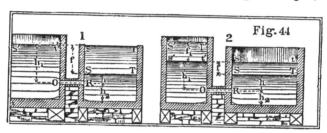

Fig. 44

discharged from an orifice at the side or bottom. The time the surface $st$, Diagram 1, Fig. 44, takes to rise to $st$, when supplied through an orifice or tube $OR$, from an upper large chamber or canal, whose surface $s't'$ remains always at the same level, is $\dfrac{2Af^{\frac{1}{2}}}{c_d a \sqrt{2g}}$,[*] and we thence get the time of rising from R to s for measures in feet.

(A.)  $$t = \frac{A}{4{\cdot}015\, c_d a} \{h_1^{\frac{1}{2}} - f^{\frac{1}{2}}\},$$

and for measures in inches

(B.)  $$t = \frac{A}{13{\cdot}9\, c_d a} \{h_1^{\frac{1}{2}} - f^{\frac{1}{2}}\},$$

in which A is the area of the horizontal section at $ST$; $a$ the sectional area of the communicating channel or orifice $OR$;

---

[*] The time of rising from s to $s$ is exactly double the time it would take if the pressure $f$ remained uniform to fill the same depth *below* R.

$c_d$ the coefficient of discharge suited to it, and $h_1$ and $f$ as shown in the diagram.

In order to find the time of filling the lower vessel to the level ST, supposing it at first empty, we have the contents of the portion below OR equal to $Ah_2$, and the time of filling it equal to

(C.) $$\frac{Ah_2}{8{\cdot}025\, c_d a h_1^{\frac{1}{2}}};$$

then the time of filling up to any level ST, for measures in feet, is equal to the sum of (A) and (C); that is,

(D.) $$T = \frac{Ah_1^{\frac{1}{2}}}{8{\cdot}025\, c_d a} \left\{ 2 + \frac{h_2}{h_1} - \frac{2f^{\frac{1}{2}}}{h_1^{\frac{1}{2}}} \right\}$$
$$= \frac{A(2h_1 + h_2 - 2f^{\frac{1}{2}} h_1^{\frac{1}{2}})}{8{\cdot}025\, c_d a h_1^{\frac{1}{2}}},$$

and for measures in inches

(E.) $$T = \frac{Ah_1^{\frac{1}{2}}}{27{\cdot}8\, c_d a} \left\{ 2 + \frac{h_2}{h_1} - \frac{2f^{\frac{1}{2}}}{h_1^{\frac{1}{2}}} \right\}$$
$$= \frac{A(2h_1 + h_2 - 2f^{\frac{1}{2}} h_1^{\frac{1}{2}})}{27{\cdot}8\, c_d a h_1^{\frac{1}{2}}}.$$

When ST coincides with $st$

(F.) $$T = \frac{A(2h_1 + h_2)}{8{\cdot}025\, c_d a h_1^{\frac{1}{2}}},$$

for measures in feet, and

(G.) $$T = \frac{A(2h_1 + h_2)}{27{\cdot}8\, c_d a h_1^{\frac{1}{2}}},$$

for measures in inches. These equations are exactly suited to the case of a closed lock-chamber filled from an adjacent canal.

When the upper level $s'T'$ is also variable, as in Diagram 2, the time which the water in both vessels takes to come to the same uniform level $s't'st$, is

(H.) $$t = \frac{2\text{AC}(h_1 + f_1 - h)^{\frac{1}{2}}}{c_d a (\text{A} + \text{C}) \sqrt{2g}} = \frac{2\text{AC}(f + f_1)^{\frac{1}{2}}}{c_d a (\text{A} + \text{C}) \sqrt{2g}};$$

in which $h_1 + f_1 - h = f + f_1$ is the difference of levels at the beginning of the flow; C the horizontal section of the upper chamber; and the other quantities as in Diagram 1. As $\text{C} f_1 = \text{A} f$, we find

$$f + f_1 = \frac{\text{C} + \text{A}}{\text{A}} f_1 = \frac{\text{C} + \text{A}}{\text{C}} f.$$

Now, in order to find the time of falling a given depth $d$ below the first level s'T', we have the head above $s't'st$ equal to $f_1 - d$ in the upper vessel, and the depth below it in the lower vessel equal to $\frac{\text{C}(f_1 - d)}{\text{A}}$; whence the difference of levels in the two vessels at the end of the fall $d$, is

$$f_1 - d + \frac{\text{C}(f_1 - d)}{\text{A}} = \frac{\text{A} + \text{C}}{\text{A}}(f_1 - d).$$

The time of falling through $d$ is, therefore, from equation (H),

(I.) $$t = \frac{2\text{AC}(f + f_1)^{\frac{1}{2}}}{c_d a (\text{A} + \text{C}) \sqrt{2g}} - \frac{2\text{AC}\left\{\left(\frac{\text{C} + \text{A}}{\text{A}}\right)(f_1 - d)\right\}^{\frac{1}{2}}}{c_d a (\text{A} + \text{C}) \sqrt{2g}}$$

$$= \frac{2\text{AC}}{c_d a (\text{A} + \text{C}) \sqrt{2g}} \left\{ (f + f_1)^{\frac{1}{2}} - \left(\frac{(\text{A} + \text{C})(f_1 - d)}{\text{A}}\right)^{\frac{1}{2}} \right\},$$

in which $\sqrt{2g} = 8\cdot 025$ for measures in feet and equal $27\cdot 8$ for measures in inches. The whole time of filling to a level the lower empty vessel, is found by adding the time of filling the portion below R, determined in a manner similar to equations (68) and (69) to be

(K.) $$\frac{2\text{C}}{c_d a \sqrt{2g}} \left\{ (h_1 + f_1)^{\frac{1}{2}} - \left(h_1 + f_1 - \frac{h_2 \text{A}}{\text{C}}\right)^{\frac{1}{2}} \right\},$$

172    THE DISCHARGE OF WATER FROM

to the time of filling above R, given in equation (H), when $h$ is taken equal to zero. Equations (H), (I), and (K) are applicable to the case of the upper and lower chambers of a double lock, after making the necessary change in the diagrams.

The above equations require further extensions when water flows into the upper vessel while also flowing from it into the lower; such extensions are, however, of little practical value, and we therefore omit them. For sluices in flood-gates with square arrises, $c_d$ may be taken at about ·545, but with rounded arrises the coefficient will rise much higher. See SECTIONS III. and VII.

---

## NOTE B.

In equations (74) and (151), the coefficient of friction $c_f$ depends on the velocity $v$, and its value can be found from an approximate value of that velocity from the small table, page 120. If, however, we use both powers of the velocity, as in equation (83), we shall get, when H is the whole head, and $h$ the head from the surface to the orifice of entry

$$(av + bv^2)\frac{l}{r} + (1 + c_r)\frac{v^2}{2g} + h = \text{H},$$

a quadratic equation from which we find

$$v = \left\{\frac{(\text{H} - h)2gr}{(1 + c_r)r + 2gbl} + \left(\frac{gal}{(1 + c_r)r + 2gbl}\right)^2\right\}^{\frac{1}{2}} - \frac{gal}{(1 + c_r)r + 2gbl}$$

or a more general value of the velocity than that given in equation (74). If now we put $c_s = c_r + c_b + c_c + c_e + c_x$ in equation (151) we shall find

$$v = \left\{\frac{(\text{H} - h)2gr}{(1 + c_s)r + 2gbl} + \left(\frac{gal}{(1 + c_s)r + 2gbl}\right)^2\right\}^{\frac{1}{2}} - \frac{gal}{(1 + c_s)r - 2gbl}$$

for a more general expression of equation (152), when the simple power of the velocity, as in equation (83), is taken into consideration. For measures in English feet, we may take $a = \cdot0000223$ and $b = \cdot0000854$, which correspond to those of Eytelwein, in equation (97). The value of $a$ is the same in English as in French measures, but the value of $b$ in equation (83), for measures in metres, must be divided by 3·2809 to find its corresponding value for measures in English feet. In considering the head $\frac{v^2}{2g} c_r$, due to contraction at the orifice of entry as not implicitly comprised in the primary values of $a$ and $b$, equation (83), Eytelwein is certainly more correct than D'Aubuisson, Traité d'Hydraulique, pp. 223 et 224, as this head varies with the nature of the junction, and should be considered in connection with the head due to the velocity, or separately. It can never be correctly considered as a portion of the head due to friction. In all Du Buât's experiments, this head was considered as a portion of that due to the velocity, and the whole head $(1 + c_r)\frac{v^2}{2g}$ deducted to find the hydraulic inclination.

VALUES OF $a$ AND $b$ FOR MEASURES IN ENGLISH FEET.

|  |  | $a$. | $b$. |
|---|---|---|---|
| Equation | (88.) | ·0000445 | ·0000944 |
| ,, | (90.) | ·0000173 | ·0001061 |
| ,, | (94.) | ·0000243 | ·0001114 |
| ,, | (98.) | ·0000223 | ·0000854 |
| ,, | (109.) | ·0000189 | ·0001044 |
| ,, | (111.) | ·0000241 | ·0001114 |
| ,, | (114.) | ·0000035 | ·000115 |
| Mean values for all straight channels, pipes, or rivers | | ·0000221 | ·0000892 |

TABLE I.—COEFFICIENTS OF DISCHARGE FROM SQUARE AND DIFFERENTLY PROPORTIONED RECTANGULAR LATERAL ORIFICES IN THIN VERTICAL PLATES.

| Heads of water measured to the upper sides of the orifices, in English inches. | Ratio of the head to the length of the orifice. | Square orifice 8″ × 8″. Ratio of the sides 1 to 1. | | Rectangular orifice 8″ × 4″. Ratio of the sides 2 to 1. | | Rectangular orifice 8″ × 2″. Ratio of the sides 4 to 1. | |
|---|---|---|---|---|---|---|---|
| | | Heads taken back from the orifice. | Heads taken at the orifice. | Heads taken back from the orifice. | Heads taken at the orifice. | Heads taken back from the orifice. | Heads taken at the orifice. |
| 0·000 | | | ·619 | | ·667 | | ·713 |
| 0·197 | ·025 | | ·597 | | ·630 | | ·668 |
| 0·394 | ·050 | | ·595 | | ·618 | ·607 | ·642 |
| 0·591 | ·075 | | ·594 | ·593 | ·615 | ·612 | ·639 |
| 0·787 | ·100 | ·572 | ·594* | ·596 | ·614 | ·615 | ·638 |
| 1·181 | ·150 | ·578 | ·593 | ·600 | ·613 | ·620 | ·637 |
| 1·575 | ·200 | ·582 | ·593* | ·603 | ·612 | ·623 | ·636 |
| 1·969 | ·250 | ·585 | ·593 | ·605 | ·612* | ·625 | ·636 |
| 2·362 | ·300 | ·587 | ·594 | ·607 | ·613 | ·627 | ·635 |
| 2·756 | ·350 | ·588 | ·594 | ·609 | ·613 | ·628 | ·635 |
| 3·150 | ·400 | ·589 | ·594 | ·610 | ·613 | ·629 | ·635 |
| 3·545 | ·450 | ·591 | ·595 | ·610 | ·614 | ·629 | ·634 |
| 3·937 | ·500 | ·592 | ·595 | ·611 | ·614 | ·630 | ·634 |
| 4·724 | ·600 | ·593 | ·596 | ·612 | ·614 | ·630 | ·633 |
| 5·512 | ·700 | ·595 | ·597 | ·613 | ·614 | ·630 | ·632 |
| 6·299 | ·800 | ·596 | ·597 | ·614 | ·615 | ·631* | ·631 |
| 7·087 | ·900 | ·597 | ·598 | ·615 | ·615 | ·630 | ·631 |
| 7·874 | 1·000 | ·598 | ·599 | ·615 | ·615 | ·630 | ·630 |
| 9·843 | 1·250 | ·599 | ·600 | ·616 | ·616 | ·630 | ·630 |
| 11·811 | 1·500 | ·600 | ·601 | ·616 | ·616 | ·629 | ·629 |
| 15·748 | 2·000 | ·602 | ·602 | ·617 | ·617 | ·628 | ·629 |
| 19·685 | 2·500 | ·603 | ·603 | ·617* | ·617* | ·628 | ·628 |
| 23·622 | 3·000 | ·604 | ·604 | ·617 | 617 | ·627 | 627 |
| 27·560 | 3·500 | ·604 | ·604 | ·616 | ·616 | ·627 | ·627 |
| 31·497 | 4·000 | ·605 | ·605 | ·616 | ·616 | ·627 | ·627 |
| 35·434 | 4·500 | ·605* | ·605* | ·615 | ·615 | ·626 | ·626 |
| 39·371 | 5·000 | ·605 | ·605 | ·615 | ·615 | ·626 | ·626 |
| 43·307 | 5·500 | ·604 | ·604 | ·614 | ·614 | ·625 | ·625 |
| 47·245 | 6·000 | ·604 | ·604 | ·614 | ·614 | ·624 | ·624 |
| 51·182 | 6·500 | ·603 | ·603 | ·613 | ·613 | ·622 | ·622 |
| 55·119 | 7·000 | ·603 | ·603 | ·612 | ·612 | ·621 | ·621 |
| 59·056 | 7·500 | ·602 | ·602 | ·611 | ·611 | ·620 | ·620 |
| 62·993 | 8·000 | ·602 | ·602 | ·611 | ·611 | ·618 | ·618 |
| 66·930 | 8·500 | ·602 | ·602 | ·610 | ·610 | ·617 | ·617 |
| 70·867 | 9·000 | ·601 | ·601 | ·609 | ·609 | ·615 | ·615 |
| 74·805 | 9·500 | ·601 | ·601 | ·608 | ·608 | ·614 | ·614 |
| 78·742 | 10·000 | ·601 | ·601 | ·607 | ·607 | ·613 | ·614 |
| 118·112 | 15·000 | ·601 | ·601 | .603 | ·603 | ·606 | ·606 |

TABLE I.—COEFFICIENTS OF DISCHARGE FROM SQUARE AND DIFFERENTLY PROPORTIONED RECTANGULAR LATERAL ORIFICES IN THIN VERTICAL PLATES.

| Rectangular orifice 8″ × 1·18″. Ratio of the sides 7 to 1 nearly. | | Rectangular orifice 8″ × 0·8″. Ratio of the sides 10 to 1. | | Rectangular orifice 8″ × 0·4″. Ratio of the sides 20 to 1. | | Ratio of the head to the length of the orifice. | Heads of water measured to the upper sides of the orifices, in English inches. |
|---|---|---|---|---|---|---|---|
| Heads taken back from the orifice. | Heads taken at the orifice. | Heads taken back from the orifice. | Heads taken at the orifice. | Heads taken back from the orifice. | Heads taken at the orifice. | | |
|  | ·766 |  | ·783 |  | ·795 |  |  |
|  | ·725 |  | ·750 | ·705 | ·778 | ·025 | 0·197 |
| ·630 | ·687 | ·660 | ·720 | ·701 | ·762 | ·050 | 0·394 |
| ·632 | ·674 | ·660 | ·707 | ·697 | ·745 | ·075 | 0·591 |
| ·634 | ·668 | ·659 | ·697 | ·694 | ·729 | ·100 | 0·787 |
| ·638 | ·659 | ·659 | ·685 | ·688 | ·708 | ·150 | 1·181 |
| ·640 | ·654 | ·658 | ·678 | ·683 | ·695 | ·200 | 1·575 |
| ·640* | ·651 | ·658 | ·672 | ·679 | ·686 | ·250 | 1·969 |
| ·640 | ·647 | ·657 | ·668 | ·676 | ·681 | ·300 | 2·362 |
| ·639 | ·645 | ·656 | ·665 | ·673 | ·677 | ·350 | 2·756 |
| ·638 | ·643 | ·656 | ·662 | ·670 | ·675 | ·400 | 3·150 |
| ·637 | ·641 | ·655 | ·659 | ·668 | ·672 | ·450 | 3·543 |
| ·637 | ·640 | ·654 | ·657 | ·666 | ·669 | ·500 | 3·937 |
| ·636 | ·637 | ·653 | ·655 | ·663 | ·665 | ·600 | 4·724 |
| ·635 | ·636 | ·651 | ·653 | ·660 | ·661 | ·700 | 5·512 |
| ·634 | ·635 | ·650 | ·651 | ·658 | ·659 | ·800 | 6·299 |
| ·634 | ·634 | ·649 | ·650 | ·657 | ·657 | ·900 | 7·087 |
| ·633 | ·633 | ·648 | ·649 | ·655 | ·656 | 1·000 | 7·874 |
| ·632 | ·632 | ·646 | ·646 | ·653 | ·653 | 1·250 | 9·843 |
| ·632 | ·632 | ·644 | ·644 | ·650 | ·651 | 1·500 | 11·811 |
| ·631 | ·631 | ·642 | ·642 | ·647 | ·647 | 2·000 | 15·748 |
| ·630 | ·630 | ·640 | ·640 | ·644 | ·645 | 2·500 | 19·685 |
| ·630 | ·630 | ·638 | ·638 | ·642 | ·643 | 3·000 | 23·622 |
| ·629 | ·629 | ·637 | ·637 | ·640 | ·640 | 3·500 | 27·560 |
| ·629 | ·629 | ·636 | ·636 | ·637 | ·637 | 4·000 | 31·497 |
| ·628 | ·628 | ·634 | ·634 | ·635 | ·635 | 4·500 | 35·434 |
| ·628 | ·628 | ·633 | ·633 | ·632 | ·632 | 5·000 | 39·371 |
| ·627 | ·627 | ·631 | ·631 | ·629 | ·629 | 5·500 | 43·307 |
| ·626 | ·626 | ·628 | ·628 | ·626 | ·626 | 6·000 | 47·245 |
| ·624 | ·624 | ·625 | ·625 | ·622 | ·622 | 6·500 | 51·182 |
| ·622 | ·622 | ·622 | ·622 | ·618 | ·618 | 7·000 | 55·119 |
| ·620 | ·620 | ·619 | ·619 | ·615 | ·615 | 7·500 | 59·056 |
| ·618 | ·618 | ·617 | ·617 | ·613 | ·613 | 8·000 | 62·993 |
| ·616 | ·616 | ·615 | ·615 | ·612 | ·612 | 8·500 | 66·930 |
| ·615 | ·615 | ·614 | ·614 | ·612 | ·612 | 9·000 | 70·867 |
| ·613 | ·613 | ·612 | ·612 | ·611 | ·611 | 9·500 | 74·805 |
| ·612 | ·612 | ·612 | ·612 | ·611 | ·611 | 10·000 | 78·742 |
| ·608 | ·608 | ·610 | ·610 | ·609 | ·609 | 15·000 | 118·112 |

TABLE II.—FOR FINDING THE VELOCITIES FROM THE ALTITUDES
AND THE ALTITUDES FROM THE VELOCITIES.

*Altitudes 0 feet $0\frac{1}{100}$ inch to 0 feet $3\frac{7}{8}$ inches.*

| Altitudes $h$ in feet and inches. | Coefficients of velocity, and the corresponding velocities of discharge in inches per second. | | | | | |
|---|---|---|---|---|---|---|
| | 1. Values of $v = 27.8\sqrt{h}$, the theoretical velocity. | 2. Values of $v = 27.077 h^{\frac{1}{2}}$ Coefficient ·974. | 3. Values of $v = 26.577 h^{\frac{1}{2}}$ Coefficient ·956. | 4. Values of $v = 23.908 h^{\frac{1}{2}}$ Coefficient ·860. | 5. Values of $v = 22.657 h^{\frac{1}{2}}$ Coefficient ·815. | 6. Values of $v = 22.24 h^{\frac{1}{2}}$ Coefficient ·800. |
| 0  0"$\frac{1}{100}$ | 2·78 | 2·71 | 2·66 | 2·39 | 2·27 | 2·22 |
| 0  0$\frac{1}{64}$ | 3·48 | 3·38 | 3·32 | 2·99 | 2·83 | 2·78 |
| 0  0$\frac{1}{16}$ | 6·95 | 6·77 | 6·64 | 5·98 | 5·66 | 5·56 |
| 0  0$\frac{1}{8}$ | 9·829 | 9·57 | 9·40 | 8·45 | 8·01 | 7·86 |
| 0  0$\frac{3}{16}$ | 12·038 | 11·72 | 11·51 | 10·35 | 9·81 | 9·63 |
| 0  0$\frac{1}{4}$ | 13·900 | 13·54 | 13·29 | 11·95 | 11·33 | 11·12 |
| 0  0$\frac{5}{16}$ | 15·541 | 15·14 | 14·86 | 13·36 | 12·67 | 12·43 |
| 0  0$\frac{3}{8}$ | 17·024 | 16·58 | 16·27 | 14·64 | 13·87 | 13·62 |
| 0  0$\frac{7}{16}$ | 18·388 | 17·91 | 17·58 | 15·81 | 14·99 | 14·71 |
| 0  0$\frac{1}{2}$ | 19·658 | 19·15 | 18·79 | 16·91 | 16·02 | 15·73 |
| 0  0$\frac{9}{16}$ | 20·850 | 20·31 | 19·93 | 17·93 | 16·99 | 16·68 |
| 0  0$\frac{5}{8}$ | 21·978 | 21·41 | 21·01 | 18·90 | 17·91 | 17·58 |
| 0  0$\frac{11}{16}$ | 23·051 | 22·45 | 22·04 | 19·82 | 18·79 | 18·44 |
| 0  0$\frac{3}{4}$ | 24·076 | 23·45 | 23·02 | 20·70 | 19·62 | 19·26 |
| 0  0$\frac{13}{16}$ | 25·059 | 24·41 | 24·00 | 21·55 | 20·42 | 20·05 |
| 0  0$\frac{7}{8}$ | 26·005 | 25·33 | 24·86 | 22·36 | 21·19 | 20·80 |
| 0  0$\frac{15}{16}$ | 26·917 | 26·22 | 25·73 | 23·15 | 21·94 | 21·53 |
| 0  1 | 27·800 | 27·08 | 26·58 | 23·91 | 22·66 | 22·24 |
| 0  1$\frac{1}{8}$ | 29·486 | 28·72 | 28·19 | 25·36 | 24·03 | 23·59 |
| 0  1$\frac{1}{4}$ | 31·081 | 30·27 | 29·71 | 26·73 | 25·33 | 24·87 |
| 0  1$\frac{3}{8}$ | 32·598 | 31·75 | 31·16 | 28·03 | 26·57 | 26·08 |
| 0  1$\frac{1}{2}$ | 34·048 | 33·19 | 32·58 | 29·30 | 27·75 | 27·26 |
| 0  1$\frac{5}{8}$ | 35·438 | 34·52 | 33·88 | 30·48 | 28·88 | 28·35 |
| 0  1$\frac{3}{4}$ | 36·776 | 35·82 | 35·16 | 31·63 | 29·97 | 29·42 |
| 0  1$\frac{7}{8}$ | 38·067 | 37·08 | 36·39 | 32·74 | 31·02 | 30·45 |
| 0  2 | 39·315 | 38·29 | 37·59 | 33·81 | 32·04 | 31·45 |
| 0  2$\frac{1}{8}$ | 40·525 | 39·47 | 38·74 | 34·85 | 33·03 | 32·42 |
| 0  2$\frac{1}{4}$ | 41·700 | 40·62 | 39·87 | 35·86 | 33·99 | 33·36 |
| 0  2$\frac{3}{8}$ | 42·843 | 41·73 | 40·96 | 36·84 | 34·92 | 34·27 |
| 0  2$\frac{1}{2}$ | 43·956 | 42·81 | 42·02 | 37·80 | 35·82 | 35·16 |
| 0  2$\frac{5}{8}$ | 45·041 | 43·87 | 43·06 | 38·74 | 36·71 | 36·03 |
| 0  2$\frac{3}{4}$ | 46·101 | 44·90 | 44·07 | 39·65 | 37·57 | 36·88 |
| 0  2$\frac{7}{8}$ | 47·137 | 45·90 | 45·06 | 40·54 | 38·42 | 37·71 |
| 0  3 | 48·151 | 46·90 | 46·03 | 41·41 | 39·24 | 38·52 |
| 0  3$\frac{1}{8}$ | 49·144 | 47·87 | 46·98 | 42·26 | 40·05 | 39·32 |
| 0  3$\frac{1}{4}$ | 50·117 | 48·81 | 47·91 | 43·10 | 40·85 | 40·09 |
| 0  3$\frac{3}{8}$ | 51·072 | 49·74 | 48·82 | 43·92 | 41·62 | 40·86 |
| 0  3$\frac{1}{2}$ | 52·009 | 50·66 | 49·72 | 44·73 | 42·39 | 41·61 |
| 0  3$\frac{5}{8}$ | 52·930 | 51·55 | 50·60 | 45·52 | 43·14 | 42·34 |
| 0  3$\frac{3}{4}$ | 53·834 | 52·43 | 51·47 | 46·30 | 43·88 | 43·07 |
| 0  3$\frac{7}{8}$ | 54·725 | 53·30 | 52·32 | 47·06 | 44·60 | 43·78 |

ORIFICES, WEIRS, PIPES, AND RIVERS.

TABLE II.—FOR FINDING THE VELOCITIES FROM THE ALTITUDES, AND THE ALTITUDES FROM THE VELOCITIES.

*Altitudes 0 feet $0\frac{1}{100}$ inch to 0 feet $3\frac{7}{8}$ inches.*

| Coefficients of velocity, and the corresponding velocities of discharge in inches per second. | | | | | | Altitudes $h$ in feet and inches. | |
|---|---|---|---|---|---|---|---|
| 7. Values of $v = 19.46 h^{\frac{1}{2}}$ Coefficient ·70. | 8. Values of $v = 18.515 h^{\frac{1}{2}}$ Coefficient ·666. | 9. Values of $v = 17.458 h^{\frac{1}{2}}$ Coefficient ·628. | 10. Values of $v = 17.153 h^{\frac{1}{2}}$ Coefficient ·617. | 11. Values of $v = 16.847 h^{\frac{1}{2}}$ Coefficient ·606. | 12. Values of $v = 15.935 h^{\frac{1}{2}}$ Coefficient ·584. | | |
| 1·95 | 1·85 | 1·75 | 1·72 | 1·68 | 1·62 | 0 | $0\frac{1}{100}''$ |
| 2·43 | 2·31 | 2·18 | 2·15 | 2·11 | 2·03 | 0 | $0\frac{1}{64}$ |
| 4·87 | 4·63 | 4·36 | 4·29 | 4·21 | 4·06 | 0 | $0\frac{1}{16}$ |
| 6·88 | 6·55 | 6·17 | 6·06 | 5·96 | 5·74 | 0 | $0\frac{1}{8}$ |
| 8·43 | 8·02 | 7·56 | 7·43 | 7·29 | 7·03 | 0 | $0\frac{3}{16}$ |
| 9·73 | 9·26 | 8·73 | 8·58 | 8·42 | 8·12 | 0 | $0\frac{1}{4}$ |
| 10·88 | 10·35 | 9·76 | 9·59 | 9·42 | 9·08 | 0 | $0\frac{5}{16}$ |
| 11·92 | 11·24 | 10·69 | 10·50 | 10·32 | 9·94 | 0 | $0\frac{3}{8}$ |
| 12·87 | 12·25 | 11·55 | 11·35 | 11·14 | 10·74 | 0 | $0\frac{7}{16}$ |
| 13·76 | 12·97 | 12·34 | 12·13 | 11·91 | 11·48 | 0 | $0\frac{1}{2}$ |
| 14·60 | 13·89 | 13·09 | 12·86 | 12·64 | 12·18 | 0 | $0\frac{9}{16}$ |
| 15·38 | 14·64 | 13·80 | 13·56 | 13·32 | 12·84 | 0 | $0\frac{5}{8}$ |
| 16·14 | 15·35 | 14·48 | 14·22 | 13·97 | 13·46 | 0 | $0\frac{11}{16}$ |
| 16·85 | 16·03 | 15·12 | 14·85 | 14·59 | 14·06 | 0 | $0\frac{3}{4}$ |
| 17·54 | 16·69 | 15·74 | 15·46 | 15·19 | 14·63 | 0 | $0\frac{13}{16}$ |
| 18·20 | 17·32 | 16·33 | 16·04 | 15·76 | 15·09 | 0 | $0\frac{7}{8}$ |
| 18·84 | 17·93 | 16·90 | 16·61 | 16·31 | 15·72 | 0 | $0\frac{15}{16}$ |
| 19·46 | 18·51 | 17·46 | 17·15 | 16·85 | 16·24 | 0 | 1 |
| 20·64 | 19·64 | 18·52 | 18·19 | 17·87 | 17·22 | 0 | $1\frac{1}{8}$ |
| 21·76 | 20·70 | 19·52 | 19·18 | 18·84 | 18·15 | 0 | $1\frac{1}{4}$ |
| 22·82 | 21·71 | 20·47 | 20·11 | 19·75 | 19·04 | 0 | $1\frac{3}{8}$ |
| 23·85 | 22·69 | 21·38 | 21·01 | 20·63 | 19·88 | 0 | $1\frac{1}{2}$ |
| 24·81 | 23·60 | 22·26 | 21·87 | 21·48 | 20·70 | 0 | $1\frac{5}{8}$ |
| 25·74 | 24·49 | 23·10 | 22·69 | 22·29 | 21·48 | 0 | $1\frac{3}{4}$ |
| 26·65 | 25·35 | 23·91 | 23·49 | 23·07 | 22·23 | 0 | $1\frac{7}{8}$ |
| 27·52 | 26·18 | 24·69 | 24·26 | 23·82 | 22·96 | 0 | 2 |
| 28·37 | 26·99 | 25·45 | 25·00 | 24·56 | 23·67 | 0 | $2\frac{1}{8}$ |
| 29·19 | 27·77 | 26·19 | 25·73 | 25·27 | 24·35 | 0 | $2\frac{1}{4}$ |
| 29·99 | 28·53 | 26·91 | 26·43 | 25·96 | 25·02 | 0 | $2\frac{3}{8}$ |
| 30·77 | 29·27 | 27·60 | 27·12 | 26·64 | 25·67 | 0 | $2\frac{1}{2}$ |
| 31·53 | 30·00 | 28·29 | 27·79 | 27·29 | 26·30 | 0 | $2\frac{5}{8}$ |
| 32·27 | 30·70 | 28·95 | 28·44 | 27·94 | 26·92 | 0 | $2\frac{3}{4}$ |
| 33·00 | 31·39 | 29·60 | 29·08 | 28·57 | 27·53 | 0 | $2\frac{7}{8}$ |
| 33·71 | 32·07 | 30·24 | 29·71 | 29·18 | 28·12 | 0 | 3 |
| 34·40 | 32·73 | 30·86 | 30·32 | 29·78 | 28·70 | 0 | $3\frac{1}{8}$ |
| 35·08 | 33·38 | 31·47 | 30·92 | 30·37 | 29·27 | 0 | $3\frac{1}{4}$ |
| 35·75 | 34·01 | 32·07 | 31·51 | 30·95 | 29·83 | 0 | $3\frac{3}{8}$ |
| 36·41 | 34·64 | 32·66 | 32·09 | 31·52 | 30·37 | 0 | $3\frac{1}{2}$ |
| 37·05 | 35·25 | 33·24 | 32·66 | 32·08 | 30·91 | 0 | $3\frac{5}{8}$ |
| 37·68 | 35·85 | 33·81 | 33·22 | 32·62 | 31·44 | 0 | $3\frac{3}{4}$ |
| 38·31 | 36·45 | 34·37 | 33·77 | 33·16 | 31·96 | 0 | $3\frac{7}{8}$ |

178 THE DISCHARGE OF WATER FROM

TABLE II.—FOR FINDING THE VELOCITIES FROM THE ALTITUDES, AND THE ALTITUDES FROM THE VELOCITIES.

*Altitudes 0 feet 4 inches to 1 foot.*

| Altitudes $h$ in feet and inches. | 1. Values of $v = 27.8 \sqrt{h}$, the theoretical velocity. | 2. Coeffic$^t$. ·974. | 3. Coeffic$^t$. ·956. | 4. Coeffit$^t$. ·86. | 5. Coeffic$^t$. ·815. | 6. Coeffic$^t$. ·8. |
|---|---|---|---|---|---|---|
| 0′ 4″ | 55·60 | 54·15 | 53·15 | 47·82 | 45·31 | 44·48 |
| 0 4⅛ | 56·462 | 54·99 | 53·98 | 48·56 | 46·02 | 45·17 |
| 0 4¼ | 57·311 | 55·82 | 54·79 | 49·29 | 46·71 | 45·85 |
| 0 4⅜ | 58·148 | 56·64 | 55·59 | 50·01 | 47·39 | 46·52 |
| 0 4½ | 58·973 | 57·44 | 56·38 | 50·72 | 48·06 | 47·18 |
| 0 4⅝ | 59·786 | 58·23 | 57·16 | 51·42 | 48·73 | 47·83 |
| 0 4¾ | 60·589 | 59·01 | 57·92 | 52·11 | 49·38 | 48·47 |
| 0 4⅞ | 61·368 | 59·77 | 58·67 | 52·78 | 50·02 | 49·09 |
| 0 5 | 62·163 | 60·55 | 59·43 | 53·46 | 50·66 | 49·73 |
| 0 5⅛ | 62·935 | 61·30 | 60·17 | 54·12 | 51·29 | 50·35 |
| 0 5¼ | 63·698 | 62·04 | 60·90 | 54·78 | 51·91 | 50·96 |
| 0 5⅜ | 64·452 | 62·78 | 61·62 | 55·43 | 52·53 | 51·56 |
| 0 5½ | 65·197 | 63·50 | 62·33 | 56·07 | 53·14 | 52·16 |
| 0 5⅝ | 65·933 | 64·22 | 63·03 | 56·70 | 53·74 | 52·75 |
| 0 5¾ | 66·662 | 64·93 | 63·73 | 57·33 | 54·33 | 53·33 |
| 0 5⅞ | 67·383 | 65·63 | 64·42 | 57·95 | 54·92 | 53·91 |
| 0 6 | 68·096 | 66·33 | 65·10 | 58·56 | 55·50 | 54·48 |
| 0 6¼ | 69·50 | 67·69 | 66·44 | 59·77 | 56·64 | 55·60 |
| 0 6½ | 70·876 | 69·03 | 67·76 | 60·95 | 57·24 | 56·70 |
| 0 6¾ | 72·227 | 70·35 | 69·05 | 62·11 | 58·86 | 57·78 |
| 0 7 | 73·552 | 71·64 | 70·32 | 63·25 | 59·95 | 58·84 |
| 0 7¼ | 74·854 | 72·91 | 71·56 | 64·37 | 61·01 | 59·88 |
| 0 7½ | 76·133 | 74·15 | 72·78 | 65·47 | 62·05 | 60·91 |
| 0 7¾ | 77·392 | 75·38 | 73·99 | 66·56 | 63·07 | 61·91 |
| 0 8 | 78·630 | 76·59 | 75·17 | 67·62 | 64·08 | 62·90 |
| 0 8¼ | 79·849 | 77·77 | 76·34 | 68·67 | 65·08 | 63·88 |
| 0 8½ | 81·050 | 78·94 | 77·48 | 69·70 | 66·06 | 64·84 |
| 0 8¾ | 82·234 | 80·10 | 78·62 | 70 72 | 67·02 | 65·79 |
| 0 9 | 83·40 | 81·23 | 79·73 | 71·72 | 67·97 | 66·72 |
| 0 9¼ | 84·550 | 82·35 | 80·83 | 72·71 | 68·91 | 67·64 |
| 0 9½ | 85·685 | 83·46 | 81·92 | 73·69 | 69·83 | 68·55 |
| 0 9¾ | 86·805 | 84·55 | 82·99 | 74·65 | 70·75 | 69 44 |
| 0 10 | 87·911 | 85·63 | 84·04 | 75·60 | 71·65 | 70·33 |
| 0 10¼ | 89·004 | 86·69 | 85·09 | 76·54 | 72·54 | 71·20 |
| 0 10½ | 90·082 | 87·74 | 86·12 | 77·47 | 73·42 | 72·07 |
| 0 10¾ | 91·148 | 88·79 | 87·14 | 78·39 | 74·29 | 72·92 |
| 0 11 | 92·202 | 89·80 | 88·15 | 79·29 | 75·14 | 73·76 |
| 0 11¼ | 93·244 | 90·82 | 89·14 | 80·19 | 75·99 | 74·59 |
| 0 11½ | 94·274 | 91·82 | 90·13 | 81·08 | 76·83 | 75·42 |
| 0 11¾ | 95·294 | 92·82 | 91·10 | 81·95 | 77 66 | 76·23 |
| 1 0 | 96·302 | 93·80 | 92·06 | 82·82 | 78·49 | 77·04 |

ORIFICES, WEIRS, PIPES, AND RIVERS.

TABLE II.—FOR FINDING THE VELOCITIES FROM THE ALTITUDES, AND THE ALTITUDES FROM THE VELOCITIES.

*Altitudes 0 feet 4 inches to 1 foot.*

| Coefficients of velocity and the corresponding velocities of discharge in inches per second. | | | | | | |
|---|---|---|---|---|---|---|
| 7. Coeffic$^t$. ·7 | 8. Coeffic$^t$. ·666. | 9. Coeffic$^t$. ·628. | 10. Coeffic$^t$. ·617. | 11. Coeffic$^t$. ·606. | 12. Coeffic$^t$. ·584. | Altitudes $h$ in feet and inches. |
| 38·92 | 37·03 | 34·92 | 34·31 | 33·69 | 32·47 | 0   4″ |
| 39·52 | 37·60 | 35·46 | 34·84 | 34·22 | 32·97 | 0   4⅛ |
| 40·12 | 38·17 | 35·99 | 35·36 | 34·73 | 33·47 | 0   4¼ |
| 40·70 | 38·73 | 36·52 | 35·88 | 35·24 | 33·96 | 0   4⅜ |
| 41·28 | 39·28 | 37·03 | 36·39 | 35·74 | 34·44 | 0   4½ |
| 41·85 | 39·82 | 37·55 | 36·89 | 36·23 | 34·92 | 0   4⅝ |
| 42·41 | 40·35 | 38·05 | 37·38 | 36·72 | 35·38 | 0   4¾ |
| 42·96 | 40·87 | 38·54 | 37·86 | 37·19 | 35·84 | 0   4⅞ |
| 43·51 | 41·40 | 39·04 | 38·35 | 37·67 | 36·30 | 0   5 |
| 44·05 | 41·91 | 39·52 | 38·83 | 38·14 | 36·75 | 0   5⅛ |
| 44·59 | 42·42 | 40·00 | 39·30 | 38·60 | 37·20 | 0   5¼ |
| 45·12 | 42·92 | 40·48 | 39·77 | 39·06 | 37·64 | 0   5⅜ |
| 45·64 | 43·42 | 40·94 | 40·23 | 39·51 | 38·07 | 0   5½ |
| 46·15 | 43·91 | 41·41 | 40·68 | 39·96 | 38·51 | 0   5⅝ |
| 46·66 | 44·40 | 41·86 | 41·13 | 40·40 | 38·93 | 0   5¾ |
| 47·17 | 44·88 | 42·32 | 41·58 | 40·83 | 39·35 | 0   5⅞ |
| 47·67 | 45·35 | 42·76 | 42·02 | 41·27 | 39·77 | 0   6 |
| 48·65 | 46·29 | 43·65 | 42·88 | 42·12 | 40·59 | 0   6¼ |
| 49·61 | 47·20 | 44·51 | 43·73 | 42·95 | 41·39 | 0   6½ |
| 50·56 | 48·10 | 45·36 | 44·56 | 43·77 | 42·18 | 0   6¾ |
| 51·49 | 48·99 | 46·19 | 45·38 | 44·57 | 42·95 | 0   7 |
| 52·40 | 49·85 | 47·01 | 46·18 | 45·36 | 43·71 | 0   7¼ |
| 53·29 | 50·70 | 47·81 | 46·97 | 46·14 | 44·46 | 0   7½ |
| 54·17 | 51·54 | 48·60 | 47·75 | 46·90 | 45·20 | 0   7¾ |
| 55·04 | 52·37 | 49·38 | 48·51 | 47·65 | 45·92 | 0   8 |
| 55·89 | 53·18 | 50·15 | 49·27 | 48·39 | 46·63 | 0   8¼ |
| 56·74 | 53·98 | 50·90 | 50·01 | 49·12 | 47·33 | 0   8½ |
| 57·56 | 54·77 | 51·64 | 50·74 | 49·83 | 48·02 | 0   8¾ |
| 58·38 | 55·54 | 52·38 | 51·46 | 50·54 | 48·71 | 0   9 |
| 59·19 | 56·31 | 53·10 | 52·17 | 51·24 | 49·38 | 0   9¼ |
| 59·98 | 57·07 | 53·81 | 52·87 | 51·93 | 50·04 | 0   9½ |
| 60·76 | 57·81 | 54·51 | 53·56 | 52·60 | 50·69 | 0   9¾ |
| 61·54 | 58·55 | 55·22 | 54·24 | 53·27 | 51·34 | 0  10 |
| 62·30 | 59·28 | 55·89 | 54·92 | 53·94 | 51·98 | 0  10¼ |
| 63·06 | 60·00 | 56·57 | 55·58 | 54·59 | 52·61 | 0  10½ |
| 63·80 | 60·70 | 57·24 | 56·24 | 55·24 | 53·23 | 0  10¾ |
| 64·54 | 61·41 | 57·90 | 56·89 | 55·87 | 53·85 | 0  11 |
| 65·27 | 62·10 | 58·56 | 57·53 | 56·51 | 54·45 | 0  11¼ |
| 65·99 | 62·79 | 59·70 | 58·17 | 57·13 | 55·06 | 0  11½ |
| 66·71 | 63·47 | 59·84 | 58·80 | 57·75 | 55·65 | 0  11¾ |
| 67·41 | 64·14 | 60·48 | 59·42 | 58·36 | 56·24 | 1   0 |

THE DISCHARGE OF WATER FROM

TABLE II.—FOR FINDING THE VELOCITIES FROM THE ALTITUDES, AND THE ALTITUDES FROM THE VELOCITIES.

*Altitudes 1 foot 0½ inch to 5 feet 3 inches.*

| Altitudes $h$ in feet and inches. | Coefficients of velocity and the corresponding velocities of discharge in inches per second. | | | | | |
|---|---|---|---|---|---|---|
| | 1. Values of $v = 27·8 \sqrt{h}$, the theoretical velocity. | 2. Coeffic$^t$. ·974. | 3. Coeffic$^t$. ·956. | 4. Coeffic$^t$. ·860. | 5. Coeffic$^t$. ·815. | 6. Coeffic$^t$. ·8. |
| 1  0½ | 98·288 | 95·73 | 93·96 | 84·53 | 80·10 | 78·63 |
| 1  1 | 100·234 | 97·63 | 95·82 | 86·20 | 81·69 | 80·19 |
| 1  1½ | 102·144 | 99·49 | 97·65 | 87·84 | 83·25 | 81·71 |
| 1  2 | 104·018 | 101·31 | 99·44 | 89·46 | 84·77 | 83·21 |
| 1  2½ | 105·859 | 103·11 | 101·20 | 91·04 | 86·28 | 84·69 |
| 1  3 | 107·669 | 104·87 | 102·93 | 92·60 | 87·75 | 86·14 |
| 1  3½ | 109·449 | 106·60 | 104·63 | 94·13 | 89·20 | 87·56 |
| 1  4 | 111·200 | 108·31 | 106·31 | 95·63 | 90·63 | 88·96 |
| 1  4½ | 112·924 | 109·99 | 107·96 | 97·11 | 92·03 | 90·34 |
| 1  5 | 114·622 | 111·42 | 109·58 | 98·58 | 93·42 | 91·70 |
| 1  5½ | 116·296 | 113·27 | 111·18 | 100·01 | 94·78 | 93·04 |
| 1  6 | 117·945 | 114·78 | 112·76 | 101·43 | 96·13 | 94·36 |
| 1  7 | 121·177 | 118·03 | 115·85 | 104·21 | 98·76 | 96·94 |
| 1  8 | 124·325 | 121·09 | 118·86 | 106·92 | 101·33 | 99·46 |
| 1  9 | 127·396 | 124·08 | 121·79 | 109·56 | 103·83 | 101·92 |
| 1 10 | 130·394 | 127·00 | 124·66 | 112·14 | 106·27 | 104·31 |
| 1 11 | 133·324 | 129·86 | 127·46 | 114·66 | 108·66 | 106·66 |
| 2  0 | 136·192 | 132·65 | 130·20 | 117·12 | 111·00 | 108·95 |
| 2  1½ | 140·383 | 136·73 | 134·21 | 120·73 | 114·41 | 112·31 |
| 2  3 | 144·453 | 140·70 | 138·10 | 124·23 | 117·73 | 115·56 |
| 2  4½ | 148·411 | 144·55 | 141·88 | 127·64 | 120·96 | 118·73 |
| 2  6 | 152·267 | 148·31 | 145·57 | 130·95 | 124·10 | 121·81 |
| 2  7½ | 156·027 | 151·97 | 149·16 | 134·18 | 127·16 | 124·82 |
| 2  9 | 159·699 | 155·55 | 152·67 | 137·34 | 130·15 | 127·76 |
| 2 10½ | 163·288 | 159·04 | 156·10 | 140·43 | 133·80 | 130·63 |
| 3  0 | 166·800 | 162·46 | 159·46 | 143·45 | 135·94 | 133·44 |
| 3  1½ | 170·240 | 165·81 | 162·75 | 146·41 | 138·75 | 136·19 |
| 3  3 | 173·611 | 169·10 | 165·97 | 149·31 | 141·49 | 138·89 |
| 3  4½ | 176·918 | 172·32 | 169·13 | 152·15 | 144·19 | 141·53 |
| 3  6 | 180·165 | 175·48 | 172·24 | 154·94 | 146·83 | 144·13 |
| 3  7½ | 183·354 | 178·59 | 175·29 | 157·68 | 149·43 | 146·68 |
| 3  9 | 186·488 | 181·64 | 178·28 | 160·38 | 151·99 | 149·19 |
| 3 10½ | 189·571 | 184·64 | 181·23 | 163·03 | 154·50 | 151·66 |
| 4  0 | 192·604 | 187·60 | 184·13 | 165·64 | 156·97 | 154·08 |
| 4  2 | 196·576 | 191·46 | 187·93 | 169·06 | 160·21 | 157·26 |
| 4  4 | 200·469 | 195·26 | 191·65 | 172·40 | 163·38 | 160·37 |
| 4  6 | 204·287 | 198·98 | 195·30 | 175·69 | 166·49 | 163·43 |
| 4  8 | 208·036 | 202·63 | 198·88 | 178·91 | 169·55 | 166·43 |
| 4 10 | 211·718 | 206·21 | 202·40 | 182·08 | 172·55 | 169·37 |
| 5  0 | 215·338 | 209·74 | 205·86 | 185·19 | 175·50 | 172·27 |
| 5  3 | 220·656 | 214·92 | 210·95 | 189·76 | 179·83 | 176·52 |

ORIFICES, WEIRS, PIPES, AND RIVERS.

TABLE II.—FOR FINDING THE VELOCITIES FROM THE ALTITUDES, AND THE ALTITUDES FROM THE VELOCITIES.

*Altitudes 1 foot 0½ inch to 5 feet 3 inches.*

| Coefficients of velocity and the corresponding velocities of discharge in inches per second. | | | | | | Altitudes $h$ in feet and inches. | |
|---|---|---|---|---|---|---|---|
| 7. Coefficᵗ. ·7. | 8. Coefficᵗ. ·666. | 9. Cocfficᵗ. ·628. | 10. Coefficᵗ. ·617. | 11. Coefficᵗ. ·606. | 12. Coefficᵗ. ·584. | | |
| 68·80 | 65·46 | 61·72 | 60·64 | 59·56 | 57·40 | 1 | 0½″ |
| 70·16 | 66·76 | 62·95 | 61·84 | 60·74 | 58·54 | 1 | 1 |
| 71·50 | 68·03 | 64·15 | 63·02 | 61·90 | 59·65 | 1 | 1½ |
| 72·81 | 69·28 | 65·32 | 64·18 | 63·03 | 60·75 | 1 | 2 |
| 74·10 | 70·50 | 66·48 | 65·32 | 64·15 | 61·82 | 1 | 2½ |
| 75·37 | 71·71 | 67·62 | 66·43 | 65·25 | 62·88 | 1 | 3 |
| 76·61 | 72·89 | 68·73 | 67·53 | 66·33 | 63·92 | 1 | 3½ |
| 77·84 | 74·06 | 69·83 | 68·61 | 67·34 | 64·94 | 1 | 4 |
| 79·05 | 75·21 | 70·92 | 69·67 | 68·43 | 65·95 | 1 | 4½ |
| 80·24 | 76·34 | 71·98 | 70·72 | 69·46 | 66·94 | 1 | 5 |
| 81·41 | 77·45 | 73·03 | 71·75 | 70·48 | 67·92 | 1 | 5½ |
| 82·56 | 78·55 | 74·07 | 72·77 | 71·47 | 68·88 | 1 | 6 |
| 84·82 | 80·70 | 76·10 | 74·77 | 73·43 | 70·77 | 1 | 7 |
| 87·03 | 82·80 | 78·08 | 76·71 | 75·34 | 72·61 | 1 | 8 |
| 89·18 | 84·85 | 80·00 | 78·60 | 77·20 | 74·40 | 1 | 9 |
| 91·28 | 86·84 | 81·89 | 80·45 | 79·02 | 76·15 | 1 | 10 |
| 93·33 | 88·79 | 83·73 | 82·26 | 80·79 | 77·86 | 1 | 11 |
| 95·33 | 90·70 | 85·53 | 84·03 | 82·53 | 79·54 | 2 | 0 |
| 98·27 | 93·50 | 88·16 | 86·62 | 85·07 | 81·98 | 2 | 1½ |
| 101·12 | 96·21 | 90·72 | 89·13 | 87·54 | 84·36 | 2 | 3 |
| 103·89 | 98·84 | 93·20 | 91·57 | 89·94 | 86·67 | 2 | 4½ |
| 106·59 | 101·41 | 95·62 | 93·95 | 92·27 | 88·92 | 2 | 6 |
| 109·22 | 103·91 | 97·99 | 96·27 | 94·55 | 91·12 | 2 | 7½ |
| 111·79 | 106·36 | 100·29 | 98·53 | 96·78 | 93·26 | 2 | 9 |
| 114·30 | 108·75 | 102·54 | 100·75 | 98·95 | 95·36 | 2 | 10½ |
| 116·76 | 111·09 | 104·75 | 102·92 | 101·08 | 97·41 | 3 | 0 |
| 119·17 | 113·38 | 106·91 | 105·04 | 103·17 | 99·42 | 3 | 1½ |
| 121·53 | 115·62 | 109·03 | 107·12 | 105·21 | 101·39 | 3 | 3 |
| 123·84 | 117·83 | 111·10 | 109·16 | 107·21 | 103·32 | 3 | 4½ |
| 126·12 | 119·99 | 113·14 | 111·16 | 109·18 | 105·22 | 3 | 6 |
| 128·35 | 122·11 | 115·15 | 113·13 | 111·11 | 107·08 | 3 | 7½ |
| 130·54 | 124·20 | 117·11 | 115·06 | 113·01 | 108·91 | 3 | 9 |
| 132·70 | 126·25 | 119·05 | 116·97 | 114·88 | 110·71 | 3 | 10½ |
| 134·82 | 128·27 | 120·96 | 118·84 | 116·72 | 112·48 | 4 | 0 |
| 137·60 | 130·92 | 123·45 | 121·29 | 119·12 | 114·80 | 4 | 2 |
| 140·33 | 133·51 | 125·89 | 123·69 | 121·48 | 117·07 | 4 | 4 |
| 143·00 | 136·06 | 128·29 | 126·05 | 123·80 | 119·30 | 4 | 6 |
| 145·63 | 138·55 | 130·65 | 128·36 | 126·07 | 121·49 | 4 | 8 |
| 148·20 | 141·00 | 132·96 | 130·63 | 128·30 | 123·64 | 4 | 10 |
| 150·74 | 143·42 | 135·23 | 132·86 | 130·49 | 125·76 | 5 | 0 |
| 154·46 | 146·96 | 138 57 | 136·14 | 133·72 | 128·86 | 5 | 3 |

TABLE II.—FOR FINDING THE VELOCITIES FROM THE ALTITUDES, AND THE ALTITUDES FROM THE VELOCITIES.

*Altitudes 5 feet 6 inches to 17 feet.*

| Altitudes $h$ in feet and inches. | Coefficients of velocity, and the corresponding velocities of discharge in inches per second. | | | | | |
|---|---|---|---|---|---|---|
| | 1. Values of $v = 27.8 \sqrt{h}$, the theoretical velocity. | 2. Coeffic$^t$. ·974. | 3. Coeffic$^t$. ·956. | 4. Coeffic$^t$. ·86. | 5. Coeffic$^t$. ·815. | 6. Coeffic$^t$. ·8. |
| 5  6 | 225·848 | 219·98 | 215·91 | 194·23 | 184·07 | 180·68 |
| 5  9 | 230·924 | 224·92 | 220·76 | 198·59 | 188·20 | 184·74 |
| 6  0 | 235·891 | 229·76 | 225·51 | 202·87 | 192·25 | 188·71 |
| 6  3 | 240·755 | 234·50 | 230·16 | 207·05 | 196·22 | 192·60 |
| 6  6 | 245·524 | 239·14 | 234·72 | 211·15 | 200·10 | 196·42 |
| 6  9 | 250·200 | 243·69 | 239·19 | 215·17 | 203·91 | 200·16 |
| 7  0 | 254·791 | 248·17 | 243·58 | 219·12 | 207·65 | 203·83 |
| 7  3 | 259·301 | 252·56 | 247·89 | 222·99 | 211·33 | 207·44 |
| 7  6 | 263·734 | 256·88 | 252·13 | 226·81 | 214·94 | 210·99 |
| 7  9 | 268·093 | 261·12 | 256·30 | 230·56 | 218·50 | 214·47 |
| 8  0 | 272·383 | 265·30 | 260·40 | 234·25 | 221·99 | 217·91 |
| 8  3 | 276·607 | 269·41 | 264·44 | 237·88 | 225·43 | 221·29 |
| 8  6 | 280·766 | 273·47 | 268·41 | 241·46 | 228·82 | 224·61 |
| 8  9 | 284·865 | 277·46 | 272·33 | 244·98 | 232·17 | 227·89 |
| 9  0 | 288·906 | 281·39 | 276·19 | 248·46 | 235·46 | 231·12 |
| 9  3 | 292·891 | 285·28 | 280·00 | 251·89 | 238·71 | 234·31 |
| 9  6 | 296·823 | 289·11 | 283·76 | 255·27 | 241·91 | 237·46 |
| 9  9 | 300·703 | 292·88 | 287·47 | 258·60 | 245·07 | 240·56 |
| 10  0 | 304·534 | 296·62 | 291·13 | 261·90 | 248·19 | 243·63 |
| 10  3 | 308·317 | 300·30 | 294·75 | 265·15 | 251·28 | 246·65 |
| 10  6 | 312·054 | 303·94 | 297·32 | 268·37 | 254·32 | 249·64 |
| 10  9 | 315·747 | 307·54 | 301·85 | 271·54 | 257·33 | 252·60 |
| 11  0 | 319·398 | 311·09 | 305·34 | 274·68 | 260·31 | 255·52 |
| 11  3 | 323·007 | 314·61 | 308·79 | 277·79 | 262·25 | 258·41 |
| 11  6 | 326·576 | 318·09 | 312·21 | 280·86 | 266·16 | 261·26 |
| 11  9 | 330·107 | 321·52 | 315·58 | 283·89 | 269·04 | 264·09 |
| 12  0 | 333·600 | 324·93 | 318·92 | 286·90 | 271·88 | 266·88 |
| 12  3 | 337·057 | 328·29 | 322·23 | 289·87 | 274·70 | 269·65 |
| 12  6 | 340·479 | 331·63 | 325·50 | 292·81 | 277·49 | 272·38 |
| 12  9 | 343·867 | 334·93 | 328·74 | 295·73 | 280·25 | 275·09 |
| 13  0 | 347·222 | 338·19 | 331·94 | 298·61 | 282·99 | 277·78 |
| 13  3 | 350·545 | 341·43 | 335·12 | 301·47 | 285·69 | 280·44 |
| 13  6 | 353·836 | 344·64 | 338·27 | 304·30 | 288·38 | 283·07 |
| 13  9 | 357·097 | 347·81 | 341·39 | 307·10 | 291·03 | 285·68 |
| 14  0 | 360·329 | 350·96 | 344·47 | 309·88 | 293·67 | 288·26 |
| 14  6 | 366·707 | 357·17 | 350·57 | 315·37 | 298·87 | 293·37 |
| 15  0 | 372·976 | 363·28 | 356·57 | 320·76 | 303·98 | 298·38 |
| 15  6 | 379·141 | 369·28 | 362·46 | 326·06 | 309·00 | 303·31 |
| 16  0 | 385·208 | 375·19 | 368·26 | 331·28 | 313·94 | 308·17 |
| 16  6 | 391·181 | 381·01 | 373·97 | 336·42 | 318·81 | 312·94 |
| 17  0 | 397·063 | 386·74 | 379·59 | 341·47 | 323·61 | 317·65 |

ORIFICES, WEIRS, PIPES, AND RIVERS.

TABLE II.—FOR FINDING THE VELOCITIES FROM THE ALTITUDES, AND THE ALTITUDES FROM THE VELOCITIES.

*Altitudes 5 feet 6 inches to 17 feet.*

| Coefficients of velocity, and the corresponding velocities of discharge in inches per second. | | | | | | Altitudes $h$ in feet and inches. | |
|---|---|---|---|---|---|---|---|
| 7. Coeffic$^t$. ·7. | 8. Coeffic$^t$. ·666. | 9. Coeffic$^t$. ·628. | 10. Coeffic$^t$. ·617. | 11. Coeffic$^t$. ·606. | 12. Coeffic$^t$. ·584. | ′ | ″ |
| 158·09 | 150·41 | 141·83 | 139·35 | 136·86 | 131·90 | 5 | 6 |
| 161·65 | 153·80 | 145·02 | 142·48 | 139·94 | 134·86 | 5 | 9 |
| 165·12 | 157·10 | 148·14 | 145·55 | 142·95 | 137·76 | 6 | 0 |
| 168·53 | 160·34 | 151·19 | 148·55 | 145·90 | 140·60 | 6 | 3 |
| 171·87 | 163·52 | 154·19 | 151·49 | 148·79 | 143·39 | 6 | 6 |
| 175·14 | 166·63 | 157·13 | 154·37 | 151·62 | 146·12 | 6 | 9 |
| 178·35 | 169·69 | 160·01 | 157·21 | 154·40 | 148·80 | 7 | 0 |
| 181·51 | 172·69 | 162·84 | 159·99 | 157·14 | 151·43 | 7 | 3 |
| 184·61 | 175·65 | 165·62 | 162·72 | 159·82 | 154·02 | 7 | 6 |
| 187·67 | 178·55 | 168·36 | 165·41 | 162·46 | 156·57 | 7 | 9 |
| 190·67 | 181·41 | 171·06 | 168·06 | 165·06 | 159·07 | 8 | 0 |
| 193·62 | 184·22 | 173·71 | 170·67 | 167·62 | 161·54 | 8 | 3 |
| 196·54 | 186·99 | 176·32 | 173·23 | 170·14 | 163·97 | 8 | 6 |
| 199·41 | 189·72 | 178·90 | 175·76 | 172·63 | 166·36 | 8 | 9 |
| 202·23 | 192·41 | 181·43 | 178·26 | 175·08 | 168·72 | 9 | 0 |
| 205·02 | 195·07 | 183·94 | 180·71 | 177·49 | 171·05 | 9 | 3 |
| 207·78 | 197·68 | 186·40 | 183·14 | 179·87 | 173·34 | 9 | 6 |
| 210·49 | 200·27 | 188·84 | 185·53 | 182·23 | 175·61 | 9 | 9 |
| 213·17 | 202·82 | 191·25 | 187·90 | 184·55 | 177·85 | 10 | 0 |
| 215·82 | 205·34 | 193·62 | 190·23 | 186·84 | 180·06 | 10 | 3 |
| 218·44 | 207·83 | 195·97 | 192·54 | 189·10 | 182·24 | 10 | 6 |
| 221·02 | 210·29 | 198·29 | 194·82 | 191·34 | 184·40 | 10 | 9 |
| 223·58 | 212·72 | 200·58 | 197·07 | 193·55 | 186·53 | 11 | 0 |
| 226·10 | 215·12 | 202·85 | 199·30 | 195·74 | 188·64 | 11 | 3 |
| 228·60 | 217·50 | 205·09 | 201·50 | 197·91 | 190·72 | 11 | 6 |
| 231·07 | 219·85 | 207·31 | 203·68 | 200·04 | 192·78 | 11 | 9 |
| 233·52 | 222·18 | 209·50 | 205·83 | 202·16 | 194·82 | 12 | 0 |
| 235·94 | 224·48 | 211·67 | 207·96 | 204·26 | 196·84 | 12 | 3 |
| 238·34 | 226·76 | 213·82 | 210·08 | 206·33 | 198·84 | 12 | 6 |
| 240·71 | 229·02 | 215·95 | 212·17 | 208·38 | 200·82 | 12 | 9 |
| 243·06 | 231·25 | 218·06 | 214·24 | 210·42 | 202·78 | 13 | 0 |
| 245·38 | 233·46 | 220·14 | 216·29 | 212·43 | 204·72 | 13 | 3 |
| 247·69 | 235·65 | 222·21 | 218·32 | 214·42 | 206·64 | 13 | 6 |
| 249·97 | 237·83 | 224·26 | 220·33 | 216·40 | 208·54 | 13 | 9 |
| 252·23 | 239·98 | 226·29 | 222·32 | 218·36 | 210·43 | 14 | 0 |
| 256·70 | 244·23 | 230·29 | 226·26 | 222·22 | 214·16 | 14 | 6 |
| 261·08 | 248·40 | 234·23 | 230·13 | 226·02 | 217·82 | 15 | 0 |
| 265·40 | 252·51 | 238·10 | 233·93 | 229·76 | 221·42 | 15 | 6 |
| 269·65 | 256·55 | 241·91 | 237·67 | 233·44 | 224·96 | 16 | 0 |
| 273·83 | 260·53 | 245·66 | 241·36 | 237·06 | 228·45 | 16 | 6 |
| 277·94 | 264·44 | 249·36 | 244·99 | 240·62 | 231·89 | 17 | 0 |

184   THE DISCHARGE OF WATER FROM

TABLE II.—FOR FINDING THE VELOCITIES FROM THE ALTITUDES,
AND THE ALTITUDES FROM THE VELOCITIES.

*Altitudes 17 feet 6 inches to 40 feet.*

| Altitudes $h$ in feet and inches. | Coefficients of velocity, and the corresponding velocities of discharge in inches per second. | | | | | |
|---|---|---|---|---|---|---|
| | 1. Values of $v = 27\cdot 8 \sqrt{h}$, the theoretical velocity. | 2. Coeffic$^t$. ·974. | 3. Coeffic$^t$. ·956. | 4. Coeffic$^t$. ·86. | 5. Coeffic$^t$. ·815. | 6. Coeffic$^t$. ·8. |
| 17′ 6″ | 402·860 | 392·39 | 385·13 | 346·46 | 328·33 | 322·29 |
| 18  0 | 408·575 | 397·95 | 390·60 | 351·37 | 332·99 | 326·86 |
| 18  6 | 414·211 | 403·44 | 395·99 | 356·22 | 337·58 | 331·37 |
| 19  0 | 419·772 | 408·86 | 401·30 | 361·00 | 342·11 | 335·82 |
| 19  6 | 425·258 | 414·20 | 406·55 | 365·72 | 346·59 | 340·21 |
| 20  0 | 430·676 | 419·48 | 411·73 | 370·38 | 351·00 | 344·54 |
| 20  6 | 436·026 | 424·69 | 416·84 | 374·98 | 355·36 | 348·82 |
| 21  0 | 441·311 | 429·84 | 421·89 | 379·53 | 359·59 | 353·05 |
| 21  6 | 446·534 | 434·92 | 426·89 | 384·02 | 363·93 | 357·23 |
| 22  0 | 451·697 | 439·95 | 431·82 | 388·46 | 368·13 | 361·36 |
| 22  6 | 456·801 | 444·92 | 436·70 | 392·85 | 372·29 | 365·44 |
| 23  0 | 461·848 | 449·84 | 441·53 | 397·19 | 376·41 | 369·48 |
| 23  6 | 466·841 | 450·70 | 446·30 | 401·48 | 380·48 | 373·47 |
| 24  0 | 471·782 | 459·52 | 451·02 | 405·73 | 384·50 | 377·43 |
| 24  6 | 476·671 | 464·28 | 455·70 | 409·94 | 388·49 | 381·34 |
| 25  0 | 481·510 | 468·99 | 460·32 | 414·10 | 392·43 | 385·21 |
| 25  6 | 486·301 | 473·66 | 464·90 | 418·22 | 396·34 | 389·04 |
| 26  0 | 491·046 | 478·28 | 469·44 | 422·30 | 400·20 | 392·84 |
| 26  6 | 495·745 | 482·86 | 473·93 | 426·34 | 404·03 | 396·60 |
| 27  0 | 500·40 | 487·39 | 478·38 | 430·34 | 407·83 | 400·32 |
| 27  6 | 505·012 | 491·88 | 482·79 | 434·31 | 411·58 | 404·01 |
| 28  0 | 509·582 | 496·33 | 487·16 | 438·24 | 415·31 | 407·67 |
| 28  6 | 514·112 | 500·75 | 491·49 | 442·14 | 419·00 | 411·29 |
| 29  0 | 518·602 | 505·12 | 495·78 | 446·00 | 422·66 | 414·88 |
| 29  6 | 523·054 | 509·45 | 500·04 | 449·83 | 426·29 | 418·44 |
| 30  0 | 527·468 | 513·75 | 504·26 | 453·62 | 429·89 | 421·97 |
| 30  6 | 531·845 | 518·02 | 508·44 | 457·39 | 433·45 | 425·48 |
| 31  ·0 | 536·187 | 522·25 | 512·59 | 461·12 | 436·99 | 428·95 |
| 31  6 | 540·494 | 526·44 | 516·71 | 464·82 | 440·50 | 432·40 |
| 32  0 | 544·767 | 530·60 | 520·80 | 468·50 | 443·98 | 435·81 |
| 32  6 | 549·006 | 534·73 | 524·85 | 472·15 | 447·44 | 439·20 |
| 33  0 | 553·213 | 538·83 | 528·87 | 475·76 | 450·87 | 442·57 |
| 33  6 | 557·388 | 542·90 | 532·86 | 479·35 | 454·27 | 445·91 |
| 34  0 | 561·532 | 546·93 | 536·83 | 482·92 | 457·65 | 449·23 |
| 34  6 | 565·646 | 550·94 | 540·76 | 486·46 | 461·00 | 452·52 |
| 35  0 | 569·730 | 554·92 | 544·66 | 489·97 | 464·33 | 455·78 |
| 36  0 | 577·812 | 562·79 | 552·39 | 496·92 | 470·92 | 462·25 |
| 37  0 | 585·782 | 570·55 | 560·01 | 503·77 | 477·41 | 468·63 |
| 38  0 | 593·646 | 578·21 | 567·53 | 510·54 | 483·82 | 474·92 |
| 39  0 | 601·406 | 585·77 | 574·94 | 517·21 | 490·15 | 481·12 |
| 40  0 | 609·067 | 593·23 | 582·27 | 523·80 | 496·39 | 487·25 |

ORIFICES, WEIRS, PIPES, AND RIVERS.

TABLE II.—FOR FINDING THE VELOCITIES FROM THE ALTITUDES, AND THE ALTITUDES FROM THE VELOCITIES.

*Altitudes* 17 *feet* 6 *inches to* 40 *feet.*

| Coefficients of velocity, and the corresponding velocities of discharge in inches per second. | | | | | | Altitudes $h$ in feet and inches. | |
|---|---|---|---|---|---|---|---|
| 7. Coeffic$^t$. ·7. | 8. Coeffic$^t$. ·666. | 9. Coeffic$^t$. ·628. | 10. Coeffic$^t$. ·617. | 11. Coeffic$^t$. ·606. | 12. Coeffic$^t$. ·584. | | |
| 282·00 | 268·30 | 253·00 | 248·56 | 244·13 | 235·27 | 17 | 6 |
| 286·00 | 272·11 | 256·59 | 252·09 | 247·60 | 238·61 | 18 | 0 |
| 289·95 | 275·86 | 260·12 | 255·57 | 251·01 | 241·90 | 18 | 6 |
| 293·84 | 279·57 | 263·32 | 259·00 | 254·38 | 245·14 | 19 | 0 |
| 297·68 | 283·22 | 267·06 | 262·38 | 257·71 | 248·35 | 19 | 6 |
| 301·47 | 286·83 | 270·46 | 265·73 | 260·99 | 251·51 | 20 | 0 |
| 305·22 | 290·39 | 273·82 | 269·03 | 264·23 | 254·64 | 20 | 6 |
| 308·92 | 293·91 | 277·08 | 272·23 | 267·37 | 257·67 | 21 | 0 |
| 312·57 | 297·39 | 280·42 | 275·51 | 270·60 | 260·78 | 21 | 6 |
| 316·19 | 300·83 | 283·67 | 278·70 | 273·73 | 263·79 | 22 | 0 |
| 319·76 | 304·23 | 286·87 | 281·85 | 276·82 | 266·77 | 22 | 6 |
| 323·29 | 307·59 | 290·04 | 284·96 | 279·88 | 269·72 | 23 | 0 |
| 326·79 | 310·92 | 293·18 | 288·04 | 282·91 | 272·64 | 23 | 6 |
| 330·25 | 314·21 | 296·28 | 291·09 | 285·90 | 275·52 | 24 | 0 |
| 333·67 | 317·46 | 299·35 | 294·11 | 288·86 | 278·38 | 24 | 6 |
| 337·06 | 320·69 | 302·39 | 297·09 | 291·80 | 281·20 | 25 | 0 |
| 340·41 | 323·88 | 305·40 | 300·05 | 294·70 | 284·00 | 25 | 6 |
| 343·73 | 327·04 | 308·38 | 302·98 | 297·57 | 286·77 | 26 | 0 |
| 347·02 | 330·17 | 311·33 | 305·87 | 300·42 | 289·52 | 26 | 6 |
| 350·28 | 333·13 | 314·25 | 308·75 | 303·24 | 292·23 | 27 | 0 |
| 353·51 | 336·34 | 317·15 | 311·59 | 306·04 | 294·93 | 27 | 6 |
| 356·71 | 339·38 | 320·02 | 314·41 | 308·81 | 297·60 | 28 | 0 |
| 359·88 | 342·40 | 322·86 | 317·20 | 311·55 | 300·24 | 28 | 6 |
| 363·02 | 345·39 | 325·68 | 319·98 | 314·27 | 302·86 | 29 | 0 |
| 366·14 | 348·35 | 328·48 | 322·72 | 316·97 | 305·46 | 29 | 6 |
| 369·23 | 351·29 | 331·25 | 325·45 | 319·65 | 308·04 | 30 | 0 |
| 372·29 | 354·21 | 334·00 | 328·15 | 322·30 | 310·60 | 30 | 6 |
| 375·33 | 357·10 | 336·73 | 330·83 | 324·93 | 313·13 | 31 | 0 |
| 378·35 | 359·97 | 339·43 | 333·48 | 327·54 | 315·60 | 31 | 6 |
| 381·34 | 362·81 | 342·11 | 336·12 | 330·13 | 318·14 | 32 | 0 |
| 384·30 | 365·64 | 344·78 | 338·74 | 332·70 | 320·62 | 32 | 6 |
| 387·25 | 368·44 | 347·42 | 341·33 | 335·25 | 323·08 | 33 | 0 |
| 390·17 | 371·22 | 350·04 | 343·01 | 337·78 | 325·51 | 33 | 6 |
| 393·07 | 373·98 | 352·64 | 346·47 | 340·29 | 327·93 | 34 | 0 |
| 395·95 | 376·72 | 355·23 | 349·00 | 342·78 | 330·34 | 34 | 6 |
| 398·81 | 379·44 | 357·79 | 351·52 | 345·26 | 332·72 | 35 | 0 |
| 404·47 | 384·82 | 362·87 | 356·51 | 350·15 | 337·44 | 36 | 0 |
| 410·05 | 390·13 | 367·87 | 361·43 | 354·98 | 342·10 | 37 | 0 |
| 415·55 | 395·37 | 372·81 | 366·28 | 359·75 | 346·69 | 38 | 0 |
| 420·98 | 400·54 | 377·68 | 371·11 | 364·45 | 351·22 | 39 | 0 |
| 426·35 | 405·64 | 382·49 | 375·79 | 369·09 | 355·70 | 40 | 0 |

TABLE III.—SQUARE ROOTS FOR FINDING THE EFFECTS OF THE VELOCITY OF APPROACH WHEN THE ORIFICE IS SMALL IN PROPORTION TO THE HEAD. ALSO FOR FINDING THE INCREASE IN THE DISCHARGE FROM AN INCREASE OF HEAD. (See p. 55

| No. | Square root. | No. | Square root. | No. | Square root. | No. | Square root. |
|---|---|---|---|---|---|---|---|
| 1·000 | 1·0000 | 1·115 | 1·0559 | 1·475 | 1·2141 | 1·975 | 1·4053 |
| 1·001 | 1·0005 | 1·120 | 1·0583 | 1·49 | 1·2207 | 1·99 | 1·4107 |
| 1·002 | 1·0010 | 1·125 | 1·0607 | 1·5 | 1·2247 | 2·00 | 1·4142 |
| 1·004 | 1·0020 | 1·13 | 1·0630 | 1·51 | 1·2288 | 2·01 | 1·4177 |
| 1·005 | 1·0025 | 1·135 | 1·0654 | 1·525 | 1·2349 | 2·025 | 1·4230 |
| 1·006 | 1·0030 | 1·14 | 1·0677 | 1·54 | 1·2410 | 2·04 | 1·4283 |
| 1·008 | 1·0040 | 1·145 | 1·0700 | 1·55 | 1·2450 | 2·05 | 1·4318 |
| 1·009 | 1·0044 | 1·15 | 1·0723 | 1·56 | 1·2490 | 2·06 | 1·4353 |
| 1·010 | 1·0050 | 1·155 | 1·0747 | 1·575 | 1·2550 | 2·075 | 1·4405 |
| 1·011 | 1·0055 | 1·16 | 1·0770 | 1·58 | 1·2570 | 2·09 | 1·4457 |
| 1·012 | 1·0060 | 1·165 | 1·0794 | 1·59 | 1·2610 | 2·10 | 1·4491 |
| 1·014 | 1·0070 | 1·17 | 1·0817 | 1·6 | 1·2649 | 2·11 | 1·4526 |
| 1·015 | 1·0075 | 1·175 | 1·0840 | 1·61 | 1·2689 | 2·125 | 1·4577 |
| 1·016 | 1·0080 | 1·18 | 1·0863 | 1·625 | 1·2748 | 2·14 | 1·4629 |
| 1·018 | 1·0090 | 1·185 | 1·0886 | 1·64 | 1·2806 | 2·15 | 1·4663 |
| 1·019 | 1·0095 | 1·19 | 1·0909 | 1·65 | 1·2845 | 2·16 | 1·4697 |
| 1·020 | 1·0100 | 1·195 | 1·0932 | 1·66 | 1·2884 | 2·175 | 1·4748 |
| 1·0225 | 1·0112 | 1·2 | 1·0954 | 1·675 | 1·2942 | 2·19 | 1·4799 |
| 1·025 | 1·0124 | 1·21 | 1·1000 | 1·69 | 1·3000 | 2·2 | 1·4832 |
| 1·0275 | 1·0137 | 1·22 | 1·1045 | 1·7 | 1·3038 | 2·21 | 1·4866 |
| 1·03 | 1·0149 | 1·23 | 1·1091 | 1·71 | 1·3077 | 2·225 | 1·4916 |
| 1·0325 | 1·0161 | 1·24 | 1·1136 | 1·725 | 1·3134 | 2·24 | 1·4937 |
| 1·035 | 1·0174 | 1·25 | 1·1180 | 1·74 | 1·3191 | 2·25 | 1·5000 |
| 1·0375 | 1·0186 | 1·26 | 1·1225 | 1·75 | 1·3229 | 2·26 | 1·5033 |
| 1·04 | 1·0198 | 1·27 | 1·1269 | 1·76 | 1·3267 | 2·275 | 1·5083 |
| 1·0425 | 1·0210 | 1·28 | 1·1314 | 1·775 | 1·3323 | 2·29 | 1·5133 |
| 1·045 | 1·0223 | 1·29 | 1·1358 | 1·79 | 1·3379 | 2·3 | 1·5166 |
| 1·0475 | 1·0235 | 1·30 | 1·1402 | 1·80 | 1·3416 | 2·31 | 1·5199 |
| 1·05 | 1·0247 | 1·31 | 1·1446 | 1·81 | 1·3454 | 2·325 | 1·5248 |
| 1·055 | 1·0271 | 1·325 | 1·1511 | 1·825 | 1·3509 | 2·34 | 1·5297 |
| 1·06 | 1·0296 | 1·34 | 1·1576 | 1·84 | 1·3565 | 2·35 | 1·5330 |
| 1·065 | 1·0320 | 1·35 | 1·1619 | 1·85 | 1·3601 | 2·36 | 1·5362 |
| 1·07 | 1·0344 | 1·36 | 1·1662 | 1·86 | 1·3638 | 2·375 | 1·5411 |
| 1·075 | 1·0368 | 1·375 | 1·1726 | 1·875 | 1·3693 | 2·39 | 1·5460 |
| 1·08 | 1·0392 | 1·39 | 1·1790 | 1·89 | 1·3748 | 2·4 | 1·5492 |
| 1·085 | 1·0416 | 1·40 | 1·1832 | 1·9 | 1·3784 | 2·41 | 1·5524 |
| 1·09 | 1·0440 | 1·41 | 1·1874 | 1·91 | 1·3820 | 2·425 | 1·5572 |
| 1·095 | 1·0464 | 1·425 | 1·1937 | 1·925 | 1·3875 | 2·44 | 1·5621 |
| 1·1 | 1·0488 | 1·44 | 1·2000 | 1·94 | 1·3928 | 2·45 | 1·5652 |
| 1·105 | 1·0512 | 1·45 | 1·2042 | 1·95 | 1·3964 | 2·46 | 1·5684 |
| 1·110 | 1·0536 | 1·46 | 1·2083 | 1·96 | 1·4000 | 2·475 | 1·5732 |

TABLE III.—SQUARE ROOTS FOR FINDING THE EFFECTS OF THE VELOCITY OF APPROACH WHEN THE ORIFICE IS SMALL IN PROPORTION TO THE HEAD. ALSO FOR FINDING THE INCREASE IN THE DISCHARGE FROM AN INCREASE OF HEAD. (See p. 55.)

| No. | Square root. | No. | Square root. | No. | Square root. | No. | Square root. |
|---|---|---|---|---|---|---|---|
| 2·49 | 1·5780 | 3·0000 | 1·7321 | 4·5 | 2·1213 | 26 | 5·0990 |
| 2·5 | 1·5811 | 3·025 | 1·7393 | 5·0 | 2·2361 | 27 | 5·1962 |
| 2·51 | 1·5843 | 3·05 | 1·7464 | 5·5 | 2·3452 | 28 | 5·2915 |
| 2·525 | 1·5890 | 3·075 | 1·7536 | 6·0 | 2·4495 | 29 | 5·3852 |
| 2·54 | 1·5937 | 3·1 | 1·7607 | 6·5 | 2·5495 | 30 | 5·4772 |
| 2·55 | 1·5969 | 3·125 | 1·7678 | 7·0 | 2·6458 | 31 | 5·5678 |
| 2·56 | 1·6000 | 3·15 | 1·7748 | 7·5 | 2·7386 | 32 | 5·6569 |
| 2·575 | 1·6047 | 3·175 | 1·7819 | 8·0 | 2·8284 | 33 | 5·7446 |
| 2·59 | 1·6093 | 3·2 | 1·7889 | 8·5 | 2·9155 | 34 | 5·8310 |
| 2·6 | 1·6125 | 3·225 | 1·7958 | 9·0 | 3·0000 | 35 | 5·9161 |
| 2·61 | 1·6155 | 3·25 | 1·8028 | 9·5 | 3·0822 | 36 | 6·0000 |
| 2·625 | 1·6202 | 3·275 | 1·8097 | 10·0 | 3·1623 | 37 | 6·0828 |
| 2·64 | 1·6248 | 3·3 | 1·8166 | 10·5 | 2·2404 | 38 | 6·1644 |
| 2·65 | 1·6279 | 3·325 | 1·8235 | 11·0 | 3·3166 | 39 | 6·2450 |
| 2·66 | 1·6310 | 3·35 | 1·8303 | 11·5 | 3·3912 | 40 | 6·3246 |
| 2·675 | 1·6355 | 3·375 | 1·8371 | 12·0 | 3·4641 | 41 | 6·4031 |
| 2·69 | 1·6401 | 3·4 | 1·8439 | 12·5 | 3·5355 | 42 | 6·4807 |
| 2·7 | 1·6432 | 3·425 | 1·8507 | 13·0 | 3·6056 | 43 | 6·5574 |
| 2·71 | 1·6462 | 3·45 | 1·8574 | 13·5 | 3·6742 | 44 | 6·6332 |
| 2·725 | 1·6508 | 3·475 | 1·8641 | 14·0 | 3·7417 | 45 | 6·7082 |
| 2·74 | 1·6553 | 3·5 | 1·8708 | 14·5 | 3·8079 | 46 | 6·7823 |
| 2·75 | 1·6583 | 3·525 | 1·8775 | 15·0 | 3·8730 | 47 | 6·8557 |
| 2·76 | 1·6613 | 3·55 | 1·8841 | 15·5 | 3·9370 | 48 | 6·9282 |
| 2·775 | 1·6658 | 3·575 | 1·8908 | 16·0 | 4·0000 | 49 | 7·0000 |
| 2·79 | 1·6703 | 3·6 | 1·8974 | 16·5 | 4·0620 | 50 | 7·0711 |
| 2·8 | 1·6733 | 3·625 | 1·9039 | 17·0 | 4·1231 | 51 | 7·1414 |
| 2·81 | 1·6763 | 3·65 | 1·9105 | 17·5 | 4·1833 | 52 | 7·2111 |
| 2·825 | 1·6808 | 3·675 | 1·9170 | 18·0 | 4·2426 | 53 | 7·2810 |
| 2·84 | 1·6852 | 3·7 | 1·9235 | 18·5 | 4·3012 | 54 | 7·3485 |
| 2·85 | 1·6882 | 3·725 | 1·9300 | 19·0 | 4·3589 | 55 | 7·4162 |
| 2·86 | 1·6912 | 3·75 | 1·9365 | 19·5 | 4·4159 | 56 | 7·4833 |
| 2·875 | 1·6956 | 3·775 | 1·9429 | 20·0 | 4·4721 | 57 | 7·5498 |
| 2·89 | 1·7000 | 3·8 | 1·9494 | 2·05 | 4·5277 | 58 | 7·6158 |
| 2·9 | 1·7029 | 3·825 | 1·9558 | 21·0 | 4·5826 | 59 | 7·6811 |
| 2·91 | 1·7059 | 3·85 | 1·9621 | 21·5 | 4·6368 | 60 | 7·7460 |
| 2·925 | 1·7103 | 3·875 | 1·9685 | 22·0 | 4·6904 | 61 | 7·8102 |
| 2·94 | 1·7146 | 3·9 | 1·9748 | 22·5 | 4·7434 | 62 | 7·8740 |
| 2·95 | 1·7176 | 3·925 | 1·9812 | 23·0 | 4·7958 | 63 | 7·9373 |
| 2·96 | 1·7205 | 3·95 | 1·9875 | 23·5 | 4·8477 | 64 | 8·0000 |
| 2·975 | 1·7248 | 3·975 | 1·9938 | 24·0 | 4·8990 | 65 | 8·0623 |
| 2·99 | 1·7292 | 4·0 | 2·0000 | 25·0 | 5·0000 | 66 | 8·.240 |

TABLE IV.—FOR FINDING THE DISCHARGE THROUGH RECTANGULAR ORIFICE IN WHICH $n = \dfrac{h}{d}$. ALSO FOR FINDING THE EFFECTS OF THE VELOCI OF APPROACH TO WEIRS, AND THE DEPRESSION ON THE CREST. (See p. 5

| $1+n$ | $n^{\frac{3}{2}}$ | $(1+n)^{\frac{3}{2}}$ | $(1+n)^{\frac{3}{2}}-n^{\frac{3}{2}}$ | $1+n$ | $n^{\frac{3}{2}}$ | $(1+n)^{\frac{3}{2}}$ | $(1+n)^{\frac{3}{2}}-n^{\frac{3}{2}}$ |
|---|---|---|---|---|---|---|---|
| 1·000 | ·0000 | 1·0000 | 1·0000 | 1·115 | ·0390 | 1·1774 | 1·1384 |
| 1·001 | ·0000 | 1·0015 | 1·0015 | 1·120 | ·0416 | 1·1853 | 1·1437 |
| 1·002 | ·0001 | 1·0030 | 1·0029 | 1·125 | ·0442 | 1·1932 | 1·1491 |
| 1·004 | ·0003 | 1·0060 | 1·0058 | 1·13 | ·0469 | 1·2012 | 1·1543 |
| 1·005 | ·0004 | 1·0075 | 1·0072 | 1·135 | ·0496 | 1·2092 | 1·1596 |
| 1·006 | ·0005 | 1·0090 | 1·0086 | 1·14 | ·0524 | 1·2172 | 1·1648 |
| 1·008 | ·0007 | 1·0120 | 1·0113 | 1·145 | ·0552 | 1·2251 | 1·1700 |
| 1·009 | ·0009 | 1·0135 | 1·0127 | 1·15 | ·0581 | 1·2332 | 1·1751 |
| 1·010 | ·0010 | 1·0150 | 1·0140 | 1·155 | ·0610 | 1·2413 | 1·1803 |
| 1·011 | ·0012 | 1·0165 | 1·0154 | 1·16 | ·0640 | 1·2494 | 1·1854 |
| 1·012 | ·0013 | 1·0181 | 1·0167 | 1·165 | ·0670 | 1·2574 | 1·1904 |
| 1·014 | ·0017 | 1·0211 | 1·0194 | 1·17 | ·0701 | 1·2655 | 1·1955 |
| 1·015 | ·0018 | 1·0226 | 1·0207 | 1·175 | ·0732 | 1·2737 | 1·2005 |
| 1·016 | ·0020 | 1·0241 | 1·0221 | 1·18 | ·0764 | 1·2818 | 1·2054 |
| 1·018 | ·0024 | 1·0271 | 1·0247 | 1·185 | ·0796 | 1·2900 | 1·2104 |
| 1·019 | ·0026 | 1·0286 | 1·0260 | 1·19 | ·0828 | 1·2981 | 1·2153 |
| 1·020 | ·0028 | 1·0301 | 1·0273 | 1·195 | ·0861 | 1·3063 | 1·2202 |
| 1·0225 | ·0034 | 1·0339 | 1·0306 | 1·2 | ·0894 | 1·3145 | 1·2251 |
| 1·025 | ·0040 | 1·0377 | 1·0338 | 1·21 | ·0962 | 1·3310 | 1·2348 |
| 1·0275 | ·0046 | 1·0415 | 1·0370 | 1·22 | ·1032 | 1·3475 | 1·2443 |
| 1·03 | ·0052 | 1·0453 | 1·0401 | 1·23 | ·1103 | 1·3641 | 1·2538 |
| 1·0325 | ·0059 | 1·0491 | 1·0433 | 1·24 | ·1176 | 1·3808 | 1·2632 |
| 1·035 | ·0065 | 1·0530 | 1·0464 | 1·25 | ·1250 | 1·3975 | 1·2725 |
| 1·0375 | ·0073 | 1·0568 | 1·0495 | 1·26 | ·1326 | 1·4143 | 1·2818 |
| 1·04 | ·0080 | 1·0606 | 1·0526 | 1·27 | ·1403 | 1·4312 | 1·2909 |
| 1·0425 | ·0088 | 1·0644 | 1·0557 | 1·28 | ·1482 | 1·4482 | 1·3000 |
| 1·045 | ·0095 | 1·0683 | 1·0587 | 1·29 | ·1562 | 1·4652 | 1·3090 |
| 1·0475 | ·0104 | 1·0721 | 1·0617 | 1·30 | ·1643 | 1·4822 | 1·3179 |
| 1·05 | ·0112 | 1·0759 | 1·0648 | 1·31 | ·1726 | 1·4994 | 1·3268 |
| 1·055 | ·0129 | 1·0836 | 1·0707 | 1·325 | ·1853 | 1·5252 | 1·3399 |
| 1·06 | ·0147 | 1·0913 | 1·0766 | 1·34 | ·1983 | 1·5512 | 1·3529 |
| 1·065 | ·0166 | 1·0991 | 1·0825 | 1·35 | ·2071 | 1·5686 | 1·3615 |
| 1·07 | ·0185 | 1·1068 | 1·0883 | 1·36 | ·2160 | 1·5860 | 1·3700 |
| 1·075 | ·0205 | 1·1146 | 1·0940 | 1·375 | ·2296 | 1·6123 | 1·3827 |
| 1·08 | ·0226 | 1·1224 | 1·0997 | 1·39 | ·2436 | 1·6388 | 1·3952 |
| 1·085 | ·0248 | 1·1302 | 1·1054 | 1·40 | ·2530 | 1·6565 | 1·4035 |
| 1·09 | ·0270 | 1·1380 | 1·1110 | 1·41 | ·2625 | 1·6743 | 1·4118 |
| 1·095 | ·0293 | 1·1458 | 1·1166 | 1·425 | ·2771 | 1·7011 | 1·4240 |
| 1·1 | ·0316 | 1·1537 | 1·1221 | 1·44 | ·2919 | 1·7280 | 1·4361 |
| 1·105 | ·0340 | 1·1616 | 1·1275 | 1·45 | ·3019 | 1·7460 | 1·4442 |
| 1·110 | ·0365 | 1·1695 | 1·1330 | 1·46 | ·3120 | 1·7641 | 1·4521 |

Values of $n$ from 0 to ·46.                [Continued on next pa

TABLE IV.—FOR FINDING THE DISCHARGE THROUGH RECTANGULAR ORIFICES; IN WHICH $n = \dfrac{h}{d}$. ALSO FOR FINDING THE EFFECTS OF THE VELOCITY OF APPROACH TO WEIRS, &c. (See p. 55.)

| $1+n$ | $n^{\frac{3}{2}}$ | $(1+n)^{\frac{3}{2}}$ | $(1+n)^{\frac{3}{2}} - n^{\frac{3}{2}}$ | $1+n$ | $n^{\frac{3}{2}}$ | $(1+n)^{\frac{3}{2}}$ | $(1+n)^{\frac{3}{2}} - n^{\frac{3}{2}}$ |
|---|---|---|---|---|---|---|---|
| 1·475 | ·3274 | 1·7914 | 1·4640 | 1·975 | ·9627 | 2·7756 | 1·8128 |
| 1·49  | ·3430 | 1·8188 | 1·4758 | 1·99  | ·9850 | 2·8072 | 1·8222 |
| 1·5   | ·3536 | 1·8371 | 1·4836 | 2·    | 1·0000 | 2·8284 | 1·8284 |
| 1·51  | ·3642 | 1·8555 | 1·4913 | 2·01  | 1·0150 | 2·8497 | 1·8346 |
| 1·525 | ·3804 | 1·8832 | 1·5028 | 2·025 | 1·0377 | 2·8816 | 1·8439 |
| 1·54  | ·3968 | 1·9111 | 1·5143 | 2·04  | 1·0606 | 2·9137 | 1·8531 |
| 1·55  | ·4079 | 1·9297 | 1·5218 | 2·05  | 1·0759 | 2·9352 | 1·8592 |
| 1·56  | ·4191 | 1·9484 | 1·5294 | 2·06  | 1·0913 | 2·9567 | 1·8653 |
| 1·575 | ·4360 | 1·9766 | 1·5406 | 2·075 | 1·1146 | 2·9890 | 1·8744 |
| 1·58  | ·4417 | 1·9860 | 1·5443 | 2·09  | 1·1380 | 3·0215 | 1·8835 |
| 1·59  | ·4532 | 2·0049 | 1·5517 | 2·10  | 1·1537 | 3·0432 | 1·8895 |
| 1·6   | ·4648 | 2·0239 | 1·5591 | 2·11  | 1·1695 | 3·0650 | 1·8955 |
| 1·61  | ·4764 | 2·0429 | 1·5664 | 2·125 | 1·1932 | 3·0977 | 1·9045 |
| 1·625 | ·4941 | 2·0715 | 1·5774 | 2·14  | 1·2172 | 3·1306 | 1·9134 |
| 1·64  | ·5120 | 2·1002 | 1·5882 | 2·15  | 1·2332 | 3·1525 | 1·9193 |
| 1·65  | ·5240 | 2·1195 | 1·5954 | 2·16  | 1·2494 | 3·1745 | 1·9252 |
| 1·66  | ·5362 | 2·1388 | 1·6026 | 2·175 | 1·2737 | 3·2077 | 1·9340 |
| 1·675 | ·5546 | 2·1678 | 1·6132 | 2·19  | 1·2981 | 3·2409 | 1·9428 |
| 1·69  | ·5732 | 2·1970 | 1·6238 | 2·2   | 1·3145 | 3·2631 | 1·9486 |
| 1·7   | ·5857 | 2·2165 | 1·6309 | 2·21  | 1·3310 | 3·2854 | 1·9544 |
| 1·71  | ·5983 | 2·2361 | 1·6379 | 2·225 | 1·3558 | 3·3189 | 1·9631 |
| 1·725 | ·6173 | 2·2656 | 1·6483 | 2·24  | 1·3808 | 3·3525 | 1·9717 |
| 1·74  | ·6366 | 2·2952 | 1·6586 | 2·25  | 1·3975 | 3·3750 | 1·9775 |
| 1·75  | ·6495 | 2·3150 | 1·6655 | 2·26  | 1·4143 | 3·3975 | 1·9832 |
| 1·76  | ·6626 | 2·3349 | 1·6724 | 2·275 | 1·4397 | 3·4314 | 1·9917 |
| 1·775 | ·6823 | 2·3648 | 1·6826 | 2·29  | 1·4652 | 3·4654 | 2·0002 |
| 1·79  | ·7022 | 2·3949 | 1·6927 | 2·3   | 1·4822 | 3·4881 | 2·0059 |
| 1·80  | ·7155 | 2·4150 | 1·6994 | 2·31  | 1·4994 | 3·5109 | 2·0115 |
| 1·81  | ·7290 | 2·4351 | 1·7061 | 2·325 | 1·5252 | 3·5451 | 2·0200 |
| 1·825 | ·7493 | 2·4654 | 1·7161 | 2·34  | 1·5512 | 3·5795 | 2·0284 |
| 1·84  | ·7699 | 2·4959 | 1·7260 | 2·35  | 1·5686 | 3·6025 | 2·0339 |
| 1·85  | ·7837 | 2·5163 | 1·7326 | 2·36  | 1·5860 | 3·6255 | 2·0395 |
| 1·86  | ·7975 | 2·5367 | 1·7392 | 2·375 | 1·6123 | 3·6601 | 2·0478 |
| 1·875 | ·8185 | 2·5674 | 1·7490 | 2·39  | 1·6388 | 3·6948 | 2·0561 |
| 1·89  | ·8396 | 2·5983 | 1·7587 | 2·4   | 1·6565 | 3·7181 | 2·0616 |
| 1·9   | ·8538 | 2·6190 | 1·7652 | 2·41  | 1·6743 | 3·7413 | 2·0670 |
| 1·91  | ·8681 | 2·6397 | 1·7716 | 2·425 | 1·7011 | 3·7763 | 2·0752 |
| 1·925 | ·8896 | 2·6709 | 1·7813 | 2·44  | 1·7280 | 3·8114 | 2·0834 |
| 1·94  | ·9114 | 2·7021 | 1·7907 | 2·45  | 1·7460 | 3·8349 | 2·0888 |
| 1·95  | ·9259 | 2·7230 | 1·7971 | 2·46  | 1·7641 | 3·8584 | 2·0942 |
| 1·96  | ·9406 | 2·7410 | 1·8034 | 2·475 | 1·7914 | 3·8937 | 2·1023 |

Values of $n$ from ·475 to 1·475.  [*Continued on next page.*]

TABLE IV.—FOR FINDING THE DISCHARGE THROUGH RECTANGULAR ORIFICES; IN WHICH $n = \dfrac{h}{d}$. ALSO FOR FINDING THE EFFECTS OF THE VELOCITY OF APPROACH TO WEIRS, &c. (See p. 55.)

| $1+n$. | $n^{\frac{3}{2}}$. | $(1+n)^{\frac{3}{2}}$. | $(1+n)^{\frac{3}{2}}-n^{\frac{3}{2}}$. | $1+n$. | $n^{\frac{3}{2}}$. | $(1+n)^{\frac{3}{2}}$. | $(1+n)^{\frac{3}{2}}-n^{\frac{3}{2}}$. |
|---|---|---|---|---|---|---|---|
| 2·49 | 1·8188 | 3·9292 | 2·1104 | 3· | 2·8284 | 5·1962 | 2·3677 |
| 2·5 | 1·8371 | 3·9528 | 2·1157 | 3·025 | 2·8816 | 5·2612 | 2·3796 |
| 2·51 | 1·8555 | 3·9766 | 2·1211 | 3·05 | 2·9352 | 5·3266 | 2·3914 |
| 2·525 | 1·8832 | 4·0123 | 2·1291 | 3·075 | 2·9890 | 5·3922 | 2·4032 |
| 2·54 | 1·9111 | 4·0481 | 2·1370 | 3·1 | 3·0432 | 5·4581 | 2·4149 |
| 2·55 | 1·9297 | 4·0720 | 2·1423 | 3·125 | 3·0977 | 5·5243 | 2·4266 |
| 2·56 | 1·9484 | 4·0960 | 2·1476 | 3·15 | 3·1525 | 5·5907 | 2·4382 |
| 2·575 | 1·9766 | 4·1321 | 2·1554 | 3·175 | 3·2077 | 5·6574 | 2·4497 |
| 2·59 | 2·0049 | 4·1682 | 2·1633 | 3·2 | 3·2631 | 5·7243 | 2·4612 |
| 2·6 | 2·0239 | 4·1924 | 2·1685 | 3·225 | 3·3189 | 5·7915 | 2·4726 |
| 2·61 | 2·0429 | 4·2166 | 2·1737 | 3·25 | 3·3750 | 5·8590 | 2·4840 |
| 2·625 | 2·0715 | 4·2530 | 2·1815 | 3·275 | 3·4314 | 5·9268 | 2·4953 |
| 2·64 | 2·1002 | 4·2895 | 2·1893 | 3·3 | 3·4881 | 5·9947 | 2·5066 |
| 2·65 | 2·1195 | 4·3139 | 2·1944 | 3·325 | 3·5451 | 6·0630 | 2·5179 |
| 2·66 | 2·1388 | 4·3383 | 2·1996 | 3·35 | 3·6025 | 6·1315 | 2·5290 |
| 2·675 | 2·1678 | 4·3751 | 2·2073 | 3·375 | 3·6601 | 6·2003 | 2·5401 |
| 2·69 | 2·1970 | 4·4119 | 2·2149 | 3·4 | 3·7181 | 6·2693 | 2·5512 |
| 2·7 | 2·2165 | 4·4366 | 2·2200 | 3·425 | 3·7763 | 6·3386 | 2·5623 |
| 2·71 | 2·2361 | 4·4612 | 2·2251 | 3·45 | 3·8349 | 6·4081 | 2·5732 |
| 2·725 | 2·2656 | 4·4983 | 2·2327 | 3·475 | 3·8937 | 6·4779 | 2·5842 |
| 2·74 | 2·2952 | 4·5355 | 2·2403 | 3·5 | 3·9528 | 6·5479 | 2·5951 |
| 2·75 | 2·3150 | 4·5604 | 2·2453 | 3·525 | 4·0123 | 6·6182 | 2·6059 |
| 2·76 | 2·3349 | 4·5853 | 2·2504 | 3·55 | 4·0720 | 6·6887 | 2·6167 |
| 2·775 | 2·3648 | 4·6227 | 2·2579 | 3·575 | 4·1321 | 6·7595 | 2·6274 |
| 2·79 | 2·3949 | 4·6602 | 2·2654 | 3·6 | 4·1924 | 6·8305 | 2·6381 |
| 2·8 | 2·4150 | 4·6853 | 2·2703 | 3·625 | 4·2530 | 6·9018 | 2·6488 |
| 2·81 | 2·4351 | 4·7104 | 2·2753 | 3·65 | 4·3139 | 6·9733 | 2·6594 |
| 2·825 | 2·4654 | 4·7482 | 2·2827 | 3·675 | 4·3751 | 7·0451 | 2·6700 |
| 2·84 | 2·4959 | 4·7861 | 2·2902 | 3·7 | 4·4366 | 7·1171 | 2·6805 |
| 2·85 | 2·5163 | 4·8114 | 2·2951 | 3·725 | 4·4983 | 7·1893 | 2·6910 |
| 2·86 | 2·5367 | 4·8367 | 2·3000 | 3·75 | 4·5604 | 7·2618 | 2·7015 |
| 2·875 | 2·5674 | 4·8748 | 2·3074 | 3·775 | 4·6227 | 7·3346 | 2·7119 |
| 2·89 | 2·5983 | 4·9130 | 2·3147 | 3·8 | 4·6853 | 7·4076 | 2·7223 |
| 2·9 | 2·6190 | 4·9385 | 2·3196 | 3·825 | 4·7482 | 7·4808 | 2·7326 |
| 2·91 | 2·6397 | 4·9641 | 2·3244 | 3·85 | 4·8114 | 7·5542 | 2·7429 |
| 2·925 | 2·6708 | 5·0025 | 2·3317 | 3·875 | 4·8748 | 7·6279 | 2·7531 |
| 2·94 | 2·7021 | 5·0411 | 2·3389 | 3·9 | 4·9385 | 7·7019 | 2·7634 |
| 2·95 | 2·7230 | 5·0668 | 2·3438 | 3·925 | 5·0025 | 7·7761 | 2·7735 |
| 2·96 | 2·7440 | 5·0926 | 2·3486 | 3·95 | 5·0668 | 7·8505 | 2·7837 |
| 2·975 | 2·7756 | 5·1313 | 2·3558 | 3·975 | 5·1313 | 7·9251 | 2·7938 |
| 2·99 | 2·8072 | 5·1702 | 2·3630 | 4· | 5·1962 | 8· | 2·8038 |

Values of $n$ from 1·49 to 3.     [*Continued on next page*

ORIFICES, WEIRS, PIPES, AND RIVERS.

BLE IV.—FOR FINDING THE DISCHARGE THROUGH RECTANGULAR ORIFICES; IN WHICH $n = \dfrac{h}{d}$. ALSO FOR FINDING THE EFFECTS OF THE VELOCITY OF APPROACH TO WEIRS, &c. (See p. 55.)

| $+n$. | $n^{\frac{3}{2}}$. | $(1+n)^{\frac{3}{2}}$. | $(1+n)^{\frac{3}{2}} - n^{\frac{3}{2}}$. | $1+n$. | $n^{\frac{3}{2}}$. | $(1+n)^{\frac{3}{2}}$. | $(1+n)^{\frac{3}{2}} - n^{\frac{3}{2}}$. |
|---|---|---|---|---|---|---|---|
| 4·5 | 6·5479 | 9·5459 | 2·9980 | 26· | 125·0000 | 132·5745 | 7·5745 |
| 5·0 | 8·0000 | 11·1803 | 3·1803 | 27· | 132·5745 | 140·2961 | 7·7216 |
| 5·5 | 9·5459 | 12·8986 | 3·3527 | 28· | 140·2961 | 148·1621 | 7·8660 |
| 6·0 | 11·1803 | 14·6969 | 3·5166 | 29· | 148·1621 | 156·1698 | 8·0077 |
| 6·5 | 12·8986 | 16·5718 | 3·6732 | 30· | 156·1698 | 164·3168 | 8·1470 |
| 7·0 | 14·6969 | 18·5203 | 3·8234 | 31· | 164·3168 | 172·6007 | 8·2839 |
| 7·5 | 16·5718 | 20·5396 | 3·9678 | 32· | 172·6007 | 181·0193 | 8·4186 |
| 8·0 | 18·5203 | 22·6274 | 4·1071 | 33· | 181·0193 | 189·5706 | 8·5513 |
| 8·5 | 20·5396 | 24·7815 | 4·2419 | 34· | 189·5706 | 198·2524 | 8·6818 |
| 9·0 | 22·6274 | 27·0000 | 4·3726 | 35· | 198·2524 | 207·0628 | 8·8104 |
| 9·5 | 24·7815 | 29·2810 | 4·4995 | 36· | 207·0628 | 216·0000 | 8·9372 |
| 10·0 | 27·0000 | 31·6228 | 4·6228 | 37· | 216·0000 | 225·0622 | 9·0622 |
| 10·5 | 29·2810 | 34·0239 | 4·7429 | 38· | 225·0622 | 234·2477 | 9·1855 |
| 11·0 | 31·6228 | 36·4829 | 4·8601 | 39· | 234·2477 | 243·5549 | 9·3072 |
| 11·5 | 34·0239 | 38·9984 | 4·9745 | 40· | 243·5549 | 252·9822 | 9·4273 |
| 12·0 | 36·4829 | 41·5692 | 5·0863 | 41· | 252·9822 | 262·5281 | 9·5459 |
| 12·5 | 38·9984 | 44·1942 | 5·1958 | 42· | 262·5281 | 272·1911 | 9·6630 |
| 13·0 | 41·5692 | 46·8722 | 5·3030 | 43· | 272·1911 | 281·9699 | 9·7788 |
| 13·5 | 44·1942 | 49·6022 | 5·4080 | 44· | 281·9699 | 291·8630 | 9·8931 |
| 14·0 | 46·8722 | 52·3832 | 5·5110 | 45· | 291·8630 | 301·8692 | 10·0062 |
| 14·5 | 49·6022 | 55·2144 | 5·6122 | 46· | 301·8692 | 311·9872 | 10·1180 |
| 15·0 | 52·3832 | 58·0947 | 5·7115 | 47· | 311·9872 | 322·2158 | 10·2286 |
| 15·5 | 55·2144 | 61·0236 | 5·8092 | 48· | 322·2158 | 332·5538 | 10·3380 |
| 16·0 | 58·0947 | 64· | 5·9053 | 49· | 332·5538 | 343·0000 | 10·4462 |
| 16·5 | 61·0236 | 67·0247 | 6·0011 | 50· | 343·0000 | 353·5534 | 10·5534 |
| 17·0 | 64· | 70·0928 | 6·0928 | 51· | 353·5534 | 364·2128 | 10·6594 |
| 17·5 | 67·0247 | 73·2078 | 6·1831 | 52· | 364·2128 | 374·9773 | 10·7645 |
| 18·0 | 70·0928 | 76·3675 | 6·2747 | 53· | 374·9773 | 385·8458 | 10·8685 |
| 18·5 | 73·2078 | 79·5715 | 6·3637 | 54· | 385·8458 | 396·8173 | 10·9715 |
| 19·0 | 76·3675 | 82·8191 | 6·4516 | 55· | 396·8173 | 407·8909 | 11·0736 |
| 19·5 | 79·5715 | 86·1097 | 6·5382 | 56· | 407·8909 | 419·0656 | 11·1747 |
| 20·0 | 82·8191 | 89·4427 | 6·6236 | 57· | 419·0656 | 430·3406 | 11·2750 |
| 20·5 | 86·1097 | 92·8177 | 6·7080 | 58· | 430·3406 | 441·7148 | 11·3742 |
| 21·0 | 89·4427 | 96·2341 | 6·7914 | 59· | 441·7148 | 453·1876 | 11·4728 |
| 21·5 | 92·8177 | 99·6914 | 6·8737 | 60· | 453·1876 | 464·7580 | 11·5704 |
| 22·0 | 96·2341 | 103·1892 | 6·9551 | 61· | 464·7580 | 476·4252 | 11·6672 |
| 22·5 | 99·6914 | 106·7269 | 7·0355 | 62· | 476·4252 | 488·1885 | 11·7633 |
| 23· | 103·1892 | 110·3041 | 7·1149 | 63· | 488·1885 | 500·0470 | 11·8585 |
| 23·5 | 106·7269 | 113·9205 | 7·1936 | 64· | 500·0470 | 512·0000 | 11·9530 |
| 24· | 110·3041 | 117·5755 | 7·2714 | 65· | 512·0000 | 524·0468 | 12·0468 |
| 25· | 117·5755 | 125· | 7·4245 | 66· | 524·0468 | 536·1865 | 12·1397 |

Values of $n$ from 3·5 to 66.

TABLE V.—COEFFICIENTS OF DISCHARGE FOR DIFFERENT RATIOS OF THE CHANNEL TO THE ORIFICE.

*Coefficients for heads in still water* ·550 *and* ·573.

| Ratio of the channel to the orifice. | Coefficient ·550 for heads in still water. | | | Coefficient ·573 for heads in still water. | | |
|---|---|---|---|---|---|---|
| | Ratio of the height due to the velocity of approach to the head. | Coefficients for orifices: the heads measured to the centres. | Coefficients for weirs: the heads measured the full depth. | Ratio of the height due to the velocity of approach to the head. | Coefficients for orifices: the heads measured to the centres. | Coefficients for weirs: the heads measured the full depth. |
| 30· | ·000 | ·550 | ·550 | ·000 | ·573 | ·573 |
| 20· | ·001 | ·550 | ·551 | ·001 | ·573 | ·574 |
| 15· | ·001 | ·550 | ·551 | ·001 | ·573 | ·574 |
| 10· | ·003 | ·551 | ·552 | ·003 | ·574 | ·576 |
| 9· | ·004 | ·551 | ·553 | ·004 | ·574 | ·576 |
| 8· | ·005 | ·551 | ·554 | ·005 | ·574 | ·577 |
| 7· | ·006 | ·552 | ·555 | ·007 | ·575 | ·578 |
| 6· | ·008 | ·552 | ·557 | ·009 | ·576 | ·580 |
| 5·5 | ·010 | ·553 | ·558 | ·011 | ·576 | ·582 |
| 5·0 | ·012 | ·553 | ·559 | ·013 | ·577 | ·584 |
| 4·5 | ·015 | ·554 | ·562 | ·016 | ·578 | ·586 |
| 4·0 | ·019 | ·555 | ·565 | ·021 | ·579 | ·589 |
| 3·75 | ·022 | ·556 | ·566 | ·024 | ·580 | ·592 |
| 3·50 | ·025 | ·557 | ·569 | ·028 | ·581 | ·594 |
| 3·25 | ·029 | ·558 | ·572 | ·032 | ·582 | ·598 |
| 3·0 | ·035 | ·559 | ·575 | ·038 | ·584 | ·602 |
| 2·75 | ·042 | ·561 | ·580 | ·045 | ·586 | ·607 |
| 2·50 | ·051 | ·564 | ·586 | ·055 | ·589 | ·614 |
| 2·25 | ·064 | ·567 | ·594 | ·069 | ·593 | ·623 |
| 2·00 | ·082 | ·572 | ·606 | ·089 | ·598 | ·636 |
| 1·95 | ·086 | ·573 | ·609 | ·094 | ·599 | ·639 |
| 1·90 | ·091 | ·575 | ·612 | ·100 | ·601 | ·643 |
| 1·85 | ·097 | ·576 | ·615 | ·106 | ·603 | ·647 |
| 1·80 | ·103 | ·578 | ·619 | ·113 | ·604 | ·651 |
| 1·75 | ·110 | ·579 | ·623 | ·120 | ·606 | ·655 |
| 1·70 | ·117 | ·581 | ·627 | ·128 | ·609 | ·660 |
| 1·65 | ·125 | ·583 | ·632 | ·137 | ·611 | ·666 |
| 1·60 | ·134 | ·586 | ·637 | ·147 | ·614 | ·671 |
| 1·55 | ·144 | ·588 | ·643 | ·158 | ·617 | ·678 |
| 1·50 | ·155 | ·591 | ·649 | ·171 | ·620 | ·685 |
| 1·45 | ·168 | ·594 | ·656 | ·185 | ·624 | ·694 |
| 1·40 | ·183 | ·598 | ·664 | ·201 | ·628 | ·703 |
| 1·35 | ·199 | ·602 | ·673 | ·220 | ·633 | ·713 |
| 1·30 | ·218 | ·607 | ·683 | ·241 | ·638 | ·724 |
| 1·25 | ·240 | ·612 | ·695 | ·266 | ·645 | ·737 |
| 1·20 | ·265 | ·619 | ·707 | ·295 | ·652 | ·753 |
| 1·15 | ·297 | ·626 | ·723 | ·330 | ·661 | ·770 |
| 1·10 | ·333 | ·635 | ·741 | ·372 | ·671 | ·791 |
| 1·05 | ·378 | ·646 | ·762 | ·424 | ·684 | ·816 |
| 1·00 | ·434 | ·659 | ·787 | ·489 | ·699 | ·845 |

See the auxiliary table, p. 68.

ORIFICES, WEIRS, PIPES, AND RIVERS.           193

TABLE V.—COEFFICIENTS OF DISCHARGE FOR DIFFERENT RATIOS OF THE CHANNEL TO THE ORIFICE.

*Coefficients for heads in still water ·584 and ·595.*

| Ratio of the channel to the orifice. | Coefficient ·584 for heads in still water. | | | Coefficient ·595 for heads in still water. | | |
|---|---|---|---|---|---|---|
| | Ratio of the height due to the velocity of approach to the head. | Coefficients for orifices: the heads measured to the centres. | Coefficients for weirs: the heads measured the full depth. | Ratio of the height due to the velocity of approach to the head. | Coefficients for orifices: the heads measured to the centres. | Coefficients for weirs: the heads measured the full depth. |
| 30· | ·000 | ·584 | ·584 | ·000 | ·595 | ·595 |
| 20· | ·001 | ·584 | ·585 | ·001 | ·595 | ·596 |
| 15· | ·002 | ·584 | ·585 | ·002 | ·595 | ·596 |
| 10· | ·003 | ·585 | ·587 | ·004 | ·596 | ·598 |
| 9·0 | ·004 | ·585 | ·588 | ·004 | ·596 | ·599 |
| 8·0 | ·005 | ·586 | ·588 | ·006 | ·597 | ·600 |
| 7·0 | ·007 | ·586 | ·590 | ·007 | ·597 | ·601 |
| 6·0 | ·010 | ·587 | ·592 | ·010 | ·598 | ·603 |
| 5·5 | ·011 | ·587 | ·593 | ·012 | ·599 | ·605 |
| 5·0 | ·014 | ·588 | ·595 | ·014 | ·599 | ·607 |
| 4·5 | ·017 | ·589 | ·598 | ·018 | ·600 | ·610 |
| 4·0 | ·022 | ·590 | ·601 | ·023 | ·602 | ·613 |
| 3·75 | ·025 | ·591 | ·604 | ·026 | ·603 | ·616 |
| 3·50 | ·029 | ·592 | ·606 | ·030 | ·604 | ·619 |
| 3·25 | ·033 | ·594 | ·610 | ·035 | ·605 | ·622 |
| 3·0 | ·039 | ·595 | ·614 | ·041 | ·607 | ·627 |
| 2·75 | ·047 | ·598 | ·620 | ·049 | ·609 | ·633 |
| 2·50 | ·058 | ·601 | ·627 | ·060 | ·613 | ·641 |
| 2·25 | ·072 | ·605 | ·637 | ·075 | ·617 | ·651 |
| 2·0 | ·093 | ·611 | ·651 | ·097 | ·623 | ·666 |
| 1·95 | ·099 | ·612 | ·654 | ·103 | ·625 | ·669 |
| 1·90 | ·104 | ·614 | ·660 | ·109 | ·627 | ·673 |
| 1·85 | ·111 | ·615 | ·662 | ·115 | ·628 | ·678 |
| 1·80 | ·118 | ·617 | ·666 | ·123 | ·630 | ·682 |
| 1·75 | ·125 | ·620 | ·671 | ·131 | ·633 | ·687 |
| 1·70 | ·134 | ·622 | ·676 | ·140 | ·635 | ·693 |
| 1·65 | ·143 | ·624 | ·682 | ·149 | ·638 | ·699 |
| 1·60 | ·154 | ·627 | ·689 | ·160 | ·641 | ·706 |
| 1·55 | ·166 | ·631 | ·696 | ·173 | ·644 | ·713 |
| 1·50 | ·179 | ·634 | ·703 | ·187 | ·648 | ·721 |
| 1·45 | ·194 | ·638 | ·712 | ·202 | ·652 | ·730 |
| 1·40 | ·211 | ·643 | ·722 | ·220 | ·657 | ·741 |
| 1·35 | ·230 | ·648 | ·732 | ·241 | ·663 | ·752 |
| 1·30 | ·253 | ·654 | ·745 | ·265 | ·669 | ·765 |
| 1·25 | ·279 | ·661 | ·759 | ·293 | ·677 | ·780 |
| 1·20 | ·310 | ·669 | ·775 | ·325 | ·685 | ·797 |
| 1·15 | ·348 | ·678 | ·794 | ·366 | ·695 | ·818 |
| 1·10 | ·393 | ·689 | ·816 | ·414 | ·707 | ·842 |
| 1·05 | ·448 | ·703 | ·842 | ·473 | ·722 | ·870 |
| 1·00 | ·518 | ·719 | ·874 | ·548 | ·740 | ·905 |

See the auxiliary table, p. 68.

O

TABLE V.—COEFFICIENTS OF DISCHARGE FOR DIFFERENT RATIOS
OF THE CHANNEL TO THE ORIFICE.

Coefficients for heads in still water ·606 and ·617.

| Ratio of the channel to the orifice. | Coefficient ·606 for heads in still water. | | | Coefficient ·617 for heads in still water. | | |
|---|---|---|---|---|---|---|
| | Ratio of the height due to the velocity of approach to the head. | Coefficients for orifices: the heads measured to the centres. | Coefficients for weirs: the heads measured the full depth. | Ratio of the height due to the velocity of approach to the head. | Coefficients for orifices: the heads measured to the centres. | Coefficients for weirs: the heads measured the full depth. |
| 30· | ·000 | ·606 | ·606 | ·000 | ·617 | ·617 |
| 20· | ·001 | ·606 | ·607 | ·001 | ·617 | ·618 |
| 15· | ·002 | ·607 | ·607 | ·002 | ·618 | ·619 |
| 10· | ·004 | ·607 | ·609 | ·004 | ·618 | ·620 |
| 9·0 | ·005 | ·607 | ·610 | ·005 | ·618 | ·621 |
| 8·0 | ·006 | ·608 | ·611 | ·006 | ·619 | ·622 |
| 7·0 | ·008 | ·608 | ·612 | ·008 | ·619 | ·624 |
| 6·0 | ·010 | ·609 | ·615 | ·011 | ·620 | ·626 |
| 5·5 | ·012 | ·610 | ·616 | ·013 | ·621 | ·628 |
| 5·0 | ·015 | ·611 | ·619 | ·015 | ·622 | ·630 |
| 4·5 | ·018 | ·612 | ·621 | ·019 | ·623 | ·633 |
| 4·0 | ·023 | ·613 | ·625 | ·024 | ·624 | ·637 |
| 3·75 | ·027 | ·614 | ·628 | ·028 | ·626 | ·640 |
| 3·50 | ·031 | ·615 | ·631 | ·032 | ·627 | ·643 |
| 3·25 | ·036 | ·617 | ·635 | ·037 | ·628 | ·647 |
| 3·00 | ·043 | ·619 | ·640 | ·044 | ·630 | ·653 |
| 2·75 | ·051 | ·621 | ·646 | ·053 | ·633 | ·660 |
| 2·50 | ·062 | ·625 | ·654 | ·065 | ·637 | ·668 |
| 2·25 | ·078 | ·629 | ·665 | ·081 | ·642 | ·679 |
| 2·00 | ·101 | ·636 | ·681 | ·105 | ·649 | ·696 |
| 1·95 | ·107 | ·638 | ·685 | ·111 | ·650 | ·700 |
| 1·90 | ·113 | ·639 | ·689 | ·118 | ·652 | ·704 |
| 1·85 | ·119 | ·641 | ·693 | ·125 | ·654 | ·709 |
| 1·80 | ·128 | ·644 | ·698 | ·133 | ·657 | ·714 |
| 1·75 | ·136 | ·646 | ·703 | ·142 | ·659 | ·720 |
| 1·70 | ·146 | ·649 | ·709 | ·152 | ·662 | ·726 |
| 1·65 | ·156 | ·652 | ·716 | ·163 | ·665 | ·733 |
| 1·60 | ·167 | ·655 | ·723 | ·175 | ·669 | ·741 |
| 1·55 | ·180 | ·658 | ·731 | ·188 | ·673 | ·749 |
| 1·50 | ·195 | ·662 | ·739 | ·204 | ·677 | ·759 |
| 1·45 | ·212 | ·667 | ·749 | ·221 | ·681 | ·768 |
| 1·40 | ·231 | ·672 | ·760 | ·241 | ·687 | ·780 |
| 1·35 | ·252 | ·678 | ·772 | ·264 | ·694 | ·793 |
| 1·30 | ·278 | ·685 | ·786 | ·291 | ·701 | ·808 |
| 1·25 | ·307 | ·693 | ·803 | ·322 | ·709 | ·825 |
| 1·20 | ·342 | ·702 | ·821 | ·359 | ·719 | ·845 |
| 1·15 | ·384 | ·713 | ·843 | ·404 | ·731 | ·868 |
| 1·10 | ·436 | ·726 | ·868 | ·459 | ·745 | ·895 |
| 1·05 | ·499 | ·742 | ·898 | ·527 | ·763 | ·928 |
| 1·00 | ·580 | ·762 | ·936 | ·615 | ·784 | ·969 |

See the auxiliary table, p. 68.

TABLE V.—COEFFICIENTS OF DISCHARGE FOR DIFFERENT RATIOS OF THE CHANNEL TO THE ORIFICE.

Mean Coefficient ·628.

Coefficients for heads in still water ·628 and ·639.

| Ratio of the channel to the orifice. | Coefficient ·628 for heads in still water. | | | Coefficient ·639 for heads in still water. | | |
|---|---|---|---|---|---|---|
| | Ratio of the height due to the velocity of approach to the head. | Coefficients for orifices: the heads measured to the centres. | Coefficients for weirs: the heads measured the full depth. | Ratio of the height due to the velocity of approach to the head. | Coefficients for orifices: the heads measured to the centres. | Coefficients for weirs: the heads measured the full depth. |
| 30· | ·000 | ·628 | ·628 | ·000 | ·639 | ·639 |
| 20· | ·001 | ·628 | ·629 | ·001 | ·639 | ·640 |
| 15· | ·002 | ·629 | ·630 | ·002 | ·640 | ·641 |
| 10· | ·004 | ·629 | ·632 | ·004 | ·640 | ·643 |
| 9·0 | ·005 | ·630 | ·632 | ·005 | ·641 | ·644 |
| 8·0 | ·006 | ·630 | ·634 | ·006 | ·641 | ·645 |
| 7·0 | ·008 | ·631 | ·635 | ·008 | ·642 | ·647 |
| 6·0 | ·011 | ·631 | ·638 | ·011 | ·643 | ·649 |
| 5·5 | ·013 | ·632 | ·640 | ·014 | ·643 | ·651 |
| 5·0 | ·016 | ·633 | ·642 | ·017 | ·644 | ·654 |
| 4·5 | ·020 | ·634 | ·645 | ·021 | ·646 | ·657 |
| 4·0 | ·025 | ·636 | ·649 | ·026 | ·647 | ·662 |
| 3·75 | ·029 | ·637 | ·652 | ·030 | ·648 | ·665 |
| 3·50 | ·033 | ·638 | ·656 | ·034 | ·650 | ·668 |
| 3·25 | ·039 | ·639 | ·659 | ·040 | ·652 | ·673 |
| 3·0 | ·046 | ·642 | ·666 | ·048 | ·654 | ·678 |
| 2·75 | ·055 | ·645 | ·672 | ·057 | ·657 | ·686 |
| 2·50 | ·067 | ·649 | ·682 | ·070 | ·661 | ·695 |
| 2·25 | ·084 | ·654 | ·694 | ·088 | ·666 | ·708 |
| 2·0 | ·109 | ·661 | ·711 | ·114 | ·674 | ·727 |
| 1·95 | ·116 | ·663 | ·715 | ·120 | ·676 | ·731 |
| 1·90 | ·123 | ·665 | ·720 | ·128 | ·679 | ·736 |
| 1·85 | ·130 | ·668 | ·725 | ·135 | ·681 | ·741 |
| 1·80 | ·139 | ·670 | ·731 | ·144 | ·684 | ·747 |
| 1·75 | ·148 | ·673 | ·737 | ·154 | ·686 | ·753 |
| 1·70 | ·158 | ·676 | ·743 | ·165 | ·690 | ·760 |
| 1·65 | ·169 | ·679 | ·750 | ·176 | ·693 | ·768 |
| 1·60 | ·182 | ·683 | ·758 | ·190 | ·697 | ·776 |
| 1·55 | ·196 | ·687 | ·767 | ·205 | ·701 | ·786 |
| 1·50 | ·213 | ·692 | ·777 | ·222 | ·706 | ·796 |
| 1·45 | ·231 | ·697 | ·788 | ·241 | ·712 | ·808 |
| 1·40 | ·252 | ·703 | ·800 | ·262 | ·718 | ·820 |
| 1·35 | ·276 | ·709 | ·814 | ·289 | ·725 | ·836 |
| 1·30 | ·304 | ·717 | ·830 | ·319 | ·734 | ·853 |
| 1·25 | ·338 | ·726 | ·846 | ·354 | ·743 | ·872 |
| 1·20 | ·377 | ·734 | ·866 | ·396 | ·755 | ·895 |
| 1·15 | ·425 | ·750 | ·894 | ·447 | ·769 | ·921 |
| 1·10 | ·484 | ·765 | ·924 | ·509 | ·785 | ·953 |
| 1·05 | ·557 | ·784 | ·959 | ·588 | ·805 | ·991 |
| 1·00 | ·651 | ·807 | 1·002 | ·690 | ·830 | 1·038 |

See the auxiliary table, p. 68.

TABLE V.—COEFFICIENTS OF DISCHARGE FOR DIFFERENT RATIOS OF THE CHANNEL TO THE ORIFICE.

Coefficients for heads in still water ·650 and ·667.

| Ratio of the channel to the orifice. | Coefficient ·650 for heads in still water. | | | Coefficient ·667 for heads in still water. | | |
|---|---|---|---|---|---|---|
| | Ratio of the height due to the velocity of approach to the head. | Coefficients for orifices: the heads measured to the centres. | Coefficients for weirs: the heads measured the full depth. | Ratio of the height due to the velocity of approach to the head. | Coefficients for orifices: the heads measured to the centres. | Coefficients for weirs: the heads measured the full depth. |
| 30· | ·000 | ·650 | ·650 | ·000 | ·667 | ·667 |
| 20· | ·001 | ·650 | ·651 | ·001 | ·667 | ·668 |
| 15· | ·002 | ·651 | ·652 | ·002 | ·667 | ·669 |
| 10· | ·004 | ·651 | ·654 | ·004 | ·668 | ·671 |
| 9· | ·005 | ·652 | ·655 | ·006 | ·669 | ·672 |
| 8· | ·007 | ·652 | ·656 | ·007 | ·669 | ·673 |
| 7·0 | ·009 | ·653 | ·658 | ·009 | ·670 | ·675 |
| 6·0 | ·012 | ·654 | ·661 | ·012 | ·671 | ·678 |
| 5·5 | ·014 | ·655 | ·663 | ·015 | ·672 | ·680 |
| 5·0 | ·017 | ·656 | ·665 | ·018 | ·673 | ·682 |
| 4·5 | ·021 | ·657 | ·669 | ·022 | ·674 | ·687 |
| 4·0 | ·027 | ·659 | ·674 | ·029 | ·676 | ·692 |
| 3·75 | ·031 | ·660 | ·677 | ·033 | ·678 | ·696 |
| 3·50 | ·036 | ·662 | ·681 | ·038 | ·679 | ·700 |
| 3·25 | ·042 | ·663 | ·686 | ·044 | ·681 | ·705 |
| 3·0 | ·049 | ·666 | ·692 | ·052 | ·684 | ·711 |
| 2·75 | ·059 | ·669 | ·699 | ·062 | ·687 | ·720 |
| 2·50 | ·073 | ·673 | ·709 | ·077 | ·692 | ·731 |
| 2·25 | ·091 | ·679 | ·723 | ·096 | ·698 | ·745 |
| 2·0 | ·118 | ·687 | ·742 | ·125 | ·707 | ·766 |
| 1·95 | ·125 | ·689 | ·747 | ·132 | ·709 | ·771 |
| 1·90 | ·133 | ·692 | ·752 | ·140 | ·712 | ·777 |
| 1·85 | ·141 | ·694 | ·758 | ·149 | ·715 | ·783 |
| 1·80 | ·150 | ·697 | ·764 | ·159 | ·718 | ·790 |
| 1·75 | ·160 | ·700 | ·771 | ·170 | ·721 | ·797 |
| 1·70 | ·172 | ·704 | ·779 | ·182 | ·725 | ·805 |
| 1·65 | ·184 | ·707 | ·786 | ·195 | ·729 | ·814 |
| 1·60 | ·198 | ·711 | ·795 | ·210 | ·733 | ·823 |
| 1·55 | ·213 | ·716 | ·805 | ·227 | ·738 | ·833 |
| 1·50 | ·231 | ·721 | ·816 | ·246 | ·744 | ·846 |
| 1·45 | ·251 | ·727 | ·828 | ·268 | ·751 | ·859 |
| 1·40 | ·275 | ·734 | ·842 | ·293 | ·758 | ·874 |
| 1·35 | ·302 | ·742 | ·858 | ·322 | ·764 | ·888 |
| 1·30 | ·333 | ·751 | ·876 | ·356 | ·776 | ·911 |
| 1·25 | ·371 | ·761 | ·896 | ·398 | ·788 | ·934 |
| 1·20 | ·415 | ·773 | ·920 | ·446 | ·802 | ·961 |
| 1·15 | ·469 | ·788 | ·949 | ·506 | ·818 | ·992 |
| 1·10 | ·537 | ·806 | ·983 | ·580 | ·838 | 1·030 |
| 1·05 | ·621 | ·828 | 1·024 | ·675 | ·863 | 1·076 |
| 1·00 | ·732 | ·855 | 1·074 | ·800 | ·894 | 1·133 |

See the auxiliary table, p. 68.

TABLE V.—COEFFICIENTS OF DISCHARGE FOR DIFFERENT RATIOS OF THE CHANNEL TO THE ORIFICE.

*Coefficients for heads in still water* $\sqrt{\cdot 5} = \cdot 7071$ *and* 1.

| Ratio of the channel to the orifice. | Coefficient ·7071 for heads in still water. | | | Coefficient 1·000 for heads in still water. | | |
|---|---|---|---|---|---|---|
| | Ratio of the height due to the velocity of approach to the head. | Coefficients for orifices: the heads measured to the centres. | Coefficients for weirs: the heads measured the full depth. | Ratio of the height due to the velocity of approach to the head. | Coefficients for orifices: the heads measured to the centres. | Coefficients for weirs: the heads measured the full depth. |
| 30· | ·001 | ·707 | ·708 | ·001 | 1·001 | 1·002 |
| 20· | ·001 | ·708 | ·708 | ·003 | 1·001 | 1·004 |
| 15· | ·001 | ·708 | ·709 | ·005 | 1·002 | 1·006 |
| 10· | ·005 | ·709 | ·712 | ·010 | 1·005 | 1·014 |
| 9· | ·006 | ·709 | ·713 | ·013 | 1·006 | 1·017 |
| 8· | ·008 | ·710 | ·714 | ·016 | 1·008 | 1·021 |
| 7· | ·010 | ·711 | ·717 | ·021 | 1·010 | 1·028 |
| 6· | ·014 | ·712 | ·721 | ·029 | 1·014 | 1·038 |
| 5·5 | ·017 | ·713 | ·723 | ·034 | 1·017 | 1·045 |
| 5·0 | ·020 | ·714 | ·727 | ·041 | 1·021 | 1·055 |
| 4·5 | ·025 | ·716 | ·731 | ·052 | 1·026 | 1·067 |
| 4·0 | ·032 | ·718 | ·737 | ·067 | 1·033 | 1·084 |
| 3·75 | ·037 | ·720 | ·742 | ·077 | 1·038 | 1·096 |
| 3·50 | ·043 | ·722 | ·747 | ·089 | 1·044 | 1·110 |
| 3·25 | ·050 | ·724 | ·753 | ·105 | 1·051 | 1·127 |
| 3·00 | ·059 | ·728 | ·760 | ·125 | 1·061 | 1·149 |
| 2·75 | ·071 | ·732 | ·770 | ·152 | 1·073 | 1·178 |
| 2·50 | ·087 | ·737 | ·783 | ·190 | 1·091 | 1·216 |
| 2·25 | ·110 | ·745 | ·801 | ·246 | 1·116 | 1·269 |
| 2·00 | ·143 | ·756 | ·826 | ·333 | 1·155 | 1·347 |
| 1·95 | ·151 | ·759 | ·832 | ·356 | 1·165 | 1·367 |
| 1·90 | ·161 | ·762 | ·839 | ·383 | 1·176 | 1·389 |
| 1·85 | ·171 | ·765 | ·846 | ·412 | 1·188 | 1·413 |
| 1·80 | ·182 | ·769 | ·854 | ·446 | 1·203 | 1·441 |
| 1·75 | ·195 | ·773 | ·863 | ·484 | 1·218 | 1·471 |
| 1·70 | ·209 | ·778 | ·873 | ·529 | 1·237 | 1·505 |
| 1·65 | ·225 | ·783 | ·883 | ·579 | 1·257 | 1·543 |
| 1·60 | ·243 | ·788 | ·895 | ·641 | 1·281 | 1·589 |
| 1·55 | ·263 | ·795 | ·908 | ·711 | 1·308 | 1·638 |
| 1·50 | ·286 | ·802 | ·923 | ·800 | 1·342 | 1·699 |
| 1·45 | ·312 | ·810 | ·939 | ·903 | 1·379 | 1·767 |
| 1·40 | ·342 | ·819 | ·958 | 1·042 | 1·429 | 1·854 |
| 1·35 | ·378 | ·830 | ·980 | 1·216 | 1·489 | 1·958 |
| 1·30 | ·421 | ·842 | 1·003 | 1·449 | 1·565 | 2·088 |
| 1·25 | ·471 | ·857 | 1·033 | 1·778 | 1·667 | 2·259 |
| 1·20 | ·532 | ·875 | 1·066 | 2·273 | 1·810 | 2·499 |
| 1·15 | ·608 | ·897 | 1·107 | 3·100 | 2·025 | 2·844 |
| 1·10 | ·704 | ·923 | 1·155 | 4·762 | 2·400 | 3·440 |
| 1·05 | ·830 | ·957 | 1·216 | 9·756 | 3·280 | 4·803 |
| 1·00 | 1·000 | 1·000 | 1·293 | infinite. | infinite. | infinite. |

198 THE DISCHARGE OF WATER FROM

TABLE VI.—THE DISCHARGE OVER WEIRS OR NOTCHES OF ONE FOOT IN LENGTH, IN CUBIC FEET PER MINUTE.

Depths ¼ inch to 10 inches.  Coefficients ·667 to ·617.

GREATER COEFFICIENTS.

| Heads in inches. | Theoretical discharge. | Coefficient ·667. | Coefficient ·650. | Coefficient ·639. | Coefficient ·628. | Coefficient ·617. |
|---|---|---|---|---|---|---|
| ·25 | ·965 | ·644 | ·627 | ·617 | ·606 | ·596 |
| ·5 | 2·730 | 1·821 | 1·775 | 1·744 | 1·714 | 1·684 |
| ·75 | 5·016 | 3·345 | 3·260 | 3·205 | 3·150 | 3·095 |
| ·1 | 7·722 | 5·151 | 5·019 | 4·934 | 4·849 | 4·764 |
| 1·25 | 10·792 | 7·198 | 7·015 | 6·896 | 6·777 | 6·659 |
| 1·5 | 14·186 | 9·462 | 9·221 | 9·065 | 8·909 | 8·753 |
| 1·75 | 17·877 | 11·924 | 11·620 | 11·423 | 11·227 | 11·030 |
| 2· | 21·842 | 14·569 | 14·197 | 13·957 | 13·717 | 13·477 |
| 2·25 | 26·062 | 17·383 | 16·940 | 16·654 | 16·367 | 16·080 |
| 2·5 | 30·524 | 20·360 | 19·841 | 19·505 | 19·169 | 18·833 |
| 2·75 | 35·215 | 23·489 | 22·890 | 22·503 | 22·115 | 21·728 |
| 3· | 40·125 | 26·763 | 26·081 | 25·640 | 25·199 | 24·757 |
| 3·25 | 45·244 | 30·178 | 29·408 | 28·911 | 28·413 | 27·915 |
| 3·5 | 50·563 | 33·726 | 32·866 | 32·310 | 31·754 | 31·197 |
| 3·75 | 56·077 | 37·403 | 36·450 | 35·833 | 35·216 | 34·599 |
| 4· | 61·777 | 41·205 | 40·155 | 39·476 | 38·796 | 38·116 |
| 4·25 | 67·658 | 45·128 | 43·978 | 43·233 | 42·489 | 41·745 |
| 4·5 | 73·714 | 49·167 | 47·914 | 47·103 | 46·292 | 45·482 |
| 4·75 | 79·942 | 53·321 | 51·962 | 51·083 | 50·203 | 49·324 |
| 5· | 86·335 | 57·585 | 56·118 | 55·168 | 54·218 | 53·269 |
| 5·25 | 92·891 | 61·958 | 60·379 | 59·357 | 58·335 | 57·314 |
| 5·5 | 99·604 | 66·436 | 64·743 | 63·647 | 62·551 | 61·456 |
| 5·75 | 106·472 | 71·017 | 69·207 | 68·036 | 66·864 | 65·693 |
| 6· | 113·491 | 75·698 | 73·769 | 72·521 | 71·272 | 70·024 |
| 6·25 | 120·657 | 80·478 | 78·427 | 77·100 | 75·772 | 74·445 |
| 6·5 | 127·969 | 85·355 | 83·180 | 81·772 | 80·365 | 78·957 |
| 6·75 | 135·422 | 90·326 | 88·024 | 86·535 | 85·045 | 83·555 |
| 7· | 143·015 | 95·391 | 92·960 | 91·387 | 89·813 | 88·240 |
| 7·25 | 150·744 | 100·546 | 97·983 | 96·325 | 94·667 | 93·009 |
| 7·5 | 158·608 | 105·792 | 103·095 | 101·350 | 99·606 | 97·861 |
| 7·75 | 166·604 | 111·125 | 108·292 | 106·460 | 104·627 | 102·795 |
| 8· | 174·731 | 116·546 | 113·575 | 111·653 | 109·731 | 107·809 |
| 8·25 | 182·984 | 122·051 | 118·940 | 116·927 | 114·914 | 112·901 |
| 8·5 | 191·365 | 127·640 | 124·387 | 122·282 | 120·177 | 118·072 |
| 8·75 | 199·869 | 133·313 | 129·915 | 127·716 | 125·518 | 123·319 |
| 9· | 208·496 | 139·067 | 135·522 | 133·229 | 130·935 | 128·642 |
| 9·25 | 217·243 | 144·901 | 141·207 | 138·818 | 136·428 | 134·039 |
| 9·5 | 226·111 | 150·816 | 146·972 | 144·485 | 141·997 | 139·510 |
| 9·75 | 235·093 | 156·807 | 152·810 | 150·225 | 147·639 | 145·053 |
| 10· | 244·193 | 162·877 | 158·725 | 156·039 | 153·353 | 150·666 |

ORIFICES, WEIRS, PIPES, AND RIVERS.

TABLE VI.—THE DISCHARGE OVER WEIRS OR NOTCHES OF ONE FOOT IN LENGTH, IN CUBIC FEET PER MINUTE.

*Depths* 10·25 *inches to* 32 *inches.*   *Coefficients* ·667 *to* ·617.

GREATER COEFFICIENTS.

| Heads in inches. | Theoretical discharge. | Coefficient ·667. | Coefficient ·650. | Coefficient ·639. | Coefficient ·628. | Coefficient ·617. |
|---|---|---|---|---|---|---|
| 10·25 | 253·407 | 169·023 | 164·715 | 161·927 | 159·140 | 156·352 |
| 10·5 | 262·734 | 175·244 | 170·777 | 167·887 | 164·997 | 162·107 |
| 10·75 | 272·173 | 181·540 | 176·913 | 173·919 | 170·925 | 167·931 |
| 11· | 281·723 | 187·909 | 183·120 | 180·021 | 176·922 | 173·823 |
| 11·25 | 291·382 | 194·352 | 189·398 | 186·193 | 182·988 | 179·782 |
| 11·5 | 301·148 | 200·866 | 195·746 | 192·434 | 189·121 | 185·808 |
| 11·75 | 311·024 | 207·451 | 202·164 | 198·743 | 195·321 | 191·900 |
| 12· | 321· | 214·107 | 208·650 | 205·119 | 201·588 | 198·057 |
| 12·5 | 341·275 | 227·628 | 221·826 | 218·072 | 214·318 | 210·564 |
| 13· | 361·950 | 241·421 | 235·268 | 231·286 | 227·305 | 223·323 |
| 13·5 | 383·031 | 255·482 | 248·970 | 244·757 | 240·543 | 236·330 |
| 14· | 404·507 | 269·806 | 262·930 | 258·480 | 254·030 | 249·581 |
| 14·5 | 426·368 | 284·387 | 277·139 | 272·449 | 267·759 | 263·069 |
| 15· | 448·611 | 299·223 | 291·597 | 286·662 | 281·728 | 276·793 |
| 15·5 | 471·228 | 314·309 | 306·298 | 301·115 | 295·931 | 290·748 |
| 16· | 494·212 | 329·639 | 321·238 | 315·801 | 310·365 | 304·929 |
| 16·5 | 517·558 | 345·211 | 336·413 | 330·720 | 325·026 | 319·333 |
| 17· | 541·261 | 361·021 | 351·820 | 345·866 | 339·912 | 333·958 |
| 17·5 | 565·315 | 377·065 | 367·455 | 361·236 | 355·018 | 348·799 |
| 18· | 589·715 | 393·340 | 383·315 | 376·828 | 370·341 | 363·854 |
| 18·5 | 614·443 | 409·833 | 399·388 | 392·629 | 385·870 | 379·111 |
| 19· | 639·533 | 426·569 | 415·696 | 408·662 | 401·627 | 394·592 |
| 19·5 | 664·944 | 443·518 | 432·214 | 424·899 | 417·585 | 410·270 |
| 20· | 690·682 | 460·685 | 448·943 | 441·346 | 433·748 | 426·151 |
| 20·5 | 716·737 | 478·064 | 465·879 | 457·995 | 450·111 | 442·227 |
| 21· | 743·125 | 495·664 | 483·031 | 474·857 | 466·683 | 458·508 |
| 21·5 | 769·823 | 513·472 | 500·385 | 491·917 | 483·449 | 474·981 |
| 22· | 796·832 | 531·487 | 517·941 | 509·176 | 500·410 | 491·645 |
| 22·5 | 824·151 | 549·709 | 535·698 | 526·632 | 517·567 | 508·501 |
| 23· | 851·775 | 568·134 | 553·654 | 544·284 | 534·915 | 525·545 |
| 23·5 | 879·700 | 586·760 | 571·805 | 562·128 | 552·452 | 542·775 |
| 24· | 907·925 | 605·586 | 590·151 | 580·164 | 570·177 | 560·190 |
| 25· | 965·253 | 643·824 | 627·414 | 616·797 | 606·179 | 595·561 |
| 26· | 1023·748 | 682·840 | 665·436 | 654·175 | 642·914 | 631·653 |
| 27· | 1083·375 | 722·611 | 704·194 | 692·277 | 680·360 | 668·442 |
| 28· | 1144·116 | 763·125 | 743·675 | 731·090 | 718·505 | 705·920 |
| 29· | 1205·950 | 804·369 | 783·868 | 770·602 | 757·337 | 744·071 |
| 30· | 1268·864 | 846·332 | 824·762 | 810·804 | 796·847 | 782·889 |
| 31· | 1332·833 | 889·000 | 866·341 | 851·680 | 837·019 | 822·358 |
| 32· | 1397·842 | 932·361 | 908·597 | 893·221 | 877·845 | 862·469 |

TABLE VI.—THE DISCHARGE OVER WEIRS OR NOTCHES OF ONE
FOOT IN LENGTH, IN CUBIC FEET PER MINUTE.
Depths 33 inches to 72 inches. Coefficients ·667 to ·617.

GREATER COEFFICIENTS.

| Heads in inches. | Theoretical discharge. | Coefficient ·667. | Coefficient ·650. | Coefficient ·639. | Coefficient ·628. | Coefficient ·617. |
|---|---|---|---|---|---|---|
| 33· | 1463·875 | 976·405 | 951·519 | 935·416 | 919·314 | 903·211 |
| 34· | 1530·917 | 1021·122 | 995·096 | 978·256 | 961·416 | 944·576 |
| 35· | 1598·951 | 1066·500 | 1039·318 | 1021·730 | 1004·141 | 986·553 |
| 36· | 1667·964 | 1112·532 | 1084·177 | 1065·829 | 1047·481 | 1029·134 |
| 37· | 1737·943 | 1159·208 | 1129·663 | 1110·546 | 1091·428 | 1072·311 |
| 38· | 1808·875 | 1206·520 | 1175·769 | 1155·871 | 1135·974 | 1116·076 |
| 39· | 1880·746 | 1254·458 | 1222·485 | 1201·797 | 1181·108 | 1160·420 |
| 40· | 1953·544 | 1303·014 | 1269·804 | 1248·315 | 1226·826 | 1205·337 |
| 41· | 2027·258 | 1352·181 | 1317·718 | 1295·418 | 1273·118 | 1250·818 |
| 42· | 2101·876 | 1401·951 | 1366·219 | 1343·099 | 1319·978 | 1296·857 |
| 43· | 2177·387 | 1452·317 | 1415·302 | 1391·350 | 1367·399 | 1343·448 |
| 44· | 2253·783 | 1503·273 | 1464·959 | 1440·167 | 1415·376 | 1390·584 |
| 45· | 2331·052 | 1554·812 | 1515·184 | 1489·542 | 1463·901 | 1438·259 |
| 46· | 2409·183 | 1606·925 | 1565·969 | 1539·468 | 1512·967 | 1486·466 |
| 47· | 2488·170 | 1659·609 | 1617·311 | 1589·941 | 1562·571 | 1535·201 |
| 48· | 2568· | 1712·856 | 1669·200 | 1640·952 | 1612·704 | 1584·456 |
| 49· | 2648·666 | 1766·660 | 1721·633 | 1692·498 | 1663·362 | 1634·227 |
| 50· | 2730·160 | 1821·021 | 1774·604 | 1744·572 | 1714·540 | 1684·509 |
| 51· | 2812·474 | 1875·920 | 1828·108 | 1797·171 | 1766·234 | 1735·296 |
| 52· | 2895·597 | 1931·363 | 1882·138 | 1850·286 | 1818·435 | 1786·583 |
| 53· | 2979·525 | 1987·343 | 1936·691 | 1903·916 | 1871·142 | 1838·367 |
| 54· | 3064·253 | 2043·857 | 1991·764 | 1958·058 | 1924·351 | 1890·644 |
| 55· | 3149·755 | 2100·887 | 2047·341 | 2012·693 | 1978·046 | 1943·399 |
| 56· | 3236·050 | 2158·445 | 2103·433 | 2067·836 | 2032·239 | 1996·643 |
| 57· | 3323·117 | 2216·519 | 2160·026 | 2123·472 | 2086·917 | 2050·363 |
| 58· | 3410·946 | 2275·101 | 2217·115 | 2179·594 | 2142·074 | 2104·554 |
| 59· | 3499·542 | 2334·195 | 2274·702 | 2236·207 | 2197·712 | 2159·217 |
| 60· | 3588·889 | 2393·789 | 2332·778 | 2293·300 | 2253·822 | 2214·344 |
| 61· | 3678·984 | 2453·882 | 2391·340 | 2350·871 | 2310·402 | 2269·933 |
| 62· | 3769·825 | 2514·473 | 2450·386 | 2408·918 | 2367·450 | 2325·982 |
| 63· | 3861·393 | 2575·549 | 2509·905 | 2467·430 | 2424·955 | 2382·479 |
| 64· | 3953·694 | 2637·114 | 2569·901 | 2526·410 | 2482·920 | 2439·429 |
| 65· | 4046·720 | 2699·162 | 2630·368 | 2585·854 | 2541·340 | 2496·826 |
| 66· | 4140·465 | 2761·690 | 2691·302 | 2645·757 | 2600·212 | 2554·667 |
| 67· | 4234·922 | 2824·693 | 2752·699 | 2706·115 | 2659·531 | 2612·947 |
| 68· | 4330·086 | 2888·167 | 2814·556 | 2766·925 | 2719·294 | 2671·663 |
| 69· | 4425·954 | 2952·111 | 2876·870 | 2828·185 | 2779·499 | 2730·814 |
| 70· | 4522·516 | 3016·518 | 2939·635 | 2889·888 | 2840·140 | 2790·392 |
| 71· | 4619·774 | 3081·389 | 3002·853 | 2952·036 | 2901·218 | 2850·401 |
| 72· | 4717·718 | 3146·718 | 3066·518 | 3014·622 | 2962·727 | 2910·832 |

ORIFICES, WEIRS, PIPES, AND RIVERS. 201

TABLE VI.—THE DISCHARGE OVER WEIRS OR NOTCHES OF ONE
FOOT IN LENGTH, IN CUBIC FEET PER MINUTE.

Depths ¼ inch to 10 inches.   Coefficients ·606 to ·518.

LESSER COEFFICIENTS.

| Heads in inches. | Coefficient ·606. | Coefficient ·595. | Coefficient ·584. | Coefficient ·562. | Coefficient ·540. | Coefficient ·518. |
|---|---|---|---|---|---|---|
| ·25 | ·585 | ·574 | ·564 | ·542 | ·521 | ·500 |
| ·5 | 1·654 | 1·624 | 1·594 | 1·534 | 1·474 | 1·414 |
| ·75 | 3·039 | 2·985 | 2·929 | 2·819 | 2·708 | 2·598 |
| 1· | 4·680 | 4·595 | 4·510 | 4·340 | 4·170 | 4·000 |
| 1·25 | 6·540 | 6·421 | 6·303 | 6·065 | 5·828 | 5·590 |
| 1·5 | 8·597 | 8·441 | 8·284 | 7·973 | 7·660 | 7·348 |
| 1·75 | 10·833 | 10·637 | 10·440 | 10·047 | 9·653 | 9·260 |
| 2· | 12·236 | 12·996 | 12·756 | 12·275 | 11·795 | 11·314 |
| 2·25 | 15·794 | 15·507 | 15·220 | 14·647 | 14·073 | 13·500 |
| 2·5 | 18·498 | 18·162 | 17·826 | 17·155 | 16·483 | 15·811 |
| 2·75 | 21·340 | 20·953 | 20·566 | 19·791 | 19·016 | 18·241 |
| 3· | 24·316 | 23·874 | 23·433 | 22·550 | 21·668 | 20·785 |
| 3·25 | 27·418 | 26·920 | 26·422 | 25·427 | 24·432 | 23·436 |
| 3·5 | 30·641 | 30·085 | 29·529 | 28·416 | 27·304 | 26·192 |
| 3·75 | 33·982 | 33·366 | 32·749 | 31·515 | 30·281 | 29·048 |
| 4· | 37·437 | 36·757 | 36·078 | 34·719 | 33·360 | 32·000 |
| 4·25 | 41·001 | 40·256 | 39·512 | 38·024 | 36·535 | 35·047 |
| 4·5 | 44·671 | 43·860 | 43·049 | 41·427 | 39·806 | 38·184 |
| 4·75 | 48·445 | 47·565 | 46·686 | 44·927 | 43·169 | 41·410 |
| 5· | 52·319 | 51·369 | 50·420 | 48·520 | 46·621 | 44·722 |
| 5·25 | 56·292 | 55·270 | 54·248 | 52·205 | 50·161 | 48·117 |
| 5·5 | 60·360 | 59·264 | 58·169 | 55·977 | 53·786 | 51·595 |
| 5·75 | 64·522 | 63·351 | 62·180 | 59·837 | 57·495 | 55·153 |
| 6· | 68·776 | 67·527 | 66·279 | 63·782 | 61·285 | 58·788 |
| 6·25 | 73·118 | 71·791 | 70·464 | 67·809 | 65·155 | 62·500 |
| 6·5 | 77·549 | 76·142 | 74·734 | 71·919 | 69·103 | 66·288 |
| 6·75 | 82·066 | 80·576 | 79·086 | 76·107 | 73·128 | 70·149 |
| 7· | 86·667 | 85·094 | 83·521 | 80·374 | 77·228 | 74·082 |
| 7·25 | 91·351 | 89·693 | 88·034 | 84·718 | 81·402 | 78·085 |
| 7·5 | 96·116 | 94·372 | 92·627 | 89·138 | 85·648 | 82·159 |
| 7·75 | 100·962 | 99·129 | 97·297 | 93·631 | 89·966 | 86·301 |
| 8· | 105·887 | 103·965 | 102·043 | 98·199 | 94·355 | 90·511 |
| 8·25 | 110·889 | 108·876 | 106·863 | 102·837 | 98·812 | 94·786 |
| 8·5 | 115·967 | 113·862 | 111·757 | 107·547 | 103·337 | 99·127 |
| 8·75 | 121·121 | 118·922 | 116·723 | 112·326 | 107·929 | 103·532 |
| 9· | 126·349 | 124·055 | 121·762 | 117·175 | 112·588 | 108·001 |
| 9·25 | 131·649 | 129·259 | 126·870 | 122·090 | 117·311 | 112·532 |
| 9·5 | 137·023 | 134·535 | 132·048 | 127·074 | 122·100 | 117·125 |
| 9·75 | 142·467 | 139·881 | 137·294 | 132·122 | 126·950 | 121·778 |
| 10· | 147·981 | 145·295 | 142·609 | 137·237 | 131·864 | 126·492 |

TABLE VI.—THE DISCHARGE OVER WEIRS OR NOTCHES OF ONE FOOT IN LENGTH, IN CUBIC FEET PER MINUTE.

*Depths 10·25 inches to 32 inches. Coefficients ·606 to ·518.*

LESSER COEFFICIENTS.

| Heads in inches. | Coeffict. ·606. | Coeffict. ·595. | Coefficient ·584. | Coefficient ·562. | Coefficient ·540. | Coefficient ·518. |
|---|---|---|---|---|---|---|
| 10·25 | 153·565 | 150·777 | 147·990 | 142·415 | 136·840 | 131·265 |
| 10·5 | 159·217 | 156·327 | 153·437 | 147·657 | 141·876 | 136·096 |
| 10·75 | 164·937 | 161·943 | 158·949 | 152·961 | 146·974 | 140·986 |
| 11· | 170·724 | 167·625 | 164·526 | 158·328 | 152·130 | 145·933 |
| 11·25 | 176·577 | 173·372 | 170·167 | 163·756 | 157·346 | 150·936 |
| 11·5 | 182·496 | 179·183 | 175·870 | 169·245 | 162·620 | 155·995 |
| 11·75 | 188·479 | 185·058 | 181·636 | 174·794 | 167·952 | 161·109 |
| 12· | 194·526 | 190·995 | 187·464 | 180·402 | 173·340 | 166·278 |
| 12·5 | 206·810 | 203·056 | 199·302 | 191·794 | 184·286 | 176·778 |
| 13· | 219·342 | 215·360 | 211·379 | 203·415 | 195·453 | 187·490 |
| 13·5 | 232·117 | 227·903 | 223·690 | 215·263 | 206·837 | 198·410 |
| 14· | 245·131 | 240·682 | 236·232 | 227·333 | 218·434 | 209·535 |
| 14·5 | 258·379 | 253·689 | 248·999 | 239·619 | 230·239 | 220·859 |
| 15· | 271·858 | 266·924 | 261·989 | 252·119 | 242·250 | 232·380 |
| 15·5 | 285·564 | 280·381 | 275·197 | 264·830 | 254·463 | 244·096 |
| 16· | 299·492 | 294·056 | 288·620 | 277·747 | 266·875 | 256·001 |
| 16·5 | 313·640 | 307·947 | 302·253 | 290·868 | 279·481 | 268·095 |
| 17· | 328·004 | 322·050 | 316·096 | 304·189 | 292·281 | 280·373 |
| 17·5 | 342·581 | 336·362 | 330·144 | 317·707 | 305·270 | 292·833 |
| 18· | 357·367 | 350·880 | 344·394 | 331·420 | 318·446 | 305·472 |
| 18·5 | 372·352 | 365·594 | 358·835 | 345·317 | 331·799 | 318·281 |
| 19· | 387·557 | 380·522 | 373·487 | 359·418 | 345·348 | 331·278 |
| 19·5 | 402·956 | 395·642 | 388·327 | 373·699 | 359·070 | 344·441 |
| 20· | 418·553 | 410·956 | 403·358 | 388·163 | 372·968 | 357·773 |
| 20·5 | 434·343 | 426·458 | 418·574 | 402·806 | 387·038 | 371·270 |
| 21· | 450·334 | 442·159 | 433·985 | 417·636 | 401·288 | 384·939 |
| 21·5 | 466·513 | 458·045 | 449·577 | 432·641 | 415·704 | 398·768 |
| 22· | 482·880 | 474·115 | 465·350 | 447·819 | 430·289 | 412·759 |
| 22·5 | 499·436 | 490·370 | 481·304 | 463·173 | 445·042 | 426·910 |
| 23· | 516·176 | 506·806 | 497·437 | 478·698 | 459·959 | 441·219 |
| 23·5 | 533·098 | 523·421 | 513·745 | 494·391 | 475·038 | 455·685 |
| 24· | 550·203 | 540·215 | 530·228 | 510·254 | 490·280 | 470·305 |
| 25· | 584·943 | 574·326 | 563·708 | 542·472 | 521·237 | 500·001 |
| 26· | 620·391 | 609·130 | 597·869 | 575·346 | 552·824 | 530·301 |
| 27· | 656·525 | 644·608 | 632·691 | 608·857 | 585·023 | 561·188 |
| 28· | 693·334 | 680·749 | 668·164 | 642·993 | 617·823 | 592·652 |
| 29· | 730·806 | 717·540 | 704·275 | 677·744 | 651·213 | 624·682 |
| 30· | 768·932 | 754·974 | 741·017 | 713·102 | 685·187 | 657·272 |
| 31· | 807·697 | 793·036 | 778·374 | 749·052 | 719·730 | 690·407 |
| 32· | 847·092 | 831·716 | 816·340 | 785·587 | 754·835 | 724·082 |

ORIFICES, WEIRS, PIPES, AND RIVERS. 203

TABLE VI.—THE DISCHARGE OVER WEIRS OR NOTCHES OF ONE FOOT IN LENGTH, IN CUBIC FEET PER MINUTE.

Depths 33 inches to 72 inches. Coefficients ·606 to ·518.

LESSER COEFFICIENTS.

| Heads in inches. | Coefficient ·606. | Coefficient ·595. | Coefficient ·584. | Coefficient ·562. | Coefficient ·540. | Coefficient ·518. |
|---|---|---|---|---|---|---|
| 33 | 887·108 | 871·006 | 854·903 | 822·698 | 790·493 | 758·287 |
| 34 | 927·736 | 910·896 | 894·056 | 860·375 | 826·695 | 793·015 |
| 35 | 968·964 | 951·376 | 933·787 | 898·610 | 863·434 | 828·257 |
| 36 | 1010·786 | 992·439 | 974·091 | 937·396 | 900·701 | 864·005 |
| 37 | 1053·193 | 1034·076 | 1014·959 | 976·724 | 938·489 | 900·254 |
| 38 | 1096·178 | 1076·281 | 1056·383 | 1016·588 | 976·793 | 936·997 |
| 39 | 1139·732 | 1119·044 | 1098·356 | 1056·979 | 1015·603 | 974·226 |
| 40 | 1183·848 | 1162·359 | 1140·870 | 1097·892 | 1054·914 | 1011·936 |
| 41 | 1228·518 | 1206·219 | 1183·919 | 1139·319 | 1094·719 | 1050·120 |
| 42 | 1273·737 | 1250·616 | 1227·496 | 1181·254 | 1135·013 | 1088·772 |
| 43 | 1319·497 | 1295·545 | 1271·594 | 1223·691 | 1175·789 | 1127·886 |
| 44 | 1365·792 | 1341·001 | 1316·209 | 1266·626 | 1217·043 | 1167·460 |
| 45 | 1412·618 | 1386·976 | 1361·334 | 1310·051 | 1258·768 | 1207·485 |
| 46 | 1459·965 | 1433·464 | 1406·963 | 1353·961 | 1300·959 | 1247·957 |
| 47 | 1507·831 | 1480·461 | 1453·091 | 1398·352 | 1343·612 | 1288·872 |
| 48 | 1556·208 | 1527·960 | 1499·712 | 1443·216 | 1386·720 | 1330·224 |
| 49 | 1605·092 | 1575·956 | 1546·821 | 1488·550 | 1430·280 | 1372·009 |
| 50 | 1654·477 | 1624·445 | 1594·413 | 1534·350 | 1474·286 | 1414·223 |
| 51 | 1704·359 | 1673·422 | 1642·485 | 1580·610 | 1518·736 | 1456·862 |
| 52 | 1754·732 | 1722·880 | 1691·029 | 1627·326 | 1563·622 | 1499·919 |
| 53 | 1805·592 | 1772·817 | 1740·043 | 1674·493 | 1608·944 | 1543·394 |
| 54 | 1856·937 | 1823·231 | 1789·524 | 1722·110 | 1654·697 | 1587·283 |
| 55 | 1908·751 | 1874·104 | 1839·457 | 1770·162 | 1700·868 | 1631·573 |
| 56 | 1961·046 | 1925·450 | 1889·853 | 1818·660 | 1747·467 | 1676·274 |
| 57 | 2013·809 | 1977·255 | 1940·700 | 1867·592 | 1794·483 | 1721·375 |
| 58 | 2067·033 | 2029·513 | 1991·992 | 1916·952 | 1841·911 | 1766·870 |
| 59 | 2120·722 | 2082·227 | 2043·733 | 1966·743 | 1889·753 | 1812·763 |
| 60 | 2174·867 | 2135·389 | 2095·911 | 2016·956 | 1938·000 | 1859·045 |
| 61 | 2229·464 | 2188·995 | 2148·527 | 2067·589 | 1986·651 | 1905·714 |
| 62 | 2284·514 | 2243·046 | 2201·578 | 2118·642 | 2035·706 | 1952·769 |
| 63 | 2340·004 | 2297·529 | 2255·054 | 2170·103 | 2085·152 | 2000·202 |
| 64 | 2395·939 | 2352·448 | 2308·957 | 2221·976 | 2134·995 | 2048·013 |
| 65 | 2452·312 | 2407·798 | 2363·284 | 2274·257 | 2185·229 | 2096·201 |
| 66 | 2509·122 | 2463·577 | 2418·032 | 2326·941 | 2235·851 | 2144·761 |
| 67 | 2566·363 | 2519·779 | 2473·194 | 2380·026 | 2286·858 | 2193·690 |
| 68 | 2624·032 | 2576·401 | 2528·770 | 2433·508 | 2338·246 | 2242·985 |
| 69 | 2682·128 | 2633·443 | 2584·757 | 2487·386 | 2390·015 | 2292·644 |
| 70 | 2740·645 | 2690·897 | 2611·149 | 2541·654 | 2442·159 | 2342·663 |
| 71 | 2799·583 | 2748·766 | 2697·948 | 2596·313 | 2494·678 | 2393·043 |
| 72 | 2858·937 | 2807·042 | 2755·147 | 2651·358 | 2547·568 | 2443·778 |

TABLE VII.—FOR FINDING THE MEAN VELOCITY FROM THE MAXIMUM VELOCITY AT THE SURFACE, IN MILL RACES, STREAMS, AND RIVERS WITH UNIFORM CHANNELS; AND THE MAXIMUM VELOCITY FROM THE MEAN VELOCITY. (See p. 101.)

*For the velocity in feet per minute, multiply by 5.*

| Maximum velocity at the surface in inches per second. | Mean velocity in large channels in inches per second. | Mean velocity in smaller channels in inches per second. | Maximum velocity at the surface in inches per second. | Mean velocity in large channels in inches per second. | Mean velocity in smaller channels in inches per second. | Maximum velocity at the surface in inches per second. | Mean velocity in large channels in inches per second. | Mean velocity in smaller channels in inches per second. |
|---|---|---|---|---|---|---|---|---|
| 1 | ·84 | ·75 | 41 | 34·24 | 33·37 | 81 | 67·64 | 68·86 |
| 2 | 1·67 | 1·51 | 42 | 35·07 | 34·23 | 82 | 68·47 | 69·77 |
| 3 | 2·51 | 2·27 | 43 | 35·91 | 35·09 | 83 | 69·31 | 70·68 |
| 4 | 3·34 | 3·04 | 44 | 36·74 | 35·95 | 84 | 70·14 | 71·59 |
| 5 | 4·18 | 3·81 | 45 | 37·58 | 36·82 | 85 | 70·98 | 72·50 |
| 6 | 5·01 | 4·58 | 46 | 38·41 | 37·69 | 86 | 71·81 | 73·42 |
| 7 | 5·85 | 5·36 | 47 | 39·25 | 38·56 | 87 | 72·65 | 74·33 |
| 8 | 6·68 | 6·14 | 48 | 40·08 | 39·43 | 88 | 73·48 | 75·24 |
| 9 | 7·52 | 6·92 | 49 | 40·92 | 40·30 | 89 | 74·32 | 76·16 |
| 10 | 8·35 | 7·71 | 50 | 41·75 | 41·17 | 90 | 75·15 | 77·08 |
| 11 | 9·19 | 8·50 | 51 | 42·59 | 42·05 | 91 | 75·99 | 77·99 |
| 12 | 10·02 | 9·29 | 52 | 43·42 | 42·92 | 92 | 76·82 | 78·91 |
| 13 | 10·86 | 10·09 | 53 | 44·26 | 43·80 | 93 | 77·66 | 79·83 |
| 14 | 11·69 | 10·88 | 54 | 45·09 | 44·68 | 94 | 78·49 | 80·75 |
| 15 | 12·53 | 11·69 | 55 | 45·93 | 45·56 | 95 | 79·33 | 81·67 |
| 16 | 13·36 | 12·49 | 56 | 46·76 | 46·44 | 96 | 80·16 | 82·59 |
| 17 | 14·20 | 13·30 | 57 | 47·60 | 47·32 | 97 | 81·00 | 83·51 |
| 18 | 15·03 | 14·11 | 58 | 48·43 | 48·21 | 98 | 81·83 | 84·43 |
| 19 | 15·87 | 14·92 | 59 | 49·27 | 49·09 | 99 | 82·67 | 85·36 |
| 20 | 16·70 | 15·73 | 60 | 50·10 | 49·98 | 100 | 83·50 | 86·28 |
| 21 | 17·54 | 16·55 | 61 | 50·94 | 50·87 | 101 | 84·34 | 87·20 |
| 22 | 18·37 | 17·37 | 62 | 51·77 | 51·76 | 102 | 85·17 | 88·13 |
| 23 | 19·21 | 18·19 | 63 | 52·61 | 52·65 | 103 | 86·01 | 89·06 |
| 24 | 20·04 | 19·02 | 64 | 53·44 | 53·54 | 104 | 86·84 | 89·98 |
| 25 | 20·88 | 19·85 | 65 | 54·28 | 54·43 | 105 | 87·68 | 90·91 |
| 26 | 21·71 | 20·68 | 66 | 55·11 | 55·33 | 106 | 88·51 | 91·84 |
| 27 | 22·55 | 21·51 | 67 | 55·95 | 56·22 | 107 | 89·35 | 92·77 |
| 28 | 23·38 | 22·34 | 68 | 56·78 | 57·12 | 108 | 90·18 | 93·69 |
| 29 | 24·22 | 23·18 | 69 | 57·62 | 58·02 | 109 | 91·02 | 94·62 |
| 30 | 25·05 | 24·02 | 70 | 58·45 | 58·91 | 110 | 91·85 | 95·55 |
| 31 | 25·89 | 24·86 | 71 | 59·29 | 59·81 | 111 | 92·69 | 96·49 |
| 32 | 26·72 | 25·70 | 72 | 60·12 | 60·71 | 112 | 93·52 | 97·42 |
| 33 | 27·56 | 26·54 | 73 | 60·96 | 61·61 | 113 | 94·36 | 98·35 |
| 34 | 28·39 | 27·39 | 74 | 61·79 | 62·52 | 114 | 95·19 | 99·28 |
| 35 | 29·23 | 28·24 | 75 | 62·63 | 63·42 | 115 | 96·03 | 100·21 |
| 36 | 30·06 | 29·09 | 76 | 63·46 | 64·32 | 116 | 96·86 | 101·15 |
| 37 | 30·90 | 29·94 | 77 | 64·30 | 65·23 | 117 | 97·70 | 102·08 |
| 38 | 31·73 | 30·79 | 78 | 65·13 | 66·13 | 118 | 98·53 | 103·02 |
| 39 | 32·57 | 31·65 | 79 | 65·97 | 67·04 | 119 | 99·37 | 103·95 |
| 40 | 33·40 | 32·51 | 80 | 66·80 | 67·95 | 120 | 100·20 | 104·89 |

ORIFICES, WEIRS, PIPES, AND RIVERS. 205

TABLE VIII.—FOR FINDING THE MEAN VELOCITIES OF WATER FLOWING IN PIPES, DRAINS, STREAMS, AND RIVERS.

For a full cylindrical pipe, divide the diameter by 4 to find the hydraulic mean depth.

*Diameters of pipes ¼ inch to 2 inches. Falls per mile 1 inch to 12 feet.*

| Falls per mile in feet and inches, and the hydraulic inclinations. || "Hydraulic mean depths," or "mean radii," and velocities in inches per second. |||||
|---|---|---|---|---|---|---|
| Falls. | Inclinations one in | 1/16 inch. | ⅛ inch. | ¼ inch. | ⅜ inch. | ½ inch. |
| F. I. | | | | | | |
| 0 1 | 63360 | ·14 | ·24 | ·38 | ·49 | ·57 |
| 0 2 | 31680 | ·22 | ·37 | ·59 | ·76 | ·90 |
| 0 3 | 21120 | ·28 | ·48 | ·75 | ·97 | 1·15 |
| 0 4 | 15840 | ·34 | ·57 | ·89 | 1·15 | 1·36 |
| 0 5 | 12672 | ·38 | ·65 | 1·02 | 1·30 | 1·55 |
| 0 6 | 10560 | ·42 | ·72 | 1·13 | 1·45 | 1·72 |
| 0 7 | 9051 | ·46 | ·78 | 1·24 | 1·58 | 1·88 |
| 0 8 | 7920 | ·50 | ·85 | 1·33 | 1·71 | 2·02 |
| 0 9 | 7040 | ·53 | ·90 | 1·43 | 1·83 | 2·16 |
| 0 10 | 6336 | ·57 | ·96 | 1·51 | 1·94 | 2·30 |
| 0 11 | 5760 | ·60 | 1·01 | 1·60 | 1·96 | 2·42 |
| 1 0 | 5280 | ·63 | 1·06 | 1·68 | 2·15 | 2·54 |
| 1 3 | 4224 | ·71 | 1·20 | 1·90 | 2·43 | 2·88 |
| 1 6 | 3520 | ·79 | 1·33 | 2·10 | 2·69 | 3·19 |
| 1 9 | 3017 | ·87 | 1·45 | 2·29 | 2·94 | 3·48 |
| 2 0 | 2640 | ·93 | 1·56 | 2·47 | 3·16 | 3·75 |
| 2 3 | Interpolated. | ·99 | 1·67 | 2·63 | 3·37 | 3·99 |
| 2 6 | 2112 | 1·05 | 1·77 | 2·79 | 3·58 | 4·24 |
| 2 9 | Interpolated. | 1·11 | 1·87 | 2·94 | 3·77 | 4·47 |
| 3 0 | 1760 | 1·16 | 1·96 | 3·09 | 3·96 | 4·69 |
| 3 3 | Interpolated. | 1·21 | 2·05 | 3·23 | 4·14 | 4·91 |
| 3 6 | 1508 | 1·26 | 2·14 | 3·37 | 4·32 | 5·12 |
| 3 9 | Interpolated. | 1·31 | 2·22 | 3·50 | 4·48 | 5·31 |
| 4 0 | 1320 | 1·36 | 2·30 | 3·63 | 4·65 | 5·51 |
| 4 6 | Interpolated. | 1·45 | 2·45 | 3·87 | 4·96 | 5·88 |
| 5 0 | 1056 | 1·54 | 2·61 | 4·11 | 5·27 | 6·24 |
| 5 6 | Interpolated. | 1·62 | 2·75 | 4·33 | 5·55 | 6·58 |
| 6 0 | 880 | 1·71 | 2·89 | 4·55 | 5·83 | 6·91 |
| 6 6 | Interpolated. | 1·78 | 3·02 | 4·76 | 6·10 | 7·22 |
| 7 0 | 754 | 1·86 | 3·15 | 4·97 | 6·36 | 7·54 |
| 7 6 | Interpolated. | 1·93 | 3·27 | 5·16 | 6·61 | 7·83 |
| 8 0 | 660 | 2·01 | 3·39 | 5·35 | 6·86 | 8·12 |
| 8 6 | Interpolated. | 2·07 | 3·51 | 5·53 | 7·09 | 8·40 |
| 9 0 | 587 | 2·14 | 3·62 | 5·72 | 7·32 | 8·68 |
| 9 6 | Interpolated. | 2·20 | 3·74 | 5·89 | 7·55 | 8·94 |
| 10 0 | 528 | 2·28 | 3·85 | 6·07 | 7·77 | 9·21 |
| 10 6 | Interpolated. | 2·33 | 3·95 | 6·24 | 7·99 | 9·47 |
| 11 0 | 480 | 2·40 | 4·06 | 6·40 | 8·20 | 9·72 |
| 11 6 | Interpolated. | 2·46 | 4·16 | 6·57 | 8·41 | 9·97 |
| 12 0 | 440 | 2·52 | 4·27 | 6·73 | 8·62 | 10·21 |

## THE DISCHARGE OF WATER FROM

TABLE VIII.—FOR FINDING THE MEAN VELOCITIES OF WATER FLOWING IN PIPES, DRAINS, STREAMS, AND RIVERS.

For a full cylindrical pipe, divide the diameter by 4 to find the hydraulic mean depth.

*Diameters of pipes $\frac{1}{4}$ inch to 2 inches. Falls per mile 13 feet to 5280 feet.*

| Falls per mile in feet, and the hydraulic inclinations. || "Hydraulic mean depths," or "mean radii," and velocities in inches per second. |||||
|---|---|---|---|---|---|---|
| Falls. | Inclinations one in | $\frac{1}{10}$ inch. | $\frac{1}{8}$ inch. | $\frac{1}{4}$ inch. | $\frac{3}{8}$ inch. | $\frac{1}{2}$ inch. |
| F. | | | | | | |
| 13·2 | 400 | 2·66 | 4·50 | 7·10 | 9·10 | 10·78 |
| 13·6 | Interpolated. | 2·71 | 4·59 | 7·24 | 9·27 | 10·98 |
| 14·1 | 375 | 2·76 | 4·67 | 7·37 | 9·44 | 11·18 |
| 14·6 | Interpolated. | 2·82 | 4·76 | 7·52 | 9·63 | 11·41 |
| 15·1 | 350 | 2·87 | 4·85 | 7·66 | 9·82 | 11·63 |
| 15·6 | Interpolated. | 2·94 | 4·96 | 7·83 | 10·03 | 11·88 |
| 16·2 | 325 | 3·00 | 5·07 | 7·99 | 10·24 | 12·13 |
| 17·6 | 300 | 3·14 | 5·30 | 8·37 | 10·72 | 12·70 |
| 19·2 | 275 | 3·30 | 5·58 | 8·80 | 11·27 | 13·35 |
| 21·1 | 250 | 3·48 | 5·89 | 9·39 | 11·90 | 14·10 |
| 23·5 | 225 | 3·70 | 6·26 | 9·87 | 12·65 | 14·99 |
| 26·4 | 200 | 3·96 | 6·70 | 10·57 | 13·54 | 16·04 |
| 30·2 | 175 | 4·28 | 7·24 | 11·42 | 14·63 | 17·33 |
| 35·2 | 150 | 4·68 | 7·92 | 12·49 | 16·00 | 18·96 |
| 37·7 | 140 | 4·88 | 8·24 | 13·00 | 16·66 | 19·74 |
| 42·2 | 125 | 5·21 | 8·81 | 13·90 | 17·80 | 21·09 |
| 48· | 110 | 5·62 | 9·50 | 14·98 | 19·19 | 22·74 |
| 52·8 | 100 | 5·94 | 10·05 | 15·85 | 20·30 | 24·06 |
| 58·7 | 90 | 6·33 | 10·69 | 16·87 | 21·61 | 25·60 |
| 66· | 80 | 6·78 | 11·47 | 18·10 | 23·17 | 27·46 |
| 75·4 | 70 | 7·35 | 12·42 | 19·59 | 25·09 | 29·73 |
| 88· | 60 | 8·05 | 13·61 | 21·48 | 27·51 | 32·60 |
| 105·6 | 50 | 8·99 | 15·19 | 23·96 | 30·69 | 36·37 |
| 117·3 | 45 | 9·57 | 16·18 | 25·53 | 32·70 | 38·75 |
| 132·0 | 40 | 10·28 | 17·37 | 27·41 | 35·11 | 41·60 |
| 150·8 | 35 | 11·14 | 18·84 | 29·71 | 38·06 | 45·10 |
| 176· | 30 | 12·23 | 20·68 | 32·62 | 41·78 | 49·51 |
| 212·2 | 25 | 13·66 | 23·09 | 36·43 | 46·67 | 55·30 |
| 264· | 20 | 15·64 | 26·44 | 41·71 | 53·43 | 63·30 |
| 352· | 15 | 18·61 | 31·46 | 49·63 | 63·57 | 75·33 |
| 528· | 10 | 23·73 | 40·11 | 63·28 | 81·06 | 96·05 |
| 586·7 | 9 | 25·26 | 42·70 | 67·37 | 86·29 | 102·25 |
| 660· | 8 | 27·08 | 45·78 | 72·22 | 92·51 | 109·61 |
| 754·3 | 7 | 29·29 | 49·51 | 78·10 | 100·04 | 118·54 |
| 880·0 | 6 | 32·05 | 54·15 | 85·43 | 109·43 | 129·66 |
| 1056· | 5 | 35·08 | 60·15 | 94·89 | 121·54 | 144·02 |
| 1320· | 4 | 40·40 | 68·29 | 107·73 | 137·99 | 163·51 |
| 1760· | 3 | 47·48 | 80·25 | 126·61 | 162·17 | 192·16 |
| 2640· | 2 | 59·47 | 100·53 | 158·59 | 203·14 | 240·70 |
| 5280· | 1 | 88·13 | 148·97 | 235·02 | 301·04 | 356·70 |

TABLE VIII.—For finding the mean velocities of water flowing in pipes, drains, streams, and rivers.

For a full cylindrical pipe, divide the diameter by 4 to find the hydraulic mean depth.

*Diameters of pipes 2½ inches to 5 inches. Falls per mile 1 inch to 12 feet.*

| Falls per mile in feet and inches, and the hydraulic inclinations. || "Hydraulic mean depths," or "mean radii," and velocities in inches per second. ||||| 
|---|---|---|---|---|---|---|
| Falls. || Inclinations one in | ⅝ inch. | ¾ inch. | ⅞ inch. | 1 inch. | 1¼ in. interpolated |
| F. | I. | | | | | | |
| 0 | 1 | 63360 | ·65 | ·73 | ·79 | ·85 | ·96 |
| 0 | 2 | 31680 | 1·02 | 1·13 | 1·23 | 1·33 | 1·49 |
| 0 | 3 | 21120 | 1·30 | 1·45 | 1·58 | 1·70 | 1·91 |
| 0 | 4 | 15840 | 1·54 | 1·71 | 1·87 | 2·01 | 2·26 |
| 0 | 5 | 12672 | 1·76 | 1·95 | 2·13 | 2·29 | 2·58 |
| 0 | 6 | 10560 | 1·95 | 2·17 | 2·36 | 2·55 | 2·86 |
| 0 | 7 | 9051 | 2·13 | 2·37 | 2·58 | 2·78 | 3·13 |
| 0 | 8 | 7920 | 2·30 | 2·55 | 2·78 | 3·00 | 3·37 |
| 0 | 9 | 7040 | 2·46 | 2·73 | 2·98 | 3·21 | 3·61 |
| 0 | 10 | 6336 | 2·61 | 2·90 | 3·16 | 3·40 | 3·83 |
| 0 | 11 | 5760 | 2·76 | 3·06 | 3·33 | 3·59 | 4·04 |
| 1 | 0 | 5280 | 2·89 | 3·21 | 3·50 | 3·77 | 4·24 |
| 1 | 3 | 4224 | 3·28 | 3·64 | 3·97 | 4·27 | 4·81 |
| 1 | 6 | 3520 | 3·63 | 4·03 | 4·39 | 4·73 | 5·32 |
| 1 | 9 | 3017 | 3·96 | 4·39 | 4·79 | 5·16 | 5·80 |
| 2 | 0 | 2640 | 4·26 | 4·73 | 5·16 | 5·55 | 6·25 |
| 2 | 3 | Interpolated. | 4·55 | 5·04 | 5·50 | 5·92 | 6·66 |
| 2 | 6 | 2112 | 4·83 | 5·35 | 5·84 | 6·29 | 7·07 |
| 2 | 9 | Interpolated. | 5·09 | 5·64 | 6·15 | 6·12 | 7·46 |
| 3 | 0 | 1760 | 5·34 | 5·92 | 6·46 | 6·96 | 7·83 |
| 3 | 3 | Interpolated. | 5·58 | 6·19 | 6·75 | 7·27 | 8·18 |
| 3 | 6 | 1508 | 5·82 | 6·46 | 7·04 | 7·59 | 8·53 |
| 3 | 9 | Interpolated. | 6·05 | 6·71 | 7·31 | 7·88 | 8·86 |
| 4 | 0 | 1320 | 6·27 | 6·95 | 7·58 | 8·17 | 9·19 |
| 4 | 6 | Interpolated. | 6·69 | 7·42 | 8·09 | 8·71 | 9·80 |
| 5 | 0 | 1056 | 7·10 | 7·88 | 8·59 | 9·25 | 10·41 |
| 5 | 6 | Interpolated. | 7·48 | 8·30 | 9·05 | 9·76 | 10·97 |
| 6 | 0 | 880 | 7·86 | 8·72 | 9·51 | 10·25 | 11·53 |
| 6 | 6 | Interpolated. | 8·22 | 9·12 | 9·94 | 10·71 | 12·05 |
| 7 | 0 | 754 | 8·57 | 9·51 | 10·37 | 11·17 | 12·57 |
| 7 | 6 | Interpolated. | 8·92 | 9·89 | 10·78 | 11·62 | 13·06 |
| 8 | 0 | 660 | 9·24 | 10·25 | 11·18 | 12·04 | 13·54 |
| 8 | 6 | Interpolated. | 9·55 | 10·60 | 11·56 | 12·45 | 14·01 |
| 9 | 0 | 587 | 9·87 | 10·95 | 11·94 | 12·86 | 14·47 |
| 9 | 6 | Interpolated. | 10·18 | 11·28 | 12·31 | 13·26 | 14·91 |
| 10 | 0 | 528 | 10·48 | 11·62 | 12·67 | 13·65 | 15·36 |
| 10 | 6 | Interpolated. | 10·77 | 11·95 | 13·03 | 14·03 | 15·78 |
| 11 | 0 | 480 | 11·06 | 12·27 | 13·38 | 14·41 | 16·21 |
| 11 | 6 | Interpolated. | 11·34 | 12·58 | 13·72 | 14·82 | 16·64 |
| 12 | 0 | 440 | 11·62 | 12·89 | 14·05 | 15·22 | 17·07 |

## THE DISCHARGE OF WATER FROM

TABLE VIII.—FOR FINDING THE MEAN VELOCITIES OF WATER FLOWING IN PIPES, DRAINS, STREAMS, AND RIVERS.

For a full cylindrical pipe, divide the diameter by 4 to find the hydraulic mean depth.

*Diameters of pipes 2½ inches to 5 inches. Falls per mile 13 feet to 5280 feet.*

| Falls per mile in feet, and the hydraulic inclinations. || "Hydraulic mean depths," or "mean radii," and velocities in inches per second. |||||
|---|---|---|---|---|---|---|
| Falls. | Inclinations one in | ⅝ inch. | ¾ inch. | ⅞ inch. | 1 inch. | 1¼ in. interpolated |
| F. | | | | | | |
| 13·2 | 400 | 12·26 | 13·60 | 14·83 | 15·98 | 17·98 |
| 13·6 | Interpolated. | 12·49 | 13·86 | 15·11 | 16·28 | 18·31 |
| 14·1 | 375 | 12·72 | 14·11 | 15·39 | 16·58 | 18·65 |
| 14·6 | Interpolated. | 12·98 | 14·39 | 15·70 | 16·91 | 19·02 |
| 15·1 | 350 | 13·23 | 14·68 | 16·00 | 17·24 | 19·40 |
| 15·6 | Interpolated. | 13·52 | 14·99 | 16·35 | 17·62 | 19·81 |
| 16·2 | 325 | 13·80 | 15·31 | 16·79 | 17·99 | 20·23 |
| 17·6 | 300 | 14·45 | 16·02 | 17·48 | 18·83 | 21·18 |
| 19·2 | 275 | 15·19 | 16·85 | 18·37 | 19·79 | 22·26 |
| 21·1 | 250 | 16·04 | 17·80 | 19·40 | 20·91 | 23·52 |
| 23·5 | 225 | 17·05 | 18·91 | 20·62 | 22·21 | 24·99 |
| 26·4 | 200 | 18·25 | 20·24 | 22·07 | 23·78 | 26·75 |
| 30·2 | 175 | 19·71 | 21·87 | 23·85 | 25·69 | 28·90 |
| 35·2 | 150 | 21·57 | 23·92 | 26·09 | 28·11 | 31·62 |
| 37·7 | 140 | 22·45 | 24·91 | 27·16 | 29·26 | 32·92 |
| 42·2 | 125 | 23·99 | 26·62 | 29·03 | 31·27 | 35·18 |
| 48· | 110 | 25·87 | 28·69 | 31·29 | 33·71 | 37·92 |
| 52·8 | 100 | 27·36 | 30·35 | 33·10 | 35·66 | 40·11 |
| 58·7 | 90 | 29·12 | 32·31 | 35·23 | 37·96 | 42·69 |
| 66· | 80 | 31·23 | 34·64 | 37·78 | 40·70 | 45·79 |
| 75·4 | 70 | 33·82 | 37·51 | 40·91 | 44·07 | 49·58 |
| 88·0 | 60 | 37·08 | 41·13 | 44·86 | 48·33 | 54·36 |
| 105·6 | 50 | 41·37 | 45·78 | 50·04 | 53·91 | 60·65 |
| 117·3 | 45 | 44·08 | 48·89 | 53·32 | 57·44 | 64·62 |
| 132· | 40 | 47·32 | 52·49 | 57·25 | 61·67 | 69·37 |
| 150·8 | 35 | 51·30 | 56·90 | 62·06 | 66·86 | 75·20 |
| 176· | 30 | 56·32 | 62·47 | 68·13 | 73·40 | 82·56 |
| 211·2 | 25 | 62·90 | 69·77 | 76·09 | 81·97 | 92·21 |
| 264· | 20 | 72·01 | 79·87 | 87·11 | 93·84 | 105·56 |
| 352· | 15 | 85·68 | 95·05 | 103·66 | 111·67 | 125·61 |
| 528· | 10 | 109·26 | 121·19 | 132·17 | 142·39 | 160·17 |
| 586·7 | 9 | 116·31 | 129·01 | 140·70 | 151·58 | 170·50 |
| 660· | 8 | 124·68 | 138·30 | 150·83 | 162·49 | 182·78 |
| 754·3 | 7 | 134·84 | 149·57 | 163·12 | 175·73 | 197·67 |
| 880· | 6 | 147·69 | 163·60 | 178·42 | 192·22 | 216·22 |
| 1056· | 5 | 163·82 | 181·71 | 198·17 | 213·50 | 240·15 |
| 1320· | 4 | 185·99 | 206·31 | 225·00 | 242·39 | 272·66 |
| 1760· | 3 | 218·58 | 242·46 | 264·42 | 284·86 | 320·43 |
| 2640· | 2 | 273·79 | 303·70 | 331·22 | 356·82 | 401·37 |
| 5280· | 1 | 405·74 | 450·07 | 490·84 | 528·79 | 594·82 |

TABLE VIII.—For finding the mean velocities of water flowing in pipes, drains, streams, and rivers.

For a full cylindrical pipe, divide the diameter by 4 to find the hydraulic mean depth.

OPEN DRAINS AND PIPES.

*Diameters of pipes 6 inches to 12 inches. Falls per mile 1 inch to 12 feet.*

| Falls per mile in feet and inches, and the hydraulic inclinations. || "Hydraulic mean depths," or "mean radii," and velocities in inches per second. |||||
|---|---|---|---|---|---|---|
| Falls. | Inclinations, one in | 1½ inch. | 1¾ in. interpolated | 2 inches. | 2½ inches. | 3 inches. |
| F. I. | | | | | | |
| 0 1 | 63360 | 1·07 | 1·15 | 1·24 | 1·40 | 1·55 |
| 0 2 | 31680 | 1·66 | 1·80 | 1·94 | 2·19 | 2·41 |
| 0 3 | 21120 | 2·12 | 2·30 | 2·48 | 2·80 | 3·08 |
| 0 4 | 15840 | 2·52 | 2·73 | 2·94 | 3·34 | 3·65 |
| 0 5 | 12672 | 2·86 | 3·11 | 3·35 | 3·77 | 4·16 |
| 0 6 | 10560 | 3·18 | 3·45 | 3·72 | 4·19 | 4·62 |
| 0 7 | 9051 | 3·47 | 3·77 | 4·06 | 4·58 | 5·04 |
| 0 8 | 7920 | 3·75 | 4·06 | 4·38 | 4·94 | 5·44 |
| 0 9 | 7040 | 4·01 | 4·34 | 4·68 | 5·28 | 5·81 |
| 0 10 | 6336 | 4·25 | 4·61 | 4·97 | 5·60 | 6·17 |
| 0 11 | 5760 | 4·49 | 4·86 | 5·24 | 5·91 | 6·51 |
| 1 0 | 5280 | 4·71 | 5·11 | 5·51 | 6·21 | 6·84 |
| 1 3 | 4224 | 5·34 | 5·79 | 6·24 | 7·03 | 7·75 |
| 1 6 | 3520 | 5·91 | 6·41 | 6·91 | 7·79 | 8·58 |
| 1 9 | 3017 | 6·44 | 6·99 | 7·53 | 8·49 | 9·35 |
| 2 0 | 2640 | 6·94 | 7·53 | 8·11 | 9·14 | 10·07 |
| 2 3 | Interpolated. | 7·40 | 8·03 | 8·65 | 9·74 | 10·74 |
| 2 6 | 2112 | 7·86 | 8·52 | 9·18 | 10·35 | 11·40 |
| 2 9 | Interpolated. | 8·28 | 8·98 | 9·67 | 10·90 | 12·01 |
| 3 0 | 1760 | 8·70 | 9·43 | 10·16 | 11·45 | 12·62 |
| 3 3 | Interpolated. | 9·09 | 9·85 | 10·62 | 11·97 | 13·19 |
| 3 6 | 1508 | 9·48 | 10·28 | 11·08 | 12·48 | 13·76 |
| 3 9 | Interpolated. | 9·84 | 10·67 | 11·50 | 12·96 | 14·29 |
| 4 0 | 1320 | 10·21 | 11·07 | 11·93 | 13·44 | 14·81 |
| 4 6 | Interpolated. | 10·89 | 11·80 | 12·72 | 14·34 | 15·80 |
| 5 0 | 1056 | 11·56 | 12·54 | 13·51 | 15·23 | 16·78 |
| 5 6 | Interpolated. | 12·18 | 13·21 | 14·24 | 16·04 | 17·68 |
| 6 0 | 880 | 12·80 | 13·88 | 14·96 | 16·86 | 18·58 |
| 6 6 | Interpolated. | 13·38 | 14·51 | 15·64 | 17·62 | 19·42 |
| 7 0 | 754 | 13·96 | 15·14 | 16·32 | 18·59 | 20·26 |
| 7 6 | Interpolated. | 14·51 | 15·73 | 16·95 | 19·10 | 21·05 |
| 8 0 | 660 | 15·05 | 16·32 | 17·58 | 19·82 | 21·84 |
| 8 6 | Interpolated. | 15·56 | 16·87 | 18·18 | 20·49 | 22·58 |
| 9 0 | 587 | 16·07 | 17·43 | 18·78 | 21·17 | 23·32 |
| 9 6 | Interpolated. | 16·57 | 17·97 | 19·36 | 21·82 | 24·04 |
| 10 0 | 528 | 17·06 | 18·50 | 19·94 | 22·47 | 24·76 |
| 10 6 | Interpolated. | 17·54 | 19·01 | 20·49 | 23·09 | 25·45 |
| 11 0 | 480 | 18·01 | 19·53 | 21·04 | 23·72 | 26·13 |
| 11 6 | Interpolated. | 18·47 | 20·02 | 21·57 | 24·32 | 26·79 |
| 12 0 | 440 | 18·92 | 20·51 | 22·11 | 24·91 | 27·45 |

# THE DISCHARGE OF WATER FROM

TABLE VIII.—FOR FINDING THE MEAN VELOCITIES OF WATER FLOWING IN PIPES, DRAINS, STREAMS, AND RIVERS.

For a full cylindrical pipe, divide the diameter by 4 to find the hydraulic mean depth.

PIPES.

*Diameters of pipes 6 inches to 14 inches. Falls per mile 13 feet to 5280 feet.*

| Falls per mile in feet, and the hydraulic inclinations. | | "Hydraulic mean depths," or "mean radii," and velocities in inches per second. | | | | |
|---|---|---|---|---|---|---|
| Falls. | Inclinations, one in | 1½ inch. | 2 inches. | 2½ inches. | 3 inches. | 3½ inches. |
| F. | | | | | | |
| 13·2 | 400 | 19·97 | 23·34 | 26·30 | 28·98 | 31·44 |
| 13·6 | | 20·34 | 23·77 | 26·79 | 29·52 | 32·03 |
| 14·1 | 375 | 20·72 | 24·21 | 27·28 | 30·06 | 32·62 |
| 14·6 | | 21·13 | 24·69 | 27·83 | 30·67 | 33·27 |
| 15·1 | 350 | 21·55 | 25·18 | 28·38 | 31·27 | 33·93 |
| 15·6 | | 22·01 | 25·72 | 28·99 | 31·94 | 34·66 |
| 16·2 | 325 | 22·48 | 26·27 | 29·60 | 32·62 | 35·39 |
| 17·6 | 300 | 23·53 | 27·50 | 30·99 | 34·15 | 37·05 |
| 19·2 | 275 | 24·74 | 28·90 | 32·57 | 35·89 | 38·94 |
| 21·1 | 250 | 26·13 | 30·53 | 34·41 | 37·91 | 41·14 |
| 23·5 | 225 | 27·76 | 32·44 | 36·56 | 40·28 | 43·71 |
| 26·4 | 200 | 29·72 | 34·72 | 39·13 | 43·12 | 46·79 |
| 30·2 | 175 | 32·11 | 37·52 | 42·28 | 46·59 | 50·55 |
| 35·2 | 150 | 35·13 | 41·04 | 46·26 | 50·97 | 55·30 |
| 37·7 | 140 | 36·57 | 42·73 | 48·16 | 53·07 | 57·58 |
| 42·2 | 125 | 39·08 | 45·66 | 51·46 | 56·71 | 61·53 |
| 48· | 110 | 42·13 | 49·23 | 55·48 | 61·13 | 66·33 |
| 52·8 | 100 | 44·57 | 52·07 | 58·69 | 64·67 | 70·17 |
| 58·7 | 90 | 47·43 | 55·42 | 62·46 | 68·83 | 74·68 |
| 66· | 80 | 50·87 | 59·44 | 66·99 | 73·81 | 80·09 |
| 75·4 | 70 | 55·08 | 64·36 | 72·50 | 79·92 | 86·72 |
| 88· | 60 | 60·39 | 70·57 | 79·53 | 87·63 | 95·09 |
| 105·6 | 50 | 67·38 | 78·73 | 88·73 | 97·77 | 106·08 |
| 117·3 | 45 | 71·79 | 83·88 | 94·54 | 104·17 | 113·03 |
| 132· | 40 | 77·07 | 90·06 | 101·50 | 118·84 | 121·35 |
| 150·8 | 35 | 83·55 | 97·63 | 110·03 | 121·24 | 131·55 |
| 176· | 30 | 91·72 | 107·18 | 120·79 | 133·10 | 144·41 |
| 211·2 | 25 | 102·44 | 119·70 | 134·90 | 148·65 | 161·29 |
| 264· | 20 | 117·28 | 137·03 | 154·44 | 170·18 | 184·65 |
| 352· | 15 | 139·56 | 163·06 | 183·78 | 202·50 | 219·72 |
| 528· | 10 | 177·95 | 207·92 | 234·33 | 258·21 | 280·16 |
| 586·7 | 9 | 189·43 | 221·34 | 249·45 | 274·87 | 298·24 |
| 660· | 8 | 203·07 | 237·28 | 267·42 | 294·67 | 319·72 |
| 754·3 | 7 | 219·61 | 256·61 | 289·20 | 318·67 | 345·77 |
| 880· | 6 | 240·22 | 281·36 | 316·33 | 348·57 | 378·20 |
| 1056· | 5 | 266·81 | 311·75 | 351·35 | 387·15 | 420·07 |
| 1320· | 4 | 302·92 | 353·95 | 398·91 | 439·55 | 476·93 |
| 1760· | 3 | 356·00 | 415·96 | 468·80 | 516·57 | 560·49 |
| 2640· | 2 | 445·93 | 521·04 | 587·22 | 647·06 | 702·08 |
| 5280· | 1 | 660·84 | 772·16 | 870·23 | 958·91 | 1040·44 |

TABLE VIII.—For finding the mean velocities of water flowing in pipes, drains, streams, and rivers.

For a full cylindrical pipe, divide the diameter by 4 to find the hydraulic mean depth.

*Diameters of pipes* 14 *inches to* 22 *inches.* *Falls per mile* 1 *inch to* 12 *feet.*

| Falls per mile in feet and inches, and the hydraulic inclinations. || "Hydraulic mean depths," or "mean radii," and velocities in inches per second. |||||
|---|---|---|---|---|---|---|
| Falls. | Inclinations, one in | 3½ inches. | 4 inches. | 4½ inches. | 5 inches. | 5½ inches. |
| F. I. | | | | | | |
| 0 1 | 63360 | 1·68 | 1·80 | 1·91 | 2·02 | 2·13 |
| 0 2 | 31680 | 2·61 | 2·81 | 2·98 | 3·15 | 3·32 |
| 0 3 | 21120 | 3·34 | 3·59 | 3·82 | 4·03 | 4·24 |
| 0 4 | 15840 | 3·96 | 4·25 | 4·52 | 4·78 | 5·02 |
| 0 5 | 12672 | 4·51 | 4·84 | 5·15 | 5·44 | 5·72 |
| 0 6 | 10560 | 5·01 | 5·37 | 5·72 | 6·04 | 6·35 |
| 0 7 | 9051 | 5·47 | 5·87 | 6·24 | 6·60 | 6·94 |
| 0 8 | 7920 | 5·90 | 6·33 | 6·74 | 7·12 | 7·48 |
| 0 9 | 7040 | 6·31 | 6·77 | 7·20 | 7·61 | 8·00 |
| 0 10 | 6336 | 6·70 | 7·18 | 7·64 | 8·08 | 8·49 |
| 0 11 | 5760 | 7·06 | 7·58 | 8·06 | 8·52 | 8·96 |
| 1 0 | 5280 | 7·42 | 7·96 | 8·47 | 8·95 | 9·41 |
| 1 3 | 4224 | 8·41 | 9·02 | 9·60 | 10·14 | 10·66 |
| 1 6 | 3520 | 9·31 | 9·99 | 10·63 | 11·23 | 11·80 |
| 1 9 | 3017 | 10·15 | 10·89 | 11·58 | 12·24 | 12·86 |
| 2 0 | 2640 | 10·93 | 11·73 | 12·47 | 13·18 | 13·86 |
| 2 3 | Interpolated. | 11·65 | 12·50 | 13·30 | 14·05 | 14·77 |
| 2 6 | 2112 | 12·37 | 13·28 | 14·12 | 14·93 | 15·69 |
| 2 9 | Interpolated. | 13·03 | 13·68 | 14·88 | 15·72 | 16·53 |
| 3 0 | 1760 | 13·69 | 14·69 | 15·63 | 16·52 | 17·36 |
| 3 3 | Interpolated. | 14·31 | 15·35 | 16·33 | 17·26 | 18·14 |
| 3 6 | 1508 | 14·92 | 16·01 | 17·03 | 18·00 | 18·92 |
| 3 9 | Interpolated. | 15·50 | 16·63 | 17·69 | 18·70 | 19·65 |
| 4 0 | 1320 | 16·07 | 17·25 | 18·35 | 19·39 | 20·38 |
| 4 6 | Interpolated. | 17·14 | 18·39 | 19·56 | 20·68 | 21·73 |
| 5 0 | 1056 | 18·21 | 19·53 | 20·78 | 21·96 | 23·08 |
| 5 6 | Interpolated. | 19·18 | 20·58 | 21·90 | 23·14 | 24·32 |
| 6 0 | 880 | 20·16 | 21·63 | 23·01 | 24·32 | 25·56 |
| 6 6 | Interpolated. | 21·07 | 22·61 | 24·05 | 25·42 | 26·72 |
| 7 0 | 754 | 21·98 | 23·59 | 25·09 | 26·52 | 27·87 |
| 7 6 | Interpolated. | 22·84 | 24·50 | 26·07 | 27·55 | 28·96 |
| 8 0 | 660 | 23·69 | 25·42 | 27·04 | 28·58 | 30·04 |
| 8 6 | Interpolated. | 24·50 | 26·29 | 27·97 | 29·55 | 31·06 |
| 9 0 | 587 | 25·31 | 27·54 | 28·89 | 30·53 | 32·09 |
| 9 6 | Interpolated. | 26·09 | 27·99 | 29·78 | 31·47 | 33·08 |
| 10 0 | 528 | 26·87 | 28·83 | 30·67 | 32·41 | 34·06 |
| 10 6 | Interpolated. | 27·61 | 29·62 | 31·52 | 33·31 | 35·01 |
| 11 0 | 480 | 28·35 | 30·42 | 32·37 | 34·20 | 35·95 |
| 11 6 | Interpolated. | 29·07 | 31·19 | 33·18 | 35·07 | 36·86 |
| 12 0 | 440 | 29·79 | 31·96 | 34·00 | 35·93 | 37·77 |

## 212 THE DISCHARGE OF WATER FROM

TABLE VIII.—FOR FINDING THE MEAN VELOCITIES OF WATER FLOWING IN PIPES, DRAINS, STREAMS, AND RIVERS.

For a full cylindrical pipe, divide the diameter by 4 to find the hydraulic mean depth.

*Diameters of pipes* 16 *inches to* 2 *feet. Falls per mile* 13 *feet to* 5280 *feet.*

| Falls per mile in feet, and the hydraulic inclinations. | | "Hydraulic mean depths," or "mean radii," and velocities in inches per second. | | | | |
|---|---|---|---|---|---|---|
| Falls. | Inclinations, one in | 4 inches. | 4½ inches. | 5 inches. | 5½ inches. | 6 inches. |
| F. | | | | | | |
| 13·2 | 400 | 33·74 | 35·89 | 37·93 | 39·87 | 41·72 |
| 13·6 | Interpolated. | 34·37 | 36·56 | 38·64 | 40·61 | 42·50 |
| 14·1 | 375 | 35·00 | 37·23 | 39·35 | 41·36 | 43·28 |
| 14·6 | Interpolated. | 35·70 | 37·98 | 40·14 | 42·19 | 44·15 |
| 15·1 | 350 | 36·40 | 38·73 | 40·92 | 43·02 | 45·01 |
| 15·6 | Interpolated. | 37·19 | 39·56 | 41·81 | 43·94 | 45·99 |
| 16·2 | 325 | 37·97 | 40·40 | 42·69 | 44·87 | 46·96 |
| 17·6 | 300 | 39·75 | 42·29 | 44·69 | 46·97 | 49·16 |
| 19·2 | 275 | 41·78 | 44·45 | 46·97 | 49·38 | 51·67 |
| 21·1 | 250 | 44·14 | 46·95 | 49·62 | 52·16 | 54·58 |
| 23·5 | 225 | 46·90 | 49·90 | 52·72 | 55·42 | 58·00 |
| 26·4 | 200 | 50·20 | 53·41 | 56·44 | 59·32 | 62·08 |
| 30·2 | 175 | 54·24 | 57·71 | 60·98 | 64·10 | 67·07 |
| 35·2 | 150 | 59·34 | 63·13 | 66·71 | 70·12 | 73·37 |
| 37·7 | 140 | 61·78 | 65·72 | 69·45 | 73·00 | 76·39 |
| 42·2 | 125 | 66·02 | 70·23 | 74·22 | 78·01 | 81·64 |
| 48· | 110 | 71·17 | 75·72 | 80·01 | 84·10 | 88·00 |
| 52·8 | 100 | 75·29 | 80·09 | 84·64 | 88·97 | 93·10 |
| 58·7 | 90 | 80·13 | 85·25 | 90·08 | 94·69 | 99·09 |
| 66· | 80 | 85·93 | 91·42 | 96·61 | 101·54 | 106·26 |
| 75·4 | 70 | 93·04 | 98·98 | 104·60 | 109·95 | 115·05 |
| 88· | 60 | 102·02 | 108·54 | 114·70 | 120·56 | 126·16 |
| 105·6 | 50 | 113·82 | 121·09 | 127·96 | 134·50 | 140·74 |
| 117·3 | 45 | 121·27 | 129·01 | 136·34 | 143·30 | 149·96 |
| 132· | 40 | 130·20 | 138·51 | 146·38 | 153·86 | 161·00 |
| 150·8 | 35 | 141·14 | 150·16 | 158·68 | 166·79 | 174·53 |
| 176· | 30 | 154·95 | 164·84 | 174·20 | 183·10 | 191·61 |
| 211·2 | 25 | 173·05 | 184·10 | 194·56 | 204·50 | 214·00 |
| 264· | 20 | 198·12 | 210·77 | 222·73 | 234·11 | 244·98 |
| 352· | 15 | 235·75 | 250·80 | 265·04 | 278·58 | 291·52 |
| 528· | 10 | 300·60 | 319·80 | 337·95 | 355·22 | 371·71 |
| 586·7 | 9 | 320·00 | 340·43 | 359·76 | 378·14 | 395·70 |
| 660· | 8 | 343·04 | 359·65 | 385·67 | 405·37 | 424·20 |
| 754·3 | 7 | 370·99 | 394·68 | 417·08 | 438·39 | 458·76 |
| 880· | 6 | 405·79 | 431·70 | 456·21 | 479·52 | 501·79 |
| 1056· | 5 | 450·71 | 479·49 | 506·71 | 532·60 | 557·34 |
| 1320· | 4 | 511·72 | 544·39 | 575·30 | 604·69 | 632·78 |
| 1760· | 3 | 601·38 | 639·78 | 676·10 | 710·64 | 743·65 |
| 2640· | 2 | 753·29 | 801·39 | 846·89 | 890·16 | 931·50 |
| 5280· | 1 | 1116·35 | 1187·62 | 1255·04 | 1319·17 | 1380·44 |

ORIFICES, WEIRS, PIPES, AND RIVERS. 213

TABLE VIII.—FOR FINDING THE MEAN VELOCITIES OF WATER FLOWING IN PIPES, DRAINS, STREAMS, AND RIVERS.

The hydraulic mean depth is found for all channels, by dividing the wetted perimeter into the area.

Hydraulic mean depths 6 inches to 10 inches.  Falls per mile 1 inch to 12 feet.

| Falls per mile in feet and inches, and the hydraulic inclinations. | | "Hydraulic mean depths," or "mean radii," and velocities in inches per second. | | | | |
|---|---|---|---|---|---|---|
| Falls. | Inclinations, one in | 6 inches. | 7 inches. | 8 inches. | 9 inches. | 10 inches. |
| F.  I. | | | | | | |
| 0  1 | 63360 | 2·23 | 2·41 | 2·58 | 2·75 | 2·90 |
| 0  2 | 31680 | 3·47 | 3·76 | 4·03 | 4·28 | 4·52 |
| 0  3 | 21120 | 4·43 | 4·80 | 5·15 | 5·47 | 5·78 |
| 0  4 | 15840 | 5·26 | 5·69 | 6·10 | 6·49 | 6·85 |
| 0  5 | 12672 | 5·93 | 6·48 | 6·95 | 7·39 | 7·80 |
| 0  6 | 10560 | 6·65 | 7·20 | 7·72 | 8·20 | 8·66 |
| 0  7 | 9051 | 7·26 | 7·86 | 8·43 | 8·96 | 9·46 |
| 0  8 | 7920 | 7·83 | 8·48 | 9·09 | 9·67 | 10·21 |
| 0  9 | 7040 | 8·37 | 9·07 | 9·72 | 10·33 | 10·91 |
| 0 10 | 6336 | 8·88 | 9·63 | 10·32 | 10·97 | 11·58 |
| 0 11 | 5760 | 9·37 | 10·16 | 10·89 | 11·57 | 12·22 |
| 1  0 | 5280 | 9·84 | 10·67 | 11·43 | 12·15 | 12·83 |
| 1  3 | 4224 | 11·16 | 12·09 | 12·95 | 13·77 | 14·54 |
| 1  6 | 3520 | 12·35 | 13·38 | 14·34 | 15·25 | 16·10 |
| 1  9 | 3017 | 13·46 | 14·58 | 15·63 | 16·61 | 17·54 |
| 2  0 | 2640 | 14·50 | 15·71 | 16·84 | 17·90 | 18·90 |
| 2  3 | Interpolated. | 15·45 | 16·75 | 18·24 | 19·08 | 20·15 |
| 2  6 | 2112 | 16·42 | 17·79 | 19·64 | 20·26 | 21·40 |
| 2  9 | Interpolated. | 17·29 | 18·74 | 20·37 | 21·34 | 22·54 |
| 3  0 | 1760 | 18·17 | 19·69 | 21·10 | 22·42 | 23·68 |
| 3  3 | Interpolated. | 18·99 | 20·57 | 22·05 | 23·43 | 24·75 |
| 3  6 | 1508 | 19·80 | 21·46 | 23·00 | 24·44 | 25·81 |
| 3  9 | Interpolated. | 20·56 | 22·28 | 23·88 | 25·38. | 26·80 |
| 4  0 | 1320 | 21·33 | 23·11 | 24·77 | 26·32 | 27·80 |
| 4  6 | Interpolated. | 22·74 | 24·64 | 26·41 | 28·07 | 29·64 |
| 5  0 | 1056 | 24·16 | 26·17 | 28·05 | 29·81 | 31·48 |
| 5  6 | Interpolated. | 25·45 | 27·58 | 29·56 | 31·42 | 33·17 |
| 6  0 | 880 | 26·75 | 28·98 | 31·06 | 33·02 | 34·86 |
| 6  6 | Interpolated. | 27·96 | 30·29 | 32·47 | 34·51 | 36·44 |
| 7  0 | 754 | 29·17 | 31·60 | 33·87 | 36·00 | 38·02 |
| 7  6 | Interpolated. | 30·30 | 32·83 | 35·19 | 37·40 | 39·50 |
| 8  0 | 660 | 31·43 | 34·06 | 36·50 | 38·80 | 40·97 |
| 8  6 | Interpolated. | 32·51 | 35·22 | 37·75 | 40·12 | 42·37 |
| 9  0 | 587 | 33·58 | 36·39 | 38·99 | 41·45 | 43·77 |
| 9  6 | Interpolated. | 34·61 | 37·50 | 40·20 | 42·72 | 45·11 |
| 10  0 | 528 | 35·65 | 38·63 | 41·40 | 44·00 | 46·46 |
| 10  6 | Interpolated. | 36·63 | 39·69 | 42·54 | 45·22 | 47·75 |
| 11  0 | 480 | 37·62 | 40·76 | 43·69 | 46·44 | 49·03 |
| 11  6 | Interpolated. | 38·57 | 41·79 | 44·79 | 47·61 | 50·27 |
| 12  0 | 440 | 39·52 | 42·82 | 45·90 | 48·78 | 51·51 |

TABLE VIII.—FOR FINDING THE MEAN VELOCITIES OF WATER FLOWING IN PIPES, DRAINS, STREAMS, AND RIVERS.

The hydraulic mean depth is found for all channels by dividing the wetted perimeter into the area.

*Hydraulic mean depths 11 inches to 21 inches. Falls per mile 1 inch to 12 feet.*

| Falls per mile in feet and inches, and the hydraulic inclinations. | | "Hydraulic mean depths," or "mean radii," and velocities in inches per second. | | | | |
|---|---|---|---|---|---|---|
| Falls. | Inclinations, one in | 11 inches. | 12 inches. | 15 inches. | 18 inches. | 21 inches. |
| F. I. | | | | | | |
| 0  1 | 63360 | 3·05 | 3·19 | 3·57 | 3·92 | 4·25 |
| 0  2 | 31680 | 4·75 | 4·97 | 5·57 | 6·12 | 6·62 |
| 0  3 | 21120 | 6·07 | 6·35 | 7·12 | 7·82 | 8·46 |
| 0  4 | 15840 | 7·19 | 7·53 | 8·44 | 9·27 | 10·03 |
| 0  5 | 12672 | 8·19 | 8·57 | 9·61 | 10·55 | 11·42 |
| 0  6 | 10560 | 9·10 | 9·52 | 10·67 | 11·72 | 12·68 |
| 0  7 | 9051 | 9·94 | 10·39 | 11·66 | 12·80 | 13·85 |
| 0  8 | 7920 | 10·72 | 11·21 | 12·57 | 13·81 | 14·94 |
| 0  9 | 7041 | 11·46 | 11·99 | 13·44 | 14·76 | 15·97 |
| 0 10 | 6336 | 12·16 | 12·72 | 14·27 | 15·66 | 16·95 |
| 0 11 | 5760 | 12·83 | 13·42 | 15·05 | 16·53 | 17·88 |
| 1  0 | 5280 | 13·48 | 14·09 | 15·81 | 17·36 | 18·78 |
| 1  3 | 4224 | 15·27 | 15·97 | 17·91 | 19·67 | 21·28 |
| 1  6 | 3520 | 16·91 | 17·68 | 19·83 | 21·78 | 23·56 |
| 1  9 | 3017 | 18·23 | 19·27 | 21·62 | 23·73 | 25·68 |
| 2  0 | 2640 | 19·85 | 20·76 | 23·28 | 25·63 | 27·66 |
| 2  3 | Interpolated. | 21·16 | 22·13 | 24·82 | 27·29 | 29·49 |
| 2  6 | 2112 | 22·48 | 23·51 | 26·36 | 28·95 | 31·32 |
| 2  9 | Interpolated. | 23·68 | 24·76 | 27·77 | 30·49 | 32·99 |
| 3  0 | 1760 | 24·88 | 26·02 | 29·18 | 32·04 | 34·67 |
| 3  3 | Interpolated. | 25·99 | 27·18 | 30·47 | 33·48 | 36·22 |
| 3  6 | 1508 | 27·11 | 28·35 | 31·77 | 34·92 | 37·78 |
| 3  9 | Interpolated. | 28·15 | 29·45 | 33·01 | 36·26 | 39·23 |
| 4  0 | 1320 | 29·20 | 30·54 | 34·25 | 37·60 | 40·69 |
| 4  6 | Interpolated. | 31·13 | 32·56 | 36·52 | 40·10 | 43·39 |
| 5  0 | 1056 | 33·07 | 34·59 | 38·79 | 42·59 | 46·09 |
| 5  6 | Interpolated. | 34·85 | 36·44 | 40·87 | 44·88 | 48·56 |
| 6  0 | 880 | 36·62 | 38·30 | 42·95 | 47·16 | 51·03 |
| 6  6 | Interpolated. | 38·28 | 40·03 | 44·90 | 49·30 | 53·34 |
| 7  0 | 754 | 39·93 | 41·76 | 46·84 | 51·43 | 55·65 |
| 7  6 | Interpolated. | 41·48 | 43·39 | 48·66 | 53·43 | 57·81 |
| 8  0 | 660 | 43·04 | 45·01 | 50·48 | 55·42 | 59·97 |
| 8  6 | Interpolated. | 44·50 | 46·54 | 52·20 | 57·32 | 62·02 |
| 9  0 | 587 | 45·97 | 48·08 | 53·92 | 59·21 | 64·06 |
| 9  6 | Interpolated. | 47·39 | 49·56 | 55·58 | 61·03 | 66·04 |
| 10  0 | 528 | 48·80 | 51·04 | 57·24 | 62·85 | 68·01 |
| 10  6 | Interpolated. | 50·15 | 52·45 | 58·83 | 64·59 | 69·89 |
| 11  0 | 480 | 51·51 | 53·87 | 60·41 | 66·33 | 71·78 |
| 11  6 | Interpolated. | 52·81 | 55·23 | 61·94 | 68·01 | 73·59 |
| 12  0 | 440 | 54·11 | 56·59 | 63·47 | 69·68 | 75·40 |

ORIFICES, WEIRS, PIPES, AND RIVERS.

TABLE VIII.—FOR FINDING THE MEAN VELOCITIES OF WATER FLOWING IN PIPES, DRAINS, STREAMS, AND RIVERS.

The hydraulic mean depth is found for all channels by dividing the wetted perimeter into the area.

*Hydraulic mean depths 24 inches to 4 feet. Falls per mile 1 inch to 12 feet.*

| Falls per mile in feet and inches, and the hydraulic inclinations. || "Hydraulic mean depths," or "mean radii," and velocities in inches per second. |||||
|---|---|---|---|---|---|---|
| Falls. || Inclinations, one in | 24 inches. | 30 inches. | 36 inches. | 42 inches. | 48 inches. |
| F. | I. | | | | | | |
| 0 | 1 | 63360 | 4·54 | 5·09 | 5·59 | 6·04 | 6·47 |
| 0 | 2 | 31680 | 7·09 | 7·94 | 8·71 | 9·42 | 10·08 |
| 0 | 3 | 21120 | 9·06 | 10·15 | 11·14 | 12·04 | 12·89 |
| 0 | 4 | 15840 | 10·73 | 12·03 | 13·20 | 14·27 | 15·27 |
| 0 | 5 | 12672 | 12·22 | 13·69 | 15·03 | 16·25 | 17·39 |
| 0 | 6 | 10560 | 13·57 | 15·21 | 16·69 | 18·05 | 19·31 |
| 0 | 7 | 9051 | 14·83 | 16·61 | 18·23 | 19·71 | 21·09 |
| 0 | 8 | 7920 | 15·99 | 17·92 | 19·66 | 21·27 | 22·76 |
| 0 | 9 | 7041 | 17·10 | 19·16 | 21·02 | 22·73 | 24·33 |
| 0 | 10 | 6336 | 18·15 | 20·33 | 22·31 | 24·13 | 25·82 |
| 0 | 11 | 5760 | 19·15 | 21·45 | 23·54 | 25·46 | 27·24 |
| 1 | 0 | 5280 | 20·11 | 22·53 | 24·72 | 26·73 | 28·61 |
| 1 | 3 | 4224 | 22·78 | 25·53 | 28·01 | 30·29 | 32·42 |
| 1 | 6 | 3520 | 25·23 | 28·27 | 31·02 | 33·54 | 35·90 |
| 1 | 9 | 3017 | 27·49 | 30·81 | 33·80 | 36·55 | 39·12 |
| 2 | 0 | 2640 | 29·62 | 33·18 | 36·41 | 39·38 | 42·14 |
| 2 | 3 | Interpolated. | 31·57 | 35·38 | 38·82 | 41·98 | 44·92 |
| 2 | 6 | 2112 | 33·53 | 37·57 | 41·22 | 44·58 | 47·71 |
| 2 | 9 | Interpolated. | 35·32 | 39·58 | 43·43 | 46·96 | 50·26 |
| 3 | 0 | 1760 | 37·11 | 41·58 | 45·63 | 49·34 | 52·81 |
| 3 | 3 | Interpolated. | 38·78 | 43·45 | 47·68 | 51·56 | 55·18 |
| 3 | 6 | 1508 | 40·45 | 45·32 | 49·73 | 53·78 | 57·55 |
| 3 | 9 | Interpolated. | 42·00 | 47·07 | 51·64 | 55·85 | 59·77 |
| 4 | 0 | 1320 | 43·56 | 48·81 | 53·56 | 57·92 | 61·98 |
| 4 | 6 | Interpolated. | 46·45 | 52·05 | 57·11 | 61·76 | 66·09 |
| 5 | 0 | 1056 | 49·34 | 55·28 | 60·66 | 65·60 | 70·20 |
| 5 | 6 | Interpolated. | 51·99 | 58·25 | 63·91 | 69·12 | 73·97 |
| 6 | 0 | 880 | 54·63 | 61·22 | 67·17 | 72·64 | 77·74* |
| 6 | 6 | Interpolated. | 57·11 | 63·99 | 70·21 | 75·93* | 81·25 |
| 7 | 0 | 754 | 59·58 | 66·76 | 73·25 | 79·21 | 84·77 |
| 7 | 6 | Interpolated. | 61·89 | 69·35 | 76·09* | 87·29 | 88·06 |
| 8 | 0 | 660 | 64·21 | 71·94 | 78·94 | 85·37 | 91·35 |
| 8 | 6 | Interpolated. | 66·40 | 74·40 | 81·63 | 88·26 | 94·47 |
| 9 | 0 | 587 | 68·59 | 76·85* | 84·32 | 91·19 | 97·59 |
| 9 | 6 | Interpolated. | 70·60 | 79·22 | 86·92 | 94·00 | 100·59 |
| 10 | 0 | 528 | 72·81 | 81·58 | 89·52 | 96·81 | 103·60 |
| 10 | 6 | Interpolated. | 74·83 | 83·84 | 91·99 | 99·49 | 106·47 |
| 11 | 0 | 480 | 76·84* | 86·10 | 94·47 | 102·17 | 109·33 |
| 11 | 6 | Interpolated. | 78·78 | 88·28 | 96·86 | 104·75 | 112·10 |
| 12 | 0 | 440 | 80·72 | 90·45 | 99·25 | 107·33 | 114·86 |

TABLE VIII.—FOR FINDING THE MEAN VELOCITIES OF WATER FLOWING IN PIPES, DRAINS, STREAMS, AND RIVERS.

The hydraulic mean depth is found for all channels by dividing the wetted perimeter into the area.

*Hydraulic mean depths 4 feet 6 inches to 7 feet.  Falls per mile 1 inch to 12 feet.*

| Falls per mile in feet and inches, and the hydraulic inclinations. | | "Hydraulic mean depths," or "mean radii," and velocities in inches per second. | | | | |
|---|---|---|---|---|---|---|
| Falls. | Inclinations, one in | 54 inches. | 60 inches. | 66 inches. | 72 inches. | 84 inches. |
| F. I. | | | | | | |
| 0  1 | 63360 | 6·86 | 7·24 | 7·60 | 7·94 | 8·58 |
| 0  2 | 31680 | 10·70 | 11·29 | 11·85 | 12·38 | 13·39 |
| 0  3 | 21120 | 13·68 | 14·62 | 15·14 | 15·83 | 17·11 |
| 0  4 | 15840 | 16·21 | 17·10 | 17·95 | 18·76 | 20·28 |
| 0  5 | 12672 | 18·46 | 19·47 | 20·43 | 21·35 | 23·13 |
| 0  6 | 10560 | 20·50 | 21·63 | 22·70 | 23·72 | 25·64 |
| 0  7 | 9051 | 22·39 | 23·62 | 24·79 | 25·90 | 28·00 |
| 0  8 | 7920 | 24·16 | 25·48 | 26·74 | 27·95 | 30·21 |
| 0  9 | 7041 | 25·83 | 27·24 | 28·59 | 29·88 | 32·30 |
| 0 10 | 6336 | 27·41 | 28·91 | 30·34 | 31·71 | 34·28 |
| 0 11 | 5760 | 28·92 | 30·51 | 32·01 | 33·46 | 36·17 |
| 1  0 | 5280 | 30·37 | 32·03 | 33·62 | 35·13 | 37·98 |
| 1  3 | 4224 | 34·41 | 36·30 | 38·10 | 39·81 | 43·04 |
| 1  6 | 3520 | 38·10 | 40·19 | 42·18 | 44·08 | 47·65 |
| 1  9 | 3017 | 41·52 | 43·80 | 45·97 | 48·04 | 51·93 |
| 2  0 | 2640 | 44·73 | 47·18 | 49·52 | 51·75 | 55·94 |
| 2  3 | Interpolated. | 47·69 | 50·30 | 52·79 | 55·17 | 59·64 |
| 2  6 | 2112 | 50·65 | 53·42 | 56·07 | 58·59 | 63·34 |
| 2  9 | Interpolated. | 53·35 | 56·28 | 59·06 | 61·72 | 66·72 |
| 3  0 | 1760 | 56·06 | 59·13 | 62·05 | 64·85 | 70·10 |
| 3  3 | Interpolated. | 58·57 | 61·79 | 64·84 | 67·76 | 73·25 |
| 3  6 | 1508 | 61·09 | 64·44 | 67·63 | 70·67 | 76·40* |
| 3  9 | Interpolated. | 63·44 | 66·92 | 70·23 | 73·39 | 79·35 |
| 4  0 | 1320 | 65·80 | 69·41 | 72·84 | 76·11* | 82·29 |
| 4  6 | Interpolated. | 70·16 | 74·01 | 77·67* | 81·16 | 87·74 |
| 5  0 | 1056 | 74·52 | 78·61* | 82·50 | 86·21 | 93·20 |
| 5  6 | Interpolated. | 78·52* | 82·83 | 86·92 | 90·84 | 98·20 |
| 6  0 | 880 | 82·52 | 87·05 | 91·35 | 95·46 | 103·20 |
| 6  6 | Interpolated. | 86·25 | 90·98 | 95·58 | 99·78 | 107·87 |
| 7  0 | 754 | 89·99 | 94·92 | 99·62 | 104·10 | 112·54 |
| 7  6 | Interpolated. | 93·48 | 98·61 | 103·48 | 108·14 | 116·91 |
| 8  0 | 660 | 96·98 | 102·30 | 107·35 | 112·19 | 121·28 |
| 8  6 | Interpolated. | 100·29 | 105·79 | 111·02 | 116·01 | 125·42 |
| 9  0 | 587 | 103·59 | 109·27 | 114·68 | 119·84 | 129·56 |
| 9  6 | Interpolated. | 106·78 | 112·64 | 118·21 | 123·53 | 133·55 |
| 10  0 | 528 | 109·97 | 116·01 | 121·74 | 127·22 | 137·54 |
| 10  6 | Interpolated. | 113·02 | 119·22 | 125·11 | 130·74 | 141·34 |
| 11  0 | 480 | 116·06 | 122·43 | 128·48 | 134·27 | 145·15 |
| 11  6 | Interpolated. | 119·00 | 125·52 | 131·73 | 137·66 | 148·82 |
| 12  0 | 440 | 121·93 | 128·61 | 134·97 | 141·05 | 152·49 |

ORIFICES, WEIRS, PIPES, AND RIVERS. 217

TABLE VIII.—For FINDING THE MEAN VELOCITIES OF WATER FLOWING IN PIPES, DRAINS, STREAMS, AND RIVERS.

The hydraulic mean depth is found for all channels by dividing the wetted perimeter into the area.

*Hydraulic mean depths 8 feet to 12 feet. Falls per mile 1 inch to 12 feet.*

| Falls per mile in feet and inches, and the hydraulic inclinations. | | "Hydraulic mean depths," or "mean radii," and velocities in inches per second. | | | | |
|---|---|---|---|---|---|---|
| Falls. | Inclinations, one in | 96 inches. | 108 inches. | 120 inches. | 132 inches. | 144 inches. |
| F.  I. | | | | | | |
| 0   1 | 63360 | 9·18 | 9·75 | 10·28 | 10·79 | 11·27 |
| 0   2 | 31680 | 14·32 | 15·20 | 16·03 | 16·82 | 17·57 |
| 0   3 | 21120 | 18·30 | 19·43 | 20·49 | 21·50 | 22·46 |
| 0   4 | 15840 | 21·69 | 23·02 | 24·28 | 25·47 | 26·62 |
| 0   5 | 12672 | 24·70 | 26·21 | 27·64 | 29·00 | 30·31 |
| 0   6 | 10560 | 27·43 | 29·11 | 30·70 | 32·21 | 33·66 |
| 0   7 | 9051 | 29·96 | 31·80 | 33·53 | 35·18 | 36·76 |
| 0   8 | 7920 | 32·32 | 34·30 | 36·18 | 37·96 | 39·66 |
| 0   9 | 7041 | 34·55 | 36·67 | 38·67 | 40·58 | 42·40 |
| 0  10 | 6336 | 36·67 | 38·92 | 41·04 | 43·07 | 45·00 |
| 0  11 | 5760 | 38·69 | 41·06 | 43·31 | 45·44 | 47·48 |
| 1   0 | 5280 | 40·63 | 43·12 | 45·48 | 47·72 | 49·86 |
| 1   3 | 4224 | 46·04 | 48·87 | 51·54 | 54·07 | 56·50 |
| 1   6 | 3520 | 50·98 | 54·11 | 57·06 | 59·87 | 62·56 |
| 1   9 | 3017 | 55·60 | 58·96 | 62·18 | 65·25 | 68·17 |
| 2   0 | 2640 | 59·85 | 63·52 | 66·98 | 70·28 | 73·44* |
| 2   3 | Interpolated. | 63·80 | 67·72 | 71·41 | 74·93* | 78·29 |
| 2   6 | 2112 | 67·76 | 71·91 | 75·84* | 79·58 | 83·15 |
| 2   9 | Interpolated. | 71·38 | 75·75* | 79·89 | 83·83 | 87·59 |
| 3   0 | 1760 | 75·00* | 79·59 | 83·94 | 88·08 | 92·03 |
| 3   3 | Interpolated. | 78·37 | 83·17 | 87·71 | 92·03 | 96·16 |
| 3   6 | 1508 | 81·74 | 86·75 | 91·48 | 95·99 | 100·30 |
| 3   9 | Interpolated. | 84·88 | 90·09 | 95·01 | 99·69 | 104·16 |
| 4   0 | 1320 | 88·03 | 93·43 | 98·53 | 103·38 | 108·02 |
| 4   6 | Interpolated. | 93·87 | 99·62 | 105·06 | 110·24 | 115·18 |
| 5   0 | 1056 | 99·70 | 105·82 | 111·59 | 117·09 | 122·34 |
| 5   6 | Interpolated. | 105·06 | 111·49 | 117·58 | 123·38 | 128·91 |
| 6   0 | 880 | 110·41 | 117·17 | 123·57 | 129·66 | 135·48 |
| 6   6 | Interpolated. | 115·40 | 122·47 | 129·16 | 135·53 | 141·61 |
| 7   0 | 754 | 120·40 | 127·76 | 134·75 | 141·39 | 147·73 |
| 7   6 | Interpolated. | 125·07 | 132·74 | 139·99 | 146·88 | 153·47 |
| 8   0 | 660 | 129·75 | 137·70 | 145·22 | 152·38 | 159·21 |
| 8   6 | Interpolated. | 134·18 | 142·40 | 150·18 | 157·57 | 164·64 |
| 9   0 | 587 | 138·60 | 147·10 | 155·13 | 162·77 | 170·07 |
| 9   6 | Interpolated. | 142·87 | 151·63 | 159·91 | 167·78 | 175·31 |
| 10  0 | 528 | 147·14 | 156·16 | 164·68 | 172·80 | 180·55 |
| 10  6 | Interpolated. | 151·21 | 160·48 | 169·24 | 177·58 | 185·55 |
| 11  0 | 480 | 155·29 | 164·80 | 173·80 | 182·36 | 190·54 |
| 11  6 | Interpolated. | 159·21 | 168·97 | 178·19 | 186·97 | 195·36 |
| 12  0 | 440 | 163·13 | 173·13 | 182·59 | 191·58 | 200·17 |

218    THE DISCHARGE OF WATER FROM

TABLE IX.—FOR FINDING THE DISCHARGE IN CUBIC FEET, PER MINUTE, WHEN THE DIAMETER OF A PIPE, OR ORIFICE, AND THE VELOCITY OF DISCHARGE ARE KNOWN; AND VICE VERSA.

| Diameters of pipes in inches. | Discharge in cubic feet per minute, for different velocities. | | | | |
|---|---|---|---|---|---|
| | Velocity of 100 inches per second. | Velocity of 200 inches per second. | Velocity of 300 inches per second. | Velocity of 400 inches per second. | Velocity of 500 inches per second. |
| ¼ | ·170442 | ·3409 | ·5113 | ·6818 | ·8522 |
| ½ | ·68177 | 1·3635 | 2·0453 | 2·7271 | 3·4089 |
| ¾ | 1·53398 | 3·0679 | 4·6019 | 6·1359 | 7·6699 |
| 1 | 2·727077 | 5·4541 | 8·1812 | 10·9083 | 13·6354 |
| 1¼ | 4·26106 | 8·5221 | 12·7832 | 17·0442 | 21·3053 |
| 1½ | 6·13593 | 12·2718 | 18·4080 | 24·5437 | 30·6797 |
| 1¾ | 8·35167 | 16·7033 | 25·0550 | 33·4067 | 41·7584 |
| 2 | 10·90831 | 21·1817 | 32·7249 | 43·6332 | 54·5415 |
| 2¼ | 13·80583 | 27·6117 | 41·4175 | 55·2233 | 69·0291 |
| 2½ | 17·04423 | 34·0885 | 51·1327 | 68·1769 | 85·2212 |
| 2¾ | 20·62352 | 41·2470 | 61·8706 | 82·4941 | 103·1176 |
| 3 | 24·54369 | 49·0874 | 73·6311 | 98·1748 | 121·7185 |
| 3¼ | 28·80475 | 57·6095 | 86·4143 | 115·2190 | 144·0238 |
| 3½ | 33·40669 | 66·8134 | 100·2201 | 133·6268 | 167·0335 |
| 3¾ | 38·34952 | 76·6990 | 115·0486 | 153·3981 | 191·7476 |
| 4 | 43·63323 | 87·2665 | 130·8997 | 174·5329 | 218·1662 |
| 4¼ | 49·25783 | 98·5157 | 147·7735 | 197·0313 | 246·2892 |
| 4½ | 55·22331 | 110·4466 | 165·6699 | 220·8932 | 276·1166 |
| 4¾ | 61·52968 | 123·0594 | 184·5890 | 246·1187 | 307·6484 |
| 5 | 68·17692 | 136·3539 | 204·5308 | 272·7077 | 340·8846 |
| 5¼ | 75·16506 | 150·3301 | 225·4952 | 300·6603 | 375·8253 |
| 5½ | 82·49408 | 164·9882 | 247·4822 | 329·9763 | 412·4704 |
| 5¾ | 90·16399 | 180·3280 | 270·4920 | 360·6560 | 450·8200 |
| 6 | 98·17478 | 196·3495 | 294·5243 | 392·6991 | 490·8739 |
| 6¼ | 106·52645 | 213·0529 | 319·5794 | 426·1058 | 532·6323 |
| 6½ | 115·2190 | 230·4380 | 345·6570 | 460·8760 | 576·0950 |
| 6¾ | 124·25245 | 248·5049 | 372·7574 | 497·0098 | 621·2623 |
| 7 | 133·6268 | 267·2536 | 400·8804 | 534·5072 | 668·1340 |
| 7¼ | 143·34199 | 286·6840 | 430·0260 | 573·3680 | 716·7100 |
| 7½ | 153·39809 | 306·7962 | 460·1943 | 613·5924 | 766·9905 |
| 7¾ | 163·79507 | 327·5901 | 491·3852 | 655·1803 | 818·9753 |
| 8 | 174·53293 | 349·0659 | 523·5988 | 698·1317 | 872·6647 |
| 8½ | 197·03132 | 394·0626 | 591·0940 | 788·1253 | 985·1566 |
| 9 | 220·89325 | 441·7865 | 662·6798 | 883·5730 | 1104·4663 |
| 9½ | 246·11871 | 492·2374 | 738·3561 | 984·4748 | 1230·5936 |
| 10 | 272·70771 | 345·4154 | 818·1231 | 1090·8308 | 1363·5386 |
| 10½ | 300·66025 | 601·3205 | 901·9808 | 1202·6410 | 1503·3013 |
| 11 | 329·97633 | 659·9527 | 989·9290 | 1319·9053 | 1649·8817 |
| 11½ | 360·65595 | 721·3119 | 1081·9679 | 1442·6238 | 1803·2798 |
| 12 | 392·6991 | 785·3982 | 1178·0973 | 1570·7964 | 1963·4955 |

ORIFICES, WEIRS, PIPES, AND RIVERS.

TABLE IX.—FOR FINDING THE DISCHARGE IN CUBIC FEET, PER MINUTE, WHEN THE DIAMETER OF A PIPE, OR ORIFICE, AND THE VELOCITY OF DISCHARGE ARE KNOWN; AND VICE VERSA.

| Discharge in cubic feet per minute, for different velocities. | | | | | Diameters of pipes in inches. |
|---|---|---|---|---|---|
| Velocity of 600 inches per second. | Velocity of 700 inches per second. | Velocity of 800 inches per second. | Velocity of 900 inches per second. | Velocity of 1000 inches per second. | |
| 1·0227 | 1·1931 | 1·3635 | 1·5340 | 1·7044 | $\frac{1}{4}$ |
| 4·0906 | 4·7724 | 5·4542 | 6·1359 | 6·8177 | $\frac{1}{2}$ |
| 9·2039 | 10·7379 | 12·2718 | 13·8058 | 15·3398 | $\frac{3}{4}$ |
| 16·3625 | 19·0895 | 21·8166 | 24·5437 | 27·2708 | 1 |
| 25·5664 | 29·8274 | 34·0885 | 38·3495 | 42·6106 | $1\frac{1}{4}$ |
| 36·8155 | 42·9515 | 49·0874 | 55·2234 | 61·3593 | $1\frac{1}{2}$ |
| 50·1100 | 58·4617 | 66·8134 | 75·1650 | 83·5167 | $1\frac{3}{4}$ |
| 65·4499 | 76·3582 | 87·2665 | 98·1748 | 109·0831 | 2 |
| 82·8350 | 96·6408 | 110·4466 | 124·2525 | 138·0583 | $2\frac{1}{4}$ |
| 102·2654 | 119·3096 | 136·3538 | 153·3981 | 170·4423 | $2\frac{1}{2}$ |
| 123·7411 | 144·3646 | 164·9882 | 185·6117 | 206·2352 | $2\frac{3}{4}$ |
| 147·2621 | 171·8059 | 196·3496 | 220·8933 | 245·4369 | 3 |
| 172·8285 | 201·6233 | 230·4380 | 259·2428 | 288·0475 | $3\frac{1}{4}$ |
| 200·4401 | 233·8468 | 267·2535 | 300·6602 | 334·0669 | $3\frac{1}{2}$ |
| 230·0971 | 268·4467 | 306·7962 | 345·1457 | 383·4952 | $3\frac{3}{4}$ |
| 261·7994 | 305·4326 | 349·0659 | 392·6991 | 436·3323 | 4 |
| 295·5470 | 344·8048 | 394·0626 | 443·3205 | 492·5783 | $4\frac{1}{4}$ |
| 331·3399 | 386·5632 | 441·7865 | 497·0098 | 552·2331 | $4\frac{1}{2}$ |
| 369·1781 | 430·7077 | 492·2374 | 553·7671 | 615·2968 | $4\frac{3}{4}$ |
| 409·0615 | 477·2384 | 545·4154 | 613·5923 | 681·7692 | 5 |
| 450·9904 | 526·1554 | 601·3205 | 676·4855 | 751·6506 | $5\frac{1}{4}$ |
| 494·9645 | 577·4586 | 659·9526 | 742·4467 | 824·9408 | $5\frac{1}{2}$ |
| 540·9839 | 631·1479 | 721·3119 | 811·4759 | 901·6399 | $5\frac{3}{4}$ |
| 589·0486 | 687·2235 | 785·3982 | 883·5730 | 981·7478 | 6 |
| 639·1587 | 745·6852 | 852·2116 | 958·7381 | 1065·2645 | $6\frac{1}{4}$ |
| 691·3141 | 806·5330 | 921·7520 | 1036·9710 | 1152·1900 | $6\frac{1}{2}$ |
| 745·5147 | 869·7672 | 994·0196 | 1118·2721 | 1242·5245 | $6\frac{3}{4}$ |
| 801·7608 | 935·3876 | 1069·0144 | 1202·6412 | 1336·2680 | 7 |
| 860·0519 | 1003·3939 | 1146·7359 | 1290·0779 | 1433·4199 | $7\frac{1}{4}$ |
| 920·3885 | 1073·7866 | 1227·1847 | 1380·5828 | 1533·9809 | $7\frac{1}{2}$ |
| 982·7704 | 1146·5655 | 1310·3605 | 1474·1556 | 1637·9507 | $7\frac{3}{4}$ |
| 1047·1976 | 1221·7305 | 1396·2634 | 1570·7964 | 1745·3293 | 8 |
| 1182·1879 | 1379·2192 | 1576·2506 | 1773·2819 | 1970·3132 | $8\frac{1}{2}$ |
| 1325·3595 | 1546·2528 | 1767·1460 | 1988·0393 | 2208·9325 | 9 |
| 1476·7123 | 1722·8310 | 1968·9497 | 2215·0683 | 2461·1871 | $9\frac{1}{2}$ |
| 1636·2463 | 1908·9540 | 2181·6617 | 2454·3694 | 2727·0771 | 10 |
| 1803·9615 | 2104·6218 | 2405·2820 | 2705·9423 | 3006·6025 | $10\frac{1}{2}$ |
| 1979·8580 | 2309·8343 | 2639·8106 | 2969·7870 | 3299·7633 | 11 |
| 2163·9357 | 2524·5917 | 2885·2476 | 3245·9936 | 3606·5595 | $11\frac{1}{2}$ |
| 2356·1946 | 2748·8937 | 3141·5928 | 3534·2919 | 3926·9910 | 12 |

TABLE X.—For finding the depths on weirs of different lengths, the quantity discharged over each being supposed constant. (See pages 149 and 150.)

| Ratios of lengths. | Coefficients. | Ratios of lengths. | Coefficients. | Ratios of lengths. | Coefficients. | Ratios of lengths. | Coefficients. |
|---|---|---|---|---|---|---|---|
| ·01 | ·0464 | ·405 | ·5474 | ·605 | ·7153 | ·805 | ·8654 |
| ·02 | ·0737 | ·410 | ·5519 | ·610 | ·7193 | ·810 | ·8689 |
| ·03 | ·0965 | ·415 | ·5564 | ·615 | ·7232 | ·815 | ·8725 |
| ·04 | ·1170 | ·420 | ·5608 | ·620 | ·7271 | ·820 | ·8761 |
| ·05 | ·1357 | ·425 | ·5653 | ·625 | ·7310 | ·825 | ·8796 |
| ·06 | ·1533 | ·430 | ·5697 | ·630 | ·7349 | ·830 | ·8832 |
| ·07 | ·1699 | ·435 | ·5741 | ·635 | ·7388 | ·835 | ·8867 |
| ·08 | ·1857 | ·440 | ·5785 | ·640 | ·7427 | ·840 | ·8903 |
| ·09 | ·2008 | ·445 | ·5829 | ·645 | ·7465 | ·845 | ·8938 |
| ·10 | ·2154 | ·450 | ·5872 | ·650 | ·7504 | ·850 | ·8973 |
| ·11 | ·2296 | ·455 | ·5916 | ·655 | ·7542 | ·855 | ·9008 |
| ·12 | ·2433 | ·460 | ·5959 | ·660 | ·7580 | ·860 | ·9043 |
| ·13 | ·2566 | ·465 | ·6002 | ·665 | ·7619 | ·865 | ·9078 |
| ·14 | ·2696 | ·470 | ·6045 | ·670 | ·7657 | ·870 | ·9113 |
| ·15 | ·2823 | ·475 | ·6088 | ·675 | ·7695 | ·875 | ·9148 |
| ·16 | ·2947 | ·480 | ·6130 | ·680 | ·7733 | ·880 | ·9183 |
| ·17 | ·3069 | ·485 | ·6173 | ·685 | ·7771 | ·885 | ·9218 |
| ·18 | ·3188 | ·490 | ·6215 | ·690 | ·7808 | ·890 | ·9253 |
| ·19 | ·3305 | ·495 | ·6258 | ·695 | ·7846 | ·895 | ·9287 |
| ·20 | ·3420 | ·500 | ·6300 | ·700 | ·7884 | ·900 | ·9322 |
| ·21 | ·3533 | ·505 | ·6342 | ·705 | ·7921 | ·905 | ·9356 |
| ·22 | ·3614 | ·510 | ·6383 | ·710 | ·7959 | ·910 | ·9391 |
| ·23 | ·3754 | ·515 | ·6425 | ·715 | ·7996 | ·915 | ·9425 |
| ·24 | ·3862 | ·520 | ·6466 | ·720 | ·8033 | ·920 | ·9459 |
| ·25 | ·3969 | ·525 | ·6508 | ·725 | ·8070 | ·925 | ·9494 |
| ·26 | ·4074 | ·530 | ·6549 | ·730 | ·8107 | ·930 | ·9528 |
| ·27 | ·4177 | ·535 | ·6590 | ·735 | ·8144 | ·935 | ·9562 |
| ·28 | ·4280 | ·540 | ·6631 | ·740 | ·8181 | ·940 | ·9596 |
| ·29 | ·4381 | ·545 | ·6672 | ·745 | ·8218 | ·945 | ·9630 |
| ·30 | ·4481 | ·550 | ·6713 | ·750 | ·8255 | ·950 | ·9664 |
| ·31 | ·4580 | ·555 | ·6754 | ·755 | ·8291 | ·955 | ·9698 |
| ·32 | ·4678 | ·560 | ·6794 | ·760 | ·8328 | ·960 | ·9732 |
| ·33 | ·4775 | ·565 | ·6834 | ·765 | ·8365 | ·965 | ·9762 |
| ·34 | ·4871 | ·570 | ·6875 | ·770 | ·8401 | ·970 | ·9799 |
| ·35 | ·4966 | ·575 | ·6915 | ·775 | ·8437 | ·975 | ·9833 |
| ·36 | ·5061 | ·580 | ·6955 | ·780 | ·8474 | ·980 | ·9866 |
| ·37 | ·5154 | ·585 | ·6995 | ·785 | ·8510 | ·985 | ·9900 |
| ·38 | ·5246 | ·590 | ·7035 | ·790 | ·8546 | ·990 | ·9933 |
| ·39 | ·5338 | ·595 | ·7074 | ·795 | ·8582 | ·995 | ·9967 |
| ·40 | ·5429 | ·600 | ·7114 | ·800 | ·8618 | 1·000 | 1·0000 |

ORIFICES, WEIRS, PIPES, AND RIVERS.

TABLE XI.—RELATIVE DIMENSIONS OF EQUAL DISCHARGING TRAPEZOIDAL CHANNELS, WITH SLOPES FROM 0 TO 1, UP TO 2 TO 1.
Half sum of the top and bottom is the mean width. The ratio of the slope, multiplied by the depth, subtracted from the mean width, will give the bottom; and if added, will give the top.

TABLE XII. gives the discharge in cubic feet per minute from the primary channel, 70 wide, and the corresponding depths taken in feet. For lesser or greater channels and discharges, see Rules, pp. 123, 124, 125, 126, and 135.

The mean widths are given in the top horizontal line, and the corresponding depths in the other horizontal lines. They may be taken in inches, feet, yards, fathoms, or any other measures whatever.

| 70 | 60 | 50 | 40 | 35 | 30 | 25 | 20 | 15 | 10 |
|---|---|---|---|---|---|---|---|---|---|
| ·125 | ·13 | ·15 | ·17 | ·20 | ·23 | ·26 | ·29 | ·35 | ·48 |
| ·25 | ·27 | ·30 | ·35 | ·40 | ·45 | ·52 | ·58 | ·71 | ·98 |
| ·375 | ·41 | ·46 | ·54 | ·60 | ·67 | ·76 | ·88 | 1·09 | 1·51 |
| ·5 | ·55 | ·62 | ·73 | ·80 | ·89 | 1·02 | 1·19 | 1·48 | 2·04 |
| ·625 | ·68 | ·78 | ·91 | 1·00 | 1·12 | 1·29 | 1·50 | 1·88 | 2·62 |
| ·75 | ·82 | ·94 | 1·10 | 1·20 | 1·35 | 1·56 | 1·82 | 2·28 | 3·22 |
| ·875 | ·96 | 1·10 | 1·29 | 1·41 | 1·58 | 1·83 | 2·14 | 2·69 | 3·86 |
| 1· | 1·10 | 1·26 | 1·48 | 1·62 | 1·81 | 2·10 | 2·46 | 3·11 | 4·50 |
| 1·125 | 1·24 | 1·42 | 1·67 | 1·83 | 2·04 | 2·37 | 2·79 | 3·54 | 5·19* |
| 1·25 | 1·39 | 1·58 | 1·86 | 2·04 | 2·28 | 2·65 | 3·12 | 3·98 | 5·89 |
| 1·375 | 1·53 | 1·74 | 2·05 | 2·25 | 2·51 | 2·92 | 3·46 | 4·43 | 6·60 |
| 1·5 | 1·67 | 1·90 | 2·24 | 2·46 | 2·75 | 3·20 | 3·80 | 4·88 | 7·31 |
| 1·625 | 1·81 | 2·06 | 2·43 | 2·67 | 2·99 | 3·47 | 4·15 | 5·34 | 8·08 |
| 1·75 | 1·95 | 2·22 | 2·62 | 2·88 | 3·23 | 3·75 | 4·50 | 5·80 | 8·86 |
| 1·875 | 2·09 | 2·38 | 2·81 | 3·09 | 3·47 | 4·03 | 4·86 | 6·29 | 9·68 |
| 2· | 2·23 | 2·54 | 3·00 | 3·31 | 3·72 | 4·32 | 5·22 | 6·78 | 10·50 |
| 2·125 | 2·37 | 2·70 | 3·19 | 3·52 | 3·96 | 4·61 | 5·58 | 7·29 | 11·37 |
| 2·25 | 2·51 | 2·86 | 3·38 | 3·73 | 4·21 | 4·91 | 5·95 | 7·81* | 12·25 |
| 2·375 | 2·65 | 3·02 | 3·57 | 3·94 | 4·45 | 5·20 | 6·31 | 8·32 | 13·12 |
| 2·5 | 2·79 | 3·18 | 3·76 | 4·16 | 4·70 | 5·50 | 6·68 | 8·84 | 14·00 |
| 2·625 | 2·93 | 3·34 | 3·95 | 4·38 | 4·95 | 5·79 | 7·06 | 9·38 | 14·92 |
| 2·75 | 3·07 | 3·51 | 4·15 | 4·60 | 5·21 | 6·09 | 7·45 | 9·93 | 15·84 |
| 2·875 | 3·21 | 3·67 | 4·34 | 4·82 | 5·46 | 6·39 | 7·83 | 10·48 | 16·76 |
| 3· | 3·35 | 3·84 | 4·54 | 5·04 | 5·72 | 6·69 | 8·22 | 11·03 | 17·68 |
| 3·125 | 3·49 | 4·00 | 4·73 | 5·26 | 5·97 | 7·00 | 8·62 | 11·60 | 18·68 |
| 3·25 | 3·63 | 4·17 | 4·93 | 5·49 | 6·23 | 7·31 | 9·02 | 12·17 | 19·68 |
| 3·375 | 3·77 | 4·33 | 5·13 | 5·72 | 6·49 | 7·62 | 9·42 | 12·74 | 20·68 |
| 3·5 | 3·91 | 4·50 | 5·33 | 5·95 | 6·75 | 7·93 | 9·82 | 13·32 | 21·68 |
| 3·625 | 4·05 | 4·66 | 5·53 | 6·17 | 7·01 | 8·25 | 10·23* | 13·92 | 22·76 |
| 3·75 | 4·19 | 4·82 | 5·73 | 6·40 | 7·28 | 8·57 | 10·65 | 14·53 | 23·84 |
| 3·875 | 4·33 | 4·98 | 5·93 | 6·62 | 7·54 | 8·89 | 11·06 | 15·14 | 24·92 |
| 4· | 4·48 | 5·14 | 6·13 | 6·85 | 7·81 | 9·21 | 11·48 | 15·75 | 26·00 |
| 4·25 | 4·76 | 5·46 | 6·54 | 7·30 | 8·35 | 9·85 | 12·33 | 16·98 | 28·18 |
| 4·5 | 5·05 | 5·79 | 6·95 | 7·75 | 8·90 | 10·50 | 13·19 | 18·22 | 30·36 |
| 4·75 | 5·33 | 6·12 | 7·35 | 8·20 | 9·45 | 11·14 | 14·07 | 19·50 | 32·68 |
| 5· | 5·62 | 6·45 | 7·75 | 8·66 | 10·00 | 11·79 | 14·96 | 20·80 | 35·00 |
| 5·25 | 5·90 | 6·78 | 8·16 | 9·14 | 10·55 | 12·51* | 15·86 | 22·13 | 37·40 |
| 5·5 | 6·18 | 7·12 | 8·57 | 9·62 | 11·10 | 13·24 | 16·77 | 23·47 | 39·81 |
| 5·75 | 6·46 | 7·46 | 8·98 | 10·11 | 11·66 | 13·94 | 17·71 | 24·86 | 42·33 |
| 6· | 6·75 | 7·80 | 9·40 | 10·60 | 12·22 | 14·65 | 18·65 | 26·25 | 44·86 |

TABLE XII.—DISCHARGES FROM THE PRIMARY CHANNEL IN THE FIRST COLUMN OF TABLE XI.

If the dimensions of the primary channel be in inches, divide the discharges in this table by 500; if in yards, multiply by 15·6; and if in fathoms, by 88·2, &c.: see pp. 126, 127. The final figures in the discharges may be rejected when they do not exceed one-half per cent., or 0·5 in 100. (See pages 123 to 126.)

| Depths of a channel whose mean width is 70:—in feet. | Falls, inclinations, and discharges in cubic feet per minute. Interpolate for intermediate falls; divide greater falls by 4, and double the corresponding discharges. | | | | | | |
|---|---|---|---|---|---|---|---|
| | 1 inch per mile, 1 in 63360 | 2 inches per mile, 1 in 31680 | 3 inches per mile, 1 in 21120 | 6 inches per mile, 1 in 10560 | 9 inches per mile, 1 in 7040. | 12 inches per mile, 1 in 5280. | 15 inches per mile, 1 in 4224. |
| ·125 | 47 | 72 | 93 | 139 | 175 | 205 | 233 |
| ·25 | 136 | 210 | 268 | 403 | 506 | 596 | 675 |
| ·375 | 249 | 389 | 498 | 746 | 940 | 1105 | 1252 |
| ·50 | 387 | 603 | 770 | 1155 | 1454 | 1709 | 1935 |
| ·625 | 541 | 849 | 1078 | 1617 | 2036 | 2395 | 2714 |
| ·75 | 714 | 1112 | 1420 | 2128 | 2681 | 3153 | 3573 |
| ·875 | 900 | 1401 | 1791 | 2685 | 3382 | 3978 | 4507 |
| 1· | 1100 | 1714 | 2190 | 3283 | 4134 | 4862 | 5507 |
| 1·125 | 1310 | 2042 | 2614 | 3909 | 4927 | 5792 | 6577 |
| 1·25 | 1534 | 2384 | 3058 | 4581 | 5766 | 6780 | 7690 |
| 1·375 | 1767 | 2757 | 3521 | 5279 | 6661 | 7823 | 8863 |
| 1·50 | 2013 | 3142 | 4006 | 6016 | 7588 | 8915 | 10099 |
| 1·625 | 2268 | 3540 | 4525 | 6781 | 8541 | 10044 | 11381 |
| 1·75 | 2534 | 3950 | 5053 | 7570 | 9537 | 11210 | 12703 |
| 1·875 | 2812 | 4384 | 5599 | 8386 | 10570 | 12429 | 14083 |
| 2· | 3090 | 4821 | 6161 | 9230 | 11628 | 13675 | 15513 |
| 2·125 | 3377 | 5273 | 6738 | 10092 | 12718 | 14956 | 16943 |
| 2·25 | 3674 | 5736 | 7331 | 10981 | 13833 | 16281 | 18435 |
| 2·375 | 3977 | 6210 | 7937 | 11889 | 14981 | 17645 | 19960 |
| 2·50 | 4293 | 6699 | 8563 | 12829 | 16161 | 19045 | 21534 |
| 2·625 | 4616 | 7203 | 9204 | 13800 | 17380 | 20434 | 23135 |
| 2·75 | 4947 | 7716 | 9865 | 14782 | 18624 | 21886 | 24800 |
| 2·875 | 5280 | 8233 | 10525 | 15773 | 19887 | 23360 | 26473 |
| 3· | 5621 | 8762 | 11204 | 16788 | 21165 | 24833 | 28176 |
| 3·125 | 5972 | 9310 | 11900 | 17830 | 22454 | 26410 | 29925 |
| 3·25 | 6329 | 9862 | 12614 | 18897 | 23780 | 27994 | 31714 |
| 3·375 | 6689 | 10420 | 13320 | 19963 | 25145 | 29570 | 33507 |
| 3·50 | 7049 | 10995 | 14048 | 21052 | 26509 | 31262 | 35329 |
| 3·625 | 7418 | 11574 | 14785 | 22153 | 27906 | 32860 | 37186 |
| 3·75 | 7794 | 12163 | 15526 | 23284 | 29321 | 34479 | 39080 |
| 3·875 | 8178 | 12753 | 16283 | 24416 | 30756 | 36170 | 41013 |
| 4· | 8566 | 13354 | 17070 | 25592 | 32225 | 37898 | 42954 |
| 4·25 | 9355 | 14582 | 18643 | 27936 | 35191 | 41368 | 46916 |
| 4·50 | 10173 | 15849 | 20267 | 30366 | 38254 | 44982 | 50973 |
| 4·75 | 11001 | 17140 | 21908 | 32818 | 41356 | 48630 | 55102 |
| 5· | 11833 | 18454 | 23595 | 35355 | 44546 | 52378 | 59346 |
| 5·25 | 12696 | 19802 | 25362 | 37939 | 47795 | 56209 | 63688 |
| 5·50 | 13576 | 21172 | 27248 | 40564 | 51097 | 60079 | 68097 |
| 5·75 | 14478 | 22580 | 29160 | 43253 | 54478 | 64058 | 72591 |
| 6· | 15393 | 23995 | 31122 | 45969 | 57897 | 68082 | 77154 |

TABLE XII.—DISCHARGES FROM THE PRIMARY CHANNEL IN THE FIRST COLUMN OF TABLE XI.

If the dimensions of the primary channel be in inches, divide the discharges in this table by 500; if in yards, multiply by 15·6, and if in fathoms by 88·2, &c.: see pp. 126 and 127. The final figures in the discharges may be rejected when they do not exceed one-half per cent., or 0·5 in 100. (See pages 123 to 126.)

| Falls, inclinations, and discharges in cubic feet per minute. Interpolate for intermediate falls; divide greater falls by 4, and double the corresponding discharges. | | | | | | | Depths of a channel whose mean width is 70 :—in feet. |
|---|---|---|---|---|---|---|---|
| 18 inches per mile, 1 in 3520. | 21 inches per mile, 1 in 3017. | 24 inches per mile, 1 in 2640. | 27 inches per mile, 1 in 2347. | 30 inches per mile, 1 in 2112. | 33 inches per mile, 1 in 1920. | 36 inches per mile, 1 in 1760. | |
| 258 | 281 | 303 | 323 | 343 | 362 | 380 | ·125 |
| 748 | 815 | 877 | 936 | 993 | 1049 | 1100 | ·25 |
| 1387 | 1511 | 1627 | 1736 | 1843 | 1952 | 2037 | ·375 |
| 2145 | 2336 | 2515 | 2684 | 2852 | 3023 | 3155 | ·50 |
| 3004 | 3274 | 3527 | 3753 | 4021 | 4207 | 4414 | ·625 |
| 3957 | 4311 | 4645 | 4966 | 5287 | 5553 | 5817 | ·75 |
| 4991 | 5422 | 5859 | 6274 | 6650 | 6992 | 7342 | ·875 |
| 6097 | 6622 | 7159 | 7631 | 8107 | 8540 | 8974 | 1· |
| 7266 | 7920 | 8531 | 9124 | 9660 | 10200 | 10693 | 1·125 |
| 8514 | 9284 | 9995 | 10658 | 11318 | 11923 | 12520 | 1·25 |
| 9816 | 10697 | 11539 | 12307 | 13045 | 13741 | 14479 | 1·375 |
| 11182 | 12185 | 13152 | 14007 | 14862 | 15656 | 16448 | 1·50 |
| 12601 | 13730 | 14821 | 15786 | 16750 | 17657 | 18552 | 1·625 |
| 14069 | 15331 | 16525 | 17616 | 18700 | 19698 | 20696 | 1·75 |
| 15593 | 16997 | 18306 | 19517 | 20728 | 21840 | 22944 | 1·875 |
| 17157 | 18697 | 20141 | 21469 | 22803 | 24017 | 25242 | 2· |
| 18766 | 20446 | 22030 | 23480 | 24938 | 26269 | 27601 | 2·125 |
| 20410 | 22247 | 23965 | 25547 | 27129 | 28578 | 30027 | 2·25 |
| 22104 | 24087 | 25947 | 27662 | 29395 | 30934 | 32512 | 2·375 |
| 23848 | 25988 | 27992 | 29841 | 31701 | 33381 | 35096 | 2·50 |
| 25669 | 27953 | 30100 | 32069 | 34086 | 35910 | 37725 | 2·625 |
| 27479 | 29933 | 32247 | 34384 | 36512 | 38471 | 40415 | 2·75 |
| 29318 | 31947 | 34408 | 36697 | 38958 | 41055 | 43135 | 2·875 |
| 31206 | 34002 | 36624 | 39050 | 41464 | 43680 | 45896 | 3· |
| 33141 | 36112 | 38897 | 41482 | 44048 | 46398 | 48747 | 3·125 |
| 35126 | 38266 | 41223 | 43954 | 46672 | 49174 | 51664 | 3·25 |
| 37109 | 40438 | 43556 | 46438 | 49330 | 51951 | 54586 | 3·375 |
| 39140 | 42631 | 45925 | 48963 | 51993 | 54775 | 57550 | 3·50 |
| 41184 | 44872 | 48343 | 51537 | 54728 | 57659 | 50580 | 3·625 |
| 43273 | 47158 | 50807 | 54162 | 57514 | 60585 | 63656 | 3·75 |
| 45407 | 49468 | 53300 | 56840 | 60341 | 63560 | 66784 | 3·875 |
| 47551 | 51818 | 55832 | 59514 | 63200 | 66576 | 69951 | 4· |
| 51911 | 56586 | 60973 | 64974 | 69013 | 72694 | 76383 | 4·25 |
| 56448 | 61508 | 66176 | 70623 | 75017 | 79017 | 82994 | 4·50 |
| 61014 | 66500 | 71625 | 76408 | 81097 | 85426 | 89767 | 4·75 |
| 65713 | 71628 | 77140 | 82250 | 87351 | 92015 | 96653 | 5· |
| 70509 | 76863 | 82779 | 88200 | 93731 | 98729 | 103745 | 5·25 |
| 75383 | 82159 | 88434 | 94344 | 100200 | 105550 | 110905 | 5·50 |
| 80379 | 87590 | 94348 | 100616 | 106823 | 112540 | 118254 | 5·75 |
| 85407 | 93093 | 100275 | 106911 | 113505 | 119616 | 125664 | 6· |

## 224 THE DISCHARGE OF WATER FROM PIPES.

TABLE XIII.—THE SQUARE ROOTS OF THE FIFTH POWERS OF NUMBERS FOR FINDING THE DIAMETER OF A PIPE, OR DIMENSIONS OF A CHANNEL FROM THE DISCHARGE, OR THE REVERSE; SHOWING THE RELATIVE DISCHARGING POWERS OF PIPES OF DIFFERENT DIAMETERS, AND OF ANY SIMILAR CHANNELS WHATEVER, CLOSED OR OPEN. (See pages 13, 123, 124, &c.)

If $d$ be the diameter of a pipe, in feet, and $D$ the discharge in cubic feet per minute, then for long straight pipes we shall have, for velocities of nearly 3 feet per second, $D = 2400\,(d^5 s)^{\frac{1}{2}}$, and $d = \cdot 044 \left(\dfrac{D^2}{s}\right)^{\frac{1}{5}}$; or if $D$ be the discharge per second, $D = 40\,(d^5 s)^{\frac{1}{2}}$, and $d = \cdot 228 \left(\dfrac{D^2}{s}\right)^{\frac{1}{5}}$. (See pages 112 to 121.)

| Relative dimensions or diameters of pipes. | Relative discharging powers. | Relative dimensions or diameters of pipes. | Relative discharging powers. | Relative dimensions or diameters of pipes. | Relative discharging powers. | Relative dimensions or diameters of pipes. | Relative discharging powers. |
|---|---|---|---|---|---|---|---|
| ·25 | ·031 | 10·5 | 357·2 | 30·5 | 5138· | 61· | 29062· |
| ·5 | ·177 | 11· | 401·3 | 31· | 5351· | 62· | 30268· |
| ·75 | ·485 | 11·5 | 448·5 | 31·5 | 5569· | 63· | 31503· |
| 1· | 1· | 12· | 498·8 | 32· | 5793· | 64· | 32768· |
| 1·25 | 1·747 | 12·5 | 552·4 | 32·5 | 6022· | 65· | 34063· |
| 1·5 | 2·756 | 13· | 609·3 | 33· | 6256· | 66· | 35388· |
| 1·75 | 4·051 | 13·5 | 669·6 | 33·5 | 6496· | 67· | 36744· |
| 2· | 5·657 | 14· | 733·4 | 34· | 6741· | 68· | 38131· |
| 2·25 | 7·594 | 14·5 | 800·6 | 34·5 | 6991· | 69· | 39543· |
| 2·5 | 9·882 | 15· | 871·4 | 35· | 7247· | 70· | 40996· |
| 2·75 | 12·541 | 15·5 | ·945·9 | 35·5 | 7509· | 71· | 42476· |
| 3· | 15·588 | 16· | 1024· | 36· | 7776· | 72· | 43988· |
| 3·25 | 19·042 | 16·5 | 1105·9 | 36·5 | 8049· | 73· | 45531· |
| 3·5 | 22·918 | 17· | 1191·6 | 37· | 8327· | 74· | 47106· |
| 3·75 | 27·232 | 17·5 | 1281·1 | 37·5 | 8611· | 75· | 48714· |
| 4· | 32· | 18· | 1374·6 | 38· | 8901· | 76· | 50354· |
| 4·25 | 37·24 | 18·5 | 1472·1 | 38·5 | 9197· | 77· | 52027· |
| 4·5 | 42·96 | 19· | 1573·6 | 39· | 9498· | 78· | 53732· |
| 4·75 | 49·17 | 19·5 | 1679·1 | 39·5 | 9806· | 79· | 55471· |
| 5· | 55·90 | 20· | 1788·9 | 40· | 10119· | 80· | 57243· |
| 5·25 | ·63·15 | 20·5 | 1902·8 | 41· | 10764· | 81· | 59049· |
| 5·5 | 70·94 | 21· | 2020·9 | 42· | 11432· | 82· | 60888· |
| 5·75 | 79·28 | 21·5 | 2143·4 | 43· | 12125· | 83· | 62762· |
| 6· | 88·18 | 22· | 2270·2 | 44· | 12842· | 84· | 64669· |
| 6·25 | 97·66 | 22·5 | 2401·4 | 45· | 13584· | 85· | 66611· |
| 6·5 | 107·72 | 23· | 2537· | 46· | 14351· | 86· | 68588· |
| 6·75 | 118·38 | 23·5 | 2677·1 | 47· | 15144· | 87· | 70599· |
| 7· | 129·64 | 24· | 2821·8 | 48· | 15963· | 88· | 72645· |
| 7·25 | 141·53 | 24·5 | 2971·1 | 49· | 16807· | 89· | 74727· |
| 7·5 | 154·05 | 25· | 3125· | 50· | 17678· | 90· | 76843· |
| 7·75 | 167·21 | 25·5 | 3283·6 | 51· | 18575· | 91· | 78996· |
| 8· | 181·02 | 26· | 3446·9 | 52· | 19499· | 92· | 81184· |
| 8·25 | 195·50 | 26·5 | 3615·1 | 53· | 20450· | 93· | 83408· |
| 8·5 | 210·64 | 27· | 3788· | 54· | 21428· | 94· | 85668· |
| 8·75 | 226·48 | 27·5 | 3965·8 | 55· | 22434· | 95· | 87965· |
| 9· | 243· | 28· | 4148·5 | 56· | 23468· | 96· | 90298· |
| 9·25 | 260·23 | 28·5 | 4336·2 | 57· | 24529· | 97· | 92668· |
| 9·5 | 278·17 | 29· | 4528·9 | 58· | 25620· | 98· | 95075· |
| 9·75 | 296·83 | 29·5 | 4726·7 | 59· | 26738· | 99· | 97519· |
| 10· | 316·23 | 30· | 4929·5 | 60· | 27886· | 100· | 100000· |

CPSIA information can be obtained
at www.ICGtesting.com
Printed in the USA
BVHW041816100522
636649BV00003B/36